T0215636

Advanced Courses in Mathematics
CRM Barcelona

Centre de Recerca Matemàtica

Managing Editor:
Manuel Castellet

Dario Catalano
Ronald Cramer
Ivan Damgård
Giovanni Di Crescenzo
David Pointcheval
Tsuyoshi Takagi

Contemporary Cryptology

Birkhäuser Verlag
Basel · Boston · Berlin

Authors:

Dario Catalano
Laboratoire d'Informatique
Ecole Normale Supérieure
45, rue d'Ulm
75230 Paris Cedex 05, France
e-mail: dario.catalano@ens.fr

Ronald Cramer
CWI
Kruislaan 413
P.O. Box 94079
1090 GB Amsterdam, The Netherlands
e-mail: Ronald.Cramer@cwi.nl

Giovanni Di Crescenzo
Telcordia Technologies, Inc.
445 South St.
Morristown, NJ 07960, USA
e-mail: giovanni@research.telcordia.com

Ivan Darmgård
Department of Computer Science
University of Aarhus
IT-parken, Aabogade 34
DK-8200 Aarhus N, Denmark
e-mail: ivan@daimi.au.dk

David Pointcheval
Département d'Informatique
Ecole Normale Supérieure
45, rue d'Ulm
75230 Paris Cedex 05, France
e-mail: David.Pointcheval@ens.fr

Tsuyoshi Takagi
Future University – Hakodate
School of Systems Information Science
116-2 Kamedanakano-cho Hakodate
Hokkaido, 041-8655, Japan
e-mail: takagi@fun.ac.jp

2000 Mathematical Subject Classification 68P25, 68P30

A CIP catalogue record for this book is available from the Library of Congress, Washington D.C., USA

Bibliografische Information Der Deutschen Bibliothek
Die Deutsche Bibliothek verzeichnet diese Publikation in der Deutschen Nationalbibliografie; detaillierte
bibliografische Daten sind im Internet über <http://dnb.ddb.de> abrufbar.

ISBN 3-7643-7294-X Birkhäuser Verlag, Basel – Boston – Berlin

© 2005 Birkhäuser Verlag, P.O. Box 133, CH-4010 Basel, Switzerland
Part of Springer Science+Business Media
Printed on acid-free paper produced from chlorine-free pulp. TCF ∞
Printed in Germany
ISBN-10: 3-7643-7294-X
ISBN-13: 978-3-7643-7294-1

9 8 7 6 5 4 3 2 1   www.birkhauser.ch

Contents

B   Multiparty Computation, an Introduction
Ronald Cramer and Ivan Damgård   **41**

C Foundations of Modern Cryptography
Giovanni Di Crescenzo **89**

D Provable Security for Public Key Schemes
David Pointcheval **133**

E   Efficient and Secure Public Key Cryptosystems
Tsuyoshi Takagi   **191**

Foreword

In February 2004, at the Universitat Politècnica de Catalunya, we presented a 45-hour *Advanced Course on Contemporary Cryptology*, organised by the Centre de Recerca Matemàtica. This volume is an expanded and unified version of the material presented in the lectures and the background material that we distributed among the participants.

As the title implies, our aim in the course and in this text is to treat selected topics of the subject of contemporary cryptology, structured in five quite independent but related themes: Efficient distributed computation modulo a shared secret, multiparty computation, modern cryptography, provable security for public key schemes, and efficient and secure public-key cryptosystems. The beauty and multidisciplinarity of this topic motivated the interest of the participants, to whom we are very much indebted for their helpful contributions.

Thanks are due to the Centre de Recerca Matemàtica for organising and sponsoring the Advanced Course, to the CRM administrative staff for smoothly working out innumerable details, and to Paz Morillo for the mathematical organisation of the course and for making it such a pleasant experience. Special thanks go to all the participants of the course for their interest in the event and for their many comments on the material.

Efficient Distributed Computation Modulo a Shared Secret

Dario Catalano

1. Introduction

In several cryptographic protocols a number of participants is required to have an RSA [49] modulus for which none of them knows the factorization. A typical example is the well known Fiat-Shamir identification scheme [22] on which all the players use the same modulus but none of them is supposed to know the factorization (for other examples the reader may look at [21, 28, 39, 43, 44]). In principle a simple solution to this problem would be to allow the "existence" of an external (with respect to the set of players) dealer which initialize the system by providing a modulus N to the players, without revealing them the corresponding factorization. The problem with this solution is, of course, that this dealer has to be *trusted*, in the sense that he has to be completely honest: he should not reveal the factorization and he should provide a correctly generated modulus.

In other scenarios the players are required not only to share an RSA modulus, but they need one of some special form. For instance, N is typically required to be the product of two *safe* primes, i.e. primes of the form $p = 2p' + 1$, where p' is itself a prime, (see [14, 31, 52] for example). While the need of safe primes can sometimes be avoided (as in [15, 23]) this comes often at the cost of needing additional assumptions.

Another case where shared generation of RSA moduli is very useful is threshold cryptography (see [30] for a nice survey on this topic). As a motivating example consider the case of threshold RSA signatures. Let N be an RSA modulus ($N = pq$ where p and q are both primes), e be the public verification key and d the corresponding (secret) signing key. Clearly one has that $ed \equiv 1 \bmod \phi(N)$. A threshold RSA signature is something quite similar to standard RSA signatures, but it involves n parties and has the additional property that any subset of, say, $t + 1 \leq n$ parties, can generate a valid signature but no less than $t + 1$ players can do the same. For this specific case we talk of $t + 1$ out of n threshold signature scheme. Another interesting feature of this type of signatures is that, unlike secret sharing schemes [51], the signature is produced *without* explicitly revealing the private key. To understand how this can be possible let us consider the following approach (originally presented in [24]) to obtain an n out of n threshold RSA signature scheme. To every players is given a (random) share d_i such that $\sum_{i=1}^{n} d_i \equiv d \bmod \phi(N)$. Then, to sign a message m, the player P_j computes $\sigma_i = m^{d_j} \bmod N$ and sends this value to an external party which we can call a *combiner* and which has no secrets. The combiner simply multiplies all the received contributions and gives back to the players

$$\sigma = \prod_{i=1}^{n} \sigma_i \bmod N = m^{d_1 + \cdots + d_n} \bmod N = m^d \bmod N$$

The obvious advantage of this solution is that no player has to store delicate information (such as a signing key would be) in his own private memory. Moreover this basic solution can be generalized to work in a more general scenario to provide a t-out-of-n solution (see [16, 19, 25, 48, 52] for details).

However, as already pointed out before, the above discussion suggests that a trusted dealer initializes the system for the players (by generating the RSA modulus and providing them the shares of the signing exponent). Clearly, however, if an intruder can compromise the dealer, he becomes able to forge signature without needing to access the players internal memory. Thus the external dealer should be not only completely honest but also "protected" enough to guarantee security. For these reasons, whenever possible, one would like not to rely on the assumption that such a dealer is available.

In this lecture we describe some efficient algorithms that allow a set of players to generate shared RSA keys without assuming the existence of a trusted dealer (interestingly efficient solutions were already known for the El-Gamal cryptosystem [20, 33, 45]). Specifically we present a "modular" approach to the problem: we propose several algorithms that can later be combined to perform the desired tasks. Note that, in theory, to generate a shared RSA key one can to resort generic secure circuit evaluation techniques [5, 12, 37, 55]. After all one can always take any (standard) algorithm to generate RSA keys and convert it into a boolean (or arithmetic) circuit. Then for each gate of this circuit the players perform a distributed multiplication modulo a small (publicly available) prime $\hat{p}$. As a consequence this general technique is rather inefficient and can hardly be considered practical (indeed note it requires that some distributed computation is performed for each gate in the circuit, and the circuit can be pretty big).

1.1. Previous Work

BONEH-FRANKLIN. The first to address the issue of an efficient solution for the problem of generating shared RSA keys were Boneh and Franklin who, in a breakthrough result, show how $n > 2$ parties can jointly generate an RSA key without a trusted dealer [7]. The main contribution of their paper is an efficient distributed algorithm to perform a *biprimality test*: the n parties jointly generate a candidate modulus N and then engage in a private distributed protocol to test that N is actually the product of two primes. The distributed biprimality test algorithm is $n - 1$ private, meaning with this that no coalition of at most $n - 1$ players should be able to get any information about the factors of N (beyond, of course, what is revealed by N itself). We will not present the details of this construction here (the interested reader is referred to the original paper), essentially for the sake of *modularity*: we describe (somehow) simple protocols and then we show how to combine them to address more complicate tasks.

OTHERS. Building on the Boneh-Franklin solution, Frankel, Mc Kenzie and Yung describe in [27] a way to add robustness [1] to the protocols in [7]. The FMY protocol follows the structure of [7] and allows to obtain a t-out-of-n threshold protocol (originally the Boneh-Franklin proposal allows for a n out of n solution). Moreover in order to achieve a t-out-of-n threshold, the FMY protocol uses *representation changes* for the sharing of the secret data. Namely, data which is shared in a t-out-of-n fashion is converted into a t-out-of-t fashion in order to perform computations, and then re-converted into a t-out-of-n sharing to preserve tolerance of crashing or malicious players. We will not discuss these issues here.

Some of the techniques that we present in this work originated in papers over robust and proactive RSA. In particular, working over the integers in order to overcome the difficulty of computing modulo an unknown integer was used in several previous papers [26, 32, 25, 48].

Finally we note that the main results presented in this article are essentially from the papers "Efficient Computation Modulo a Shared Secret with Application to the Generation of Shared Safe-Prime Products" by Joy Algesheimer, Jan Camenish and Victor Shoup (appeared in the proceedings of Crypto 2002) [1] and "Computing Inverses over a Shared Secret Modulus" by Dario Catalano, Rosario Gennaro and Shai Halevi (appeared in the proceedings of Eurocrypt 2000) [11]. More precisely the results presented in Sections 5, 6, 7, 8 and 9 are from [1] while the results presented in Section 11 are from [11].

1.2. Organization of this Lecture

We start by introducing some preliminaries in Section 2 (and in particular we give definitions and notations and we discuss the network model we are going to employ in the rest of this document). Then in Section 3 we describe some well known secret sharing methods. In Section 4 we discuss some basic protocols that are going to be useful as tools to "construct" the protocols we will later describe. Section 5 is devoted to describe a quite unusual approach to perform modular arithmetic. In Section 6, we describe some methods to convert between different secret sharing schemes. Then we present efficient algorithms to perform some distributed computation with respect to a shared modulus – and in particular to perform modular reductions – in Section 7. On top of this we pass discussing

---

[1] Informally a protocol is said to be robust if it maintains its security properties even in the presence of maliciously behaving players.

some important applications of the distributed modular reduction algorithm in Sections 8 and 9. In Section 10 we illustrate how to generate shared RSA keys. Finally we discuss an efficient algorithm to compute inverses over a shared secret modulus in Section 11.

2. Preliminaries

2.1. The Network Model

We consider a network of n players, that are connected by point-to-point private channels and by a broadcast channel.[2] We model failures in the network by an adversary $\mathcal{A}$, who can corrupt at most t of the players. We distinguish between the following types of "failures":

- *honest but curious:* the adversary can just read the memory of the corrupted players but not modify their behavior;
- *halting:* an "honest but curious" adversary who may also cause any of the corrupted players to crash and abort the protocol;
- *malicious:* the adversary may cause players to deviate arbitrarily from the protocol.

For the sake of simplicity we will present protocols that are secure with respect only to an honest but curious behaving adversary, which moreover is *static*, i.e. the set of corrupted players is decided at the beginning of the computation of a protocol. Note that all the above assumptions can be relaxed using standard techniques. For example it is possible to force the parties to behave honestly by having them to commit to their inputs and to prove (using the so called zero-knowledge proofs[38, 36]) that they followed the protocol correctly. However we believe that such a formulation would make the presentation more intricate, thus distracting the reader from the focus of this article, which are the protocols for efficient distributed computations modulo a shared value.

Finally we assume communication is synchronous, except that we allow *rushing* adversaries (i.e. adversaries who decide the messages of the bad players at round R after having seen the messages of the good players at the same round).

2.2. Definitions and Notations

In the following we denote with $\mathbb{N}$ the set of natural numbers and with $\mathbb{R}^+$ the set of positive real numbers. We say that a function $\mathsf{negl} : \mathbb{N} \to \mathbb{R}^+$ is *negligible* iff for every polynomial $P(n)$ there exists a $n_0 \in \mathbb{N}$ s.t. for all $n > n_0$, $\mathsf{negl}(n) \leq 1/P(n)$.

Let X_k and Y_k be two probability distributions on the set $\{0,1\}^k$ (this means that by $a \leftarrow X_k$ we intend that $a \in \{0,1\}^k$ and it is chosen according to the

---

[2]The communication assumptions allow us to focus on a high-level description of the protocols, and they can be eliminated using standard techniques for privacy, authentication, commitment and agreement.

distribution X_k). We say that X_k and Y_k are statistically indistinguishable if there exist a negligible function $\mathsf{negl}(\cdot)$ such that for sufficiently large k

$$\sum_{\forall a \in \{0,1\}^k} |Pr_{x \leftarrow X_k}[x = a] - Pr_{y \leftarrow Y_k}[y = a]| < \mathsf{negl}(k).$$

A (probabilistic) distributed protocol for a task T, running in a network of n players is a sequence of programs $R = (R_1, \ldots, R_n)$ where R_j is the program ran by the player P_j

CORRECTNESS. Let $x_1, \ldots, x_n$ be a secret sharing of some secret x, where x_j constitutes the local input of player P_j. We say that a protocol R for a task T is *correct* if its output values $d_1, \ldots, d_n$ constitute a secret sharing of $T(x) = d$.

PRIVACY. We define privacy using the usual simulation approach. That is, we consider the view of the adversary $\mathcal{A}$ during a protocol to be the set of messages sent and received by the bad players during a run of the protocol. We say that a protocol is *private* if for any adversary $\mathcal{A}$ there exists a *simulator* S that runs an execution of the protocol together with $\mathcal{A}$ and produces for it a view that is indistinguishable from the real one.

SECURITY. We say that a protocol is *secure* if it is correct and private.

Remark 1. We point out here that basically all the protocols we are going to present in this article can be proven secure with respect to a slightly different definition, proposed by Canetti [8]. Roughly speaking, Canetti suggested a model in which one shows that a protocol is secure by proving that running the protocol is just as safe as running an idealized computational process where security is inherently guaranteed. In the context of secure multiparty computation this "ideal process" can be seen as all the players handing their inputs to some trusted third party who performs the required computation and outputs back to each player the appropriate "portion" of the function. Thus, in this ideal process, the adversary controlling a minority of players is very limited, because he can only learn and possibly modify the data of the corrupted players. Next we say that a protocol securely performs the required task if it is correct and executing the protocol amounts to *emulating* the ideal process for the considered task.

Using this definition it is possible to prove that security is preserved under non concurrent, modular composition of protocols [8].

For the sole sake of simplicity, however, we preferred to *not* consider this definition here and prove our protocols secure with respect to the simpler one given before.

3. Building Blocks

In this section we will discuss some well known secret sharing methods. First, however, we introduce some terminology. The efficiency of a multiparty protocol is in general measured in terms of two parameters: the *communication* complexity and the *round* complexity. The first parameter measures the number of bits sent by each player. The round complexity, on the other hand, is the number of communication rounds that the parties have to perform in order to complete the protocol. As an additional parameter we consider the *bit-complexity* that measures the number of bit-operations performed by each player.

In the following we will assume that q is a public prime number and that n is the number of players involved in the protocol (and in particular $q > n$). All the primitives presented in this lecture require $O(1)$ rounds of communication. Furthermore, denoted with k the size of the prime q, their communication complexity is $O(kn)$ bits.

3.1. Additive Sharing over $\mathbb{Z}_q$

To share a secret a, player P_j chooses $n-1$ random elements $a_i \in \mathbb{Z}_q$ (for $i \neq j$) and sends a_i to player P_i. Finally he sets his own share a_j as

$$a_j = a - \sum_{i=1, i\neq j}^{n} a_i \bmod q.$$

Note that a player has to perform n additions to share a secret. Since adding two k-bit integers requires k bit operations the entire operation can be done with $O(kn)$ bit operations.

To (publicly) reconstruct the secret every player is required to disclose his share. The secret value is obtained as the sum of all the published contributions.

3.2. Polynomial Sharing over $\mathbb{Z}_q$

In this section we describe a method for constructing a $t+1$ out of n (with $t < n$) threshold scheme originally proposed by Shamir [51]. This method allows n players to share a secret in a way such that any subset of $t+1$ participants can later retrieve the secret but no subgroup of, at most, t participants can do so.
To share a secret a a player P_j randomly chooses t elements $b_i \in \mathbb{Z}_q$ and sets $f(z)$ as the polynomial

$$f(z) = a + \sum_{i=1}^{t} b_i z^i \bmod q.$$

Then for $i \neq j$ he sends the values $f(i)$ to player P_i. Note that the polynomial is evaluated only for small inputs (i.e. $f(i)$ is computed only for the i's denoting the indexes of the remaining players), this means that we can safely assume that $z < \log n$ in the above relations. Thus, since we can assume that multiplying a k-bit integer by a ℓ bit integer requires $O(k\ell)$ bit operations, we can conclude that the proposed method requires

1. t additions of k bit integers. This costs, of course, at most $O(tk)$ bit operations.
2. t exponentiations of a $\log n$ bit integer to a $\log n$ bit exponent. The cost of such exponentiations can be bounded by $O(tn \log^2 n)$.
3. t multiplications of a k-bit long number with a (at most) $t \log n$-bit number. This produces a cost of $O(t^2 k \log n)$ bit operations.

Thus, since $t < n$ and $\log n << k$ we have a total bit complexity of $O(n^2 k \log n)$.

Let us give a look on how any subset of $t+1$ participants can reconstruct the secret. Basically this is achieved by means of polynomial interpolation. Here we will describe a simple method to do that, based on the Lagrange interpolation formula for polynomials.

In a nutshell the Lagrange interpolation formula allows one to retrieve the unique polynomial f of degree at most t from $t+1$ points of it. Let $S = \{P_{i_1}, \ldots, P_{i_{t+1}}\}$ be any subset of $t+1$ players. The formula is

$$f(z) = \sum_{j=1}^{t+1} f(i_j) \prod_{1 \le k \le t+1, \, k \ne j} \frac{z - i_k}{i_j - i_k} \bmod q.$$

Since we are interested only in the free term of the polynomial we can rewrite the formula as

$$f(0) = \sum_{j=1}^{t+1} f(i_j) \prod_{1 \le k \le t+1, \, k \ne j} \frac{i_k}{i_k - i_j} \bmod q.$$

If we set

$$\lambda_{i_j} = \prod_{1 \le k \le t+1, \, k \ne j} \frac{i_k}{i_k - i_j} \bmod q,$$

then we have that

$$f(0) = \sum_{j=1}^{t+1} f(i_j) \lambda_{i_j} \bmod q.$$

We will refer to the λ's as to the *Lagrange interpolation coefficients*. Note that their value depends on q but is independent from the specific polynomial one wants to interpolate. For this reason the Lagrange interpolation coefficients can be precomputed and their values do not need to be kept secret.

3.3. Additive Sharing over $\mathbb{Z}$

To share a secret a, chosen in a given interval $[-A, A]$ player P_j chooses $n-1$ random elements a_i in the larger interval $[-A2^\rho, A2^\rho]$, where, as usual, $i \ne j$. Then he sets $a_j = a - \sum_{i=1, i \ne j}^{n} a_i$ and sends a_i to player P_i. The need of considering a larger interval to choose the a_i's comes from the fact that one has to make sure that the shares release no information (in a statistical sense) about the secret being shared. Note that we did not have this problem when considering additive sharing modulo a prime. The problem here is that the quantity $a_j = a - \sum_{i=1, i \ne j}^{n} a_i$, when computed over the integers, is in general *not* random and may strongly depend on the specific secret a. It goes without saying that having shares that depend too much on the secret is not a very desirable problem when designing a secure secret sharing scheme. To overcome this problem we impose to choose the a_i's in a interval that is sufficiently larger than the one where a is sampled. In this way it is possible to prove that for sufficiently large ρ (in practice one may set $\rho = 128$ for instance), the distributions of shares of distinct secrets are statistically indistinguishable, for *any* set of $n-1$ players.

A simple analysis shows that this sharing technique requires $O(n(\rho + \log A))$ bit operations.

3.4. Polynomial Sharing over $\mathbb{Z}$

In principle, to share a secret using polynomials over the integers, one may think of using the same technique described in Section 3.2 for the case of polynomials over $\mathbb{Z}_q$. There are some technical problems that need to be discussed however. First of all to share a secret a, chosen in a given interval $[-A, A]$ one has to choose the coefficients b_j of the polynomial in a larger interval $[-A2^\rho, A2^\rho]$ (for similar reasons as those seen for the case of additive sharing over $\mathbb{Z}$).

The second difficulty is a little more subtle. In order to prove that a secret sharing scheme is secure one has to prove that, unless enough players pool together their shares and become thus able to reconstruct the secret, no information (either in an information theoretic or computational sense) about the secret is revealed. When the sharing is performed via a t degree polynomial this means that $t + 1$ shares are sufficient to interpolate the secret. On the other hand no information about the secret should be obtained from up to t shares. A way to prove this may be to show that the distribution of t shares of some secret a with polynomial $f(z)$ is indistinguishable from the distribution of t shares that result from sharing another value b with polynomial $\hat{f}$ (without loss of generality we assume that the t shares are those of players $1, \ldots, t$). In other words one has to prove that, with high probability, there is a sharing of b using polynomial $\hat{f}$ with integer coefficients in the same range as f and such that $\hat{f}(j) = f(j)$ (for $j = 1, \ldots, t$). A way to achieve this is to define a polynomial $h(z)$ such that $h(0) = a - b$ and $h(1) = \ldots = h(t) = 0$. Then the desired polynomial is $\hat{f}(z) = f(z) - h(z)$.

Observe that the polynomial $h(z)$ can easily be interpolated as

$$h(z) = \sum_{i=0}^{t} h(i) \prod_{j \neq i, j=0,\ldots,t} \frac{z-j}{i-j} = (a-b) \prod_{j=1,\ldots,t} \frac{z-j}{-j}$$

where the coefficient of z^i is

$$(a-b) \sum_{B \subseteq \{1,\ldots,t\}, |B|=i} \frac{\prod_{j \in B}(-j)}{\prod_{j=1,\ldots,t}(-j)}.$$

Note, however, that the above coefficients *are not* necessarily integers (actually they are fractions).

To overcome this problem we adopt the following trick. To share a secret a one shares the related value La, where $L = n!$. In this way the polynomial $h(z)$ above can be re-defined as the one such that $h(0) = (a - b)L$ and $h(1) = \ldots = h(t) = 0$. That is

$$h(z) = \sum_{i=0}^{t} h(i) \prod_{j \neq i, j=0,\ldots,t} \frac{z-j}{i-j} = L(a-b) \prod_{j=1,\ldots,t} \frac{z-j}{-j},$$

where the coefficient of z^i is

$$L(a-b) \sum_{B \subseteq \{1,...,t\}, |B|=i} \frac{\prod_{j \in B}(-j)}{\prod_{j=1...,t}(-j)}.$$

Note that because $L = n!$ this value is an integer. Moreover it can be bounded in absolute value by

$$\sum_{B \subseteq \{1,...,t\}, |B|=i} L(a-b) \leq L(a-b)\binom{t}{i} \leq \frac{(a-b)Lt!}{i!(t-i)!} \leq (a-b)Lt! \leq L^2 A.$$

This means that the coefficients of $\hat{f}(z)$ are in the range $[-L^2 A - 2^\rho A, 2^\rho A + L^2 A]$. Thus the probability that they are outside the legal range is $t\frac{2L^2 A}{2(L^2 A + 2^\rho A)} = \frac{tL^2}{L^2 + 2^\rho}$, which for sufficiently large ρ is negligible.

4. Basic Protocols

Once we briefly described some secret sharing basics, we pass considering some important protocols to perform some basic tasks that are going to be used as underlying building blocks for the protocols presented in the following sections.

4.1. Distributed Computation Modulo q

In this paragraph we briefly discuss the problem of performing basic operations with shared secrets using the polynomial sharing technique described above. The basic operations we want to perform are essentially the following:

1. *Multiplication or addition of a constant (public) value and a polynomially shared secret.*
   This is done by having each player multiply (or add) his share to the constant. This is because, by the properties of polynomials, if $f(i)$ is a share of a, then $f(i) + c$ will be a share of $a + c$ and $cf(i)$ one of $c \cdot a$.
2. *Addition of two polynomially shared values.*
   This is done by having the players locally add their own shares. In particular denoting with $f(i)$ a share of a secret a and with $g(i)$ a share of a secret b, the value $f(i) + g(i)$ is actually a share of the sum $a + b$.
3. *Multiplication of two polynomially shared values.*
   This is just a little more complicate. In principle one can adopt the same strategy already described for addition: every player locally multiplies his own shares $f(i)$ and $g(i)$ and sets $h(i) = f(i)g(i)$ as his share of the product (note that the free coefficient of the polynomial $h(x)$ is actually $f(0)g(0)$). However there are two problems with using the polynomial $h(x)$ to encode the product of the two secrets. The first, rather obvious, is that, if f and g are polynomials of degree t their product will be a polynomial of degree $2t$. This fact creates no problems in interpolating h if n is bigger than $2t$. However it is easy to see that further multiplications raise the degree and once such

degree becomes larger than n, interpolation becomes impossible (we will not have enough points).

The second problem is more subtle: $h(z)$ is not a random polynomial of degree $2t$ (for example, being a product of two polynomials, it is not irreducible).

To solve these problems one can adopt a solution proposed by Ben-Or, Goldwasser and Widgerson [5] that allows to efficiently randomize the coefficients of the polynomial $h(x)$ and to reduce its degree, while, of course, keeping the free coefficient unaltered.

Recently a more efficient variant of the Ben-Or, Goldwasser and Widgerson protocol was proposed by Gennaro, Rabin and Rabin [34] and requires $O(k^2 n + kn^2 \log n)$ bit operations per player.

In the rest of this document we will refer to the latter protocol as to $\mathtt{MUL}(f(i), g(i))$.

4.2. Joint Random Sharing over $\mathbb{Z}_q$

In this section we describe how to generate shares of a secret chosen jointly and at random in $\mathbb{Z}_q$ by the players.

Each player chooses a random value $r_i \in \mathbb{Z}_q$, shares it according to the adopted secret sharing scheme and sends the obtained shares to the remaining players involved in the protocol. At this point each players sums up (modulo q) all the received values and sets the obtained value as his share of the jointly chosen random value.

In the following we will refer to this protocol as $\mathtt{JRS}(\mathbb{Z}_q)$ if the players get additive shares and $\mathtt{JRP}(\mathbb{Z}_q)$ if, on the other hand, they get polynomial shares. It is not hard to see that the first protocol requires $O(nk)$ bit operations per player while the second one requires $O(kn^2 \log n)$ bit operations per player.

4.3. Joint Random Sharing of 0 in $\mathbb{Z}_q$

In many protocols it is often useful to be able to generate a sharing of zero to re-randomize shares obtained from some earlier performed computation. The joint random sharing of zero protocol is pretty simple and can be described as follows. Each player performs a sharing of zero, according to the secret sharing scheme adopted, and sends the produced shares to the remaining players. Next each player sums up (modulo q) the received values and sets the result as his share for zero. As before we denote this protocol with $\mathtt{JRSZ}(\mathbb{Z}_q)$ if the the players get additive shares and with $\mathtt{JRPZ}(\mathbb{Z}_q)$ if, on the other hand, they get polynomial shares. The protocols require $O(nk)$ and $O(kn^2 \log n)$ bit operations per player, respectively.

In case one wants to get additive shares over the integers the technique is basically the same as that seen to produce an additive sharing over $\mathbb{Z}$. It is given a range $[-2^\rho A..2^\rho A]$ from which the players sample the shares they send to the other participants. We denote this protocol by $\mathtt{JRIZ}([-2^\rho A..2^\rho A])$ and it requires $O(n(\rho + \log A))$ bit operations per player.

4.4. Computing Shares of the Inverse of a Shared Secret

The protocol we are going to describe works only for polynomial sharings over $\mathbb{Z}_q$. Let a be an *invertible* element in $\mathbb{Z}_q$. We say that an element is invertible in $\mathbb{Z}_q$ if $gcd(a, q) = 1$. Since we are considering q a prime number, every non zero element is invertible in $\mathbb{Z}_q$. Note that for every invertible element a there exist a $b \in \mathbb{Z}_q$ such that $ab \equiv 1 \bmod q$ and this element is efficiently computable (using the well known extended Euclid's algorithm). Now assume that a is shared among the players and denote with a_i the share held by player P_i. The following protocol, due to Bar-Ilan and Beaver [4], allows to compute shares of b from shares of a. The idea is the following. First the players run the $\mathtt{JRP}(\mathbb{Z}_q)$ protocol to jointly generate a shared random value r, then they multiply the two shared secrets a and r by means of the $\mathtt{MUL}(a_i, r_i)$ protocol. To conclude this phase the players reveal the shares obtained after the execution of the multiplication protocol and jointly reconstruct the value $u \equiv ar \bmod q$. If $u \equiv 0 \bmod q$ the protocol is restarted. Otherwise u is invertible modulo q and every player can locally compute his share of $a^{-1} \bmod q$ by setting $b_i = r_i \cdot u^{-1} \bmod q$. We denote this protocol by $\mathtt{INV}(a_i)$. It requires an (expected) number of $O(k^2 n + k n^2 \log n)$ bit operations per player.

4.5. Joint Random Invertible Element Sharing

This protocol is a variant of the one presented in the previous section and was proposed by Bar-Ilan and Beaver [4] as well. It allows a set of players to generate a random element with the additional property that this element is invertible in $\mathbb{Z}_q$. The players start by generating shares of two random values r and s by running the $\mathtt{JRP}(\mathbb{Z}_q)$ protocol and then jointly compute their product using the $\mathtt{MUL}(s_i, r_i)$ procedure. Finally they reveal the obtained results and reconstruct the value $u \equiv r \cdot s \bmod q$. If u is not zero modulo q each player sets his share of r as the share of the desired random invertible element (otherwise they simply repeat the protocol). As before this protocol, that we call $\mathtt{JRP\text{-}INV}(\mathbb{Z}_q)$, requires an (expected) number of $O(k^2 n + k n^2 \log n)$ bit operations per player.

5. A Different Approach

For some of the protocols that we are going to present in this article, it is more useful to perform modular arithmetic in a slightly different way. So far we adopted the standard notation by which, given two integers a, b and a positive integer q, we write $a \equiv b \bmod q$ if q divides $a - b$. In particular this can be interpreted as follows. Suppose we divide a and b by q, obtaining integer quotients and remainders; we assumed that the remainders were always positive integers between 0 and $q - 1$. This means that, denoting $a = Q_1 q + R_1$ and $b = Q_2 q + R_2$ one has that $0 \leq R_1, R_2 \leq q - 1$. By this position $a \equiv b \bmod q$ if and only if $R_1 = R_2$ and the notation $a \bmod q$ denotes the remainder when a is divided by q, i.e. the value R_1 above.

However there is no need to assume that the remainder has to be a positive integer. Here we will describe a different approach by which modular arithmetic is done centered around zero. We will adopt the symbol 'rem' rather than 'mod' as the operator of modular reduction to remind the reader of this.

Let a be a real number, we denote with $\lfloor a \rfloor$ the largest integer b such that $b \le a$. Conversely we indicate with $\lceil a \rceil$ the smallest integer $b \ge a$. Finally we denote with $\lceil a \rfloor$ the largest integer $b < \frac{1}{2} + a$. We denote with $\text{trunc}(a)$ the operator

$$\text{trunc}(a) = \begin{cases} \lceil a \rceil & \text{if } a < 0 \\ \lfloor a \rfloor & \text{if } a \ge 0 \end{cases} .$$

Thus trunc actually truncates a towards zero.

Now let q be a positive integer and define $\mathcal{Z}_q$ as the set $\{x \in \mathbb{Z} \mid -q/2 < x \le q/2\}$. Clearly any integer a can be written as $c + \frac{k}{2}q$ with $c \in \mathcal{Z}_q$ and $\frac{k}{2} \in \mathbb{Z}$. Now consider the value $\lceil \frac{a}{q} \rfloor$, one has that

$$\left\lceil \frac{a}{q} \right\rfloor = \left\lceil \frac{c}{q} + \frac{k}{2} \right\rfloor .$$

Since $\frac{k}{2}$ is an integer and $|\frac{c}{q}| < 1/2$ we can conclude that $\lceil \frac{a}{q} \rfloor = \frac{k}{2}$. Thus

$$a \text{ rem } q = c = a - \left\lceil \frac{a}{q} \right\rfloor q.$$

It is not hard to see that all the protocols described in Section 3 work in this new representation setting as well (basically one simply needs to rewrite them using the 'rem' operator to replace the 'mod' one).

6. Converting among Different Secret Sharing Methods

In the protocols we are going to describe, we will need to use all the basic algorithms described in the previous sections and to adopt all the three secret sharing schemes discussed so far. For this reason we will devote this section to explain some efficient methods to convert shares from a secret sharing scheme into shares of a different one.

6.1. Converting between Additive and Polynomial Shares

Converting from additive shares in $\mathcal{Z}_q$ to polynomial shares in $\mathcal{Z}_q$ is very simple. Let a_i be the share held by player P_i. We start the conversion by allowing every player P_i to share his contribution a_i by mean of a polynomial $f_i(z)$ of degree $t < n$. In particular P_i chooses at random t coefficients β_j and sets $f_i(z) = a_i + \sum_{j=1}^{t} \beta_j z^j$ rem q. Finally he sends to every other player P_j the value $f_i(j)$ rem q. Upon having received all the contributions from the other parties, player P_i sets his polynomial share for $a = \sum a_i$ rem q as $f(i) = \sum_{j=1}^{n} f_j(i)$ rem q.

Converting from polynomial shares in $\mathcal{Z}_q$ to additive shares in $\mathcal{Z}_q$ is very easy as well. Let a be the shared secret, $\lambda_1, \ldots \lambda_n$ be the Lagrange interpolation coefficients and denote with $S = \{P_{i_1}, \ldots, P_{i_{t+1}}\}$ any subset of $t + 1$ parties. Of course the players in S can interpolate and reconstruct the secret as showed in Section 3.2. This means that, in particular,

$$a = \sum_{j=1}^{t+1} \lambda_{i_j} f(i_j)$$

where f is the sharing polynomial. So every player in S just performs an additive sharing of his own contribution and sends the shares to the respective parties, which just add them up to obtain an additive sharing of a.

Conversions between additive and polynomial sharings over the integers are done – basically – in the same way.

6.2. Converting between Integer Shares and $\mathcal{Z}_q$ Shares

Converting integer shares into shares over $\mathcal{Z}_q$ clearly requires that $q/2$ is bigger than the absolute value of the secret. If this is the case let c be an integer additively shared over the integers, and c_i be the share held by player P_i. An additive sharing over $\mathcal{Z}_q$ can be easily obtained by having each player reduce his own share modulo q. More precisely each player P_i sets $c'_i = c_i$ rem q. Clearly $\sum_{i=1}^{n} c'_i = c$ rem q.

Converting shares over $\mathcal{Z}_q$ into integer shares, however, is not as easy. The problem here is that if one simply considers the additive shares over $\mathcal{Z}_q$ as additive shares over the integers, then the resulting secret may be off by some multiple of q with respect to the actual one. For example if the c'_i's are additive shares of c in $\mathcal{Z}_q$, then one has that $\sum_{i=1}^{n} c'_i = c$ rem q. However this equation simply tells us that $\sum_{i=1}^{n} c'_i = c + kq$ where k is the quotient of $\sum_{i=1}^{n} c'_i$ and q (and in general such a quotient is not zero).

Here we describe a method that allows to determine this quotient without revealing anything about the secret c. The basic idea of the proposed solution is the following. Assume that the shared secret is much smaller than the modulus q (one may assume it is at least ρ bits smaller, where, as usual, ρ is a security parameter). If this is the case, then one can expect the shares c_i to be much larger than c. Consequently every player can reveal the high order bits of his share without compromising the secrecy of the shared value. As we will see, knowledge of these bits is sufficient to compute the desired quotient.

The formal protocol is presented in Figure 1.

Remark 2. For the sake of simplicity, we assume that (unless otherwise explicitly noted) all the protocols presented in this article use, as underlying primitive, an n out of n (additive or polynomial) sharing mechanism. This, in particular, means that we assume that no player can stop participating to the protocol before the end of the protocol itself. This may seem a very strong requirement. However we point out here that standard techniques (see [48] for instance) can be used to relax this assumption.

With the following theorem we prove that the SQ-to-Si protocol is actually secure (i.e. that is correct and private).

SQ-to-SI Protocol

Public Parameters: A value k such that $-2^{k-1} < c = \sum_i c_i$ rem $q < 2^{k-1}$.
A security parameter ρ and a truncation parameter $t = \rho + k + 2$.
Common Input: A modulus $q > 2^{\rho+k+\log n + 4}$.
Private Input (for player P_j): A share $c_j \in \mathcal{Z}_q$ of the secret c.

Player P_j does as follows:

1. Reveal $a_j = \mathtt{trunc}\left(\frac{c_j}{2^t}\right)$.

2. Publicly compute $\ell = \left\lceil \frac{2^t \sum_i a_i}{q} \right\rceil$.

3. The players run the protocol $\mathtt{JRIZ}(-2^\rho q, 2^\rho q)$ to produce an additive sharing of zero over the integers Denote with σ_j the resulting share obtained by P_j.

4. If $j \le |\ell|$ set the output to $c'_j = c_j - q + \sigma_j$ if $\ell > 0$ and to $c'_j = c_j + q + \sigma_j$ if $\ell < 0$
   If $j > |\ell|$ set the output to $c'_j = c_j + \sigma_j$.

FIGURE 1.   A protocol to convert shares over $\mathcal{Z}_q$ into integer shares

---

Theorem 1. *Let $c_1, \ldots, c_n$ a random additive sharing of $-2^{k-1} < c = \sum_i c_i$ rem $q < 2^{k-1}$. If $q > 2^{\rho+k+\log n + 4}$, then the protocol in Figure 1 securely computes additive shares of c over the integers.*

Proof. We divide the proof in two steps. First we prove that the protocol is correct and then that it is also private.

To prove that the protocol is correct we have to show that the local outputs of the players are actually shares of c. Let $\hat{\ell} = \left\lceil \frac{\sum_i c_i}{q} \right\rceil$. Clearly $c = \sum_{i=1}^n c_i - \hat{\ell} q$ where $|c| < 2^{k-1}$ by our assumption. We want to show that $\hat{\ell}$ is actually the same ℓ computed in the protocol.

Let $b_i = c_i - 2^t a_i$. Since $2^t a_i$ contains the t most significant bits of c_i, $|b_i| < 2^t$. Moreover we have $\sum_{i=1}^n b_i = \sum_{i=1}^n c_i - 2^t \sum_{i=1}^n a_i = c + \hat{\ell} q - 2^t \sum_{i=1}^n a_i$ and then $2^t \sum_{i=1}^n a_i = c + \hat{\ell} q - \sum_{i=1}^n b_i$. This means that

$$\ell = \left\lceil \frac{2^t \sum_i a_i}{q} \right\rceil = \left\lceil \frac{c}{q} + \hat{\ell} - \frac{\sum_i b_i}{q} \right\rceil.$$

Since $\hat{\ell}$ is an integer we have that $\hat{\ell} = \ell$ if $\left|\frac{c}{q}\right| < 1/4$ and $\left|\frac{\sum_i b_i}{q}\right| < 1/4$, that is if $k < \log q - 1$ and $t + \log n + 2 = \rho + k + \log n + 4 < \log q$ hold. Moreover since $c_i \in \mathcal{Z}_q$ for all i we have that $\ell = \left\lceil \frac{\sum_i c_i}{q} \right\rceil < n$.

Now we prove that the protocol is private. We do this by showing that the protocol is *simulatable*. According to our definition (see Section 2.2), we need to provide a simulator S that runs an execution of the protocol together with an adversary $\mathcal{A}$ and produces for it a view that is indistinguishable with respect to the one a *real* execution of the protocol would have produced. In the simulated scenario we assume that the simulator controls one single player (and without loss of generality we can assume this player is P_n). For such a player the simulator holds as initial value an element c'_n which result from a sharing of a secret $c' \neq c$ (in the correct range). Note that the distribution of $n - 1$ additive shares (over $\mathcal{Z}_q$) of a secret c is indistinguishable from the distribution of $n-1$ additive shares of a different secret c'. This is because if $(c_1, \ldots, c_n)$ is an additive sharing (over $\mathcal{Z}_q$) of c, an additive sharing of c' can easily be obtained as $(c_1, \ldots, c_{n-1}, c'_n = c_n - c + c'$ rem $q)$. Next for such a share c'_n the simulator computes $a'_n = \text{trunc}\left(\frac{c'_n}{2^t}\right)$ and publishes this value. For steps 2, 3 and 4 the simulator simply follows the same instructions as the protocol. Thus, to conclude the proof, we need to show that the distribution of the a'_n produced in step 1 is statistically indistinguishable from the output produced when running the protocol on a different secret. We prove this by showing that the distributions of the a_i's for different shared values c are statistically indistinguishable. In particular we consider the probability that the a_i's take different values when a different secret c is shared. Without loss of generality let us concentrate on the case on which $c_1, \ldots, c_{n-1}$ are random values and c_n is set as $c_n = c - \sum_{i=1}^{n-1} c_i$ rem q. In this case clearly $C = -\sum_{i=1}^{n-1} c_i$ rem q is uniformly distributed over $\mathcal{Z}_q$ and the values a_i cannot depend on the secret. It remains to consider a_n. By definition $c_n = a_n 2^t + b_n$ where $b_n < 2^t$. Let us consider the quantity $c = c_n - C$ rem q. This value has to be in the range $[-2^{k-1}..2^{k-1}]$. However if we focus on the quantity $c_n - C$, considered over the integers, this value may not be equal to c (i.e. a wrap around occurs). Thus two cases have to be considered, depending on whether $c_n - C$ wraps around or not.

First assume that $c_n - C$ wraps around. This means that $c = c_n - C \pm q$ and in particular one has that either $-2^{k-1} \leq c_n - (C+q) \leq 0$ or $0 \leq c_n - (C-q) \leq 2^{k-1}$. From the two relations above we get that c_n is independent of c if $|C| \leq q - 2^{k-1}$. Note that, since C is uniformly distributed over $\mathcal{Z}_q$, this happens with probability $(1 - 2^{k-1-|q|})$

By a similar argument one can prove that, when the quantity $c_n - C$ does not wrap around, if C rem $2^t < 2^t - 2^{k-1}$, then c_n gets a value that is completely independent from c. Since this second event happens with probability $1 - 2^{k-1-t}$, one has that the total probability that c_n (and thus a_n) gets a value that depends on c is bounded by

$$
\begin{aligned}
Pr[(|C| > q - 2^{k-1})] + Pr[(|C \text{ rem } 2^t| > 2^t - 2^{k-1})] &\leq \frac{2^k}{q} + \frac{2^k}{2^t} \\
&\leq \frac{2^{k+1}}{2^t} = 2^{k+1-t}.
\end{aligned}
\tag{1}
$$

Now that we have determined which are the "bad" cases it is not too hard to show that the statistical difference between the distributions of the a_i's for different shared secrets c's has to be smaller than $2 \cdot 2^{k+2-t} = 2^{-\rho}$, which by our assumption is negligible. □

The bit complexity of the proposed protocol is $O(kn^2 \log n + k^2 n)$ and its communication complexity is $O(kn)$. The round complexity is, clearly, $O(1)$. Note that one may use this protocol also to convert from polynomial shares over $\mathcal{Z}_q$ to polynomial shares over $\mathcal{Z}_{q'}$ where $q' \neq q$, if, of course, q and q' are sufficiently large with respect to the secret and the security parameter being considered.

6.3. Computing Shares of the Binary Representation of a Secret

In some situations is useful to have a secret shared bit by bit. Unfortunately the only solution we know to perform this task is not very efficient because it requires one to resort to general multiparty computation protocols. In a nutshell, the basic idea is as follows. Assume the players hold additive shares of a k bit secret b. In order to obtain shares of the bits of b, each player distributes polynomial shares, modulo some prime q' of the bits of his additive share. Then the players engage in a general multiparty computation protocol to add these bits and obtain shares of the bits of b. As noticed by [1] this multiparty computation can be done over $Z_{q'}^*$ where q' can be rather small (say $\rho + \log n$ bits). The details of this construction are omitted here, but it is possible to prove that such a solution (we will refer to it as to the ADD-to-BIN protocol) requires $O(kn^3 \log q' \log n + kn^2 (\log q')^2 + k^2 n^2 \log n)$ bit operations per player. The communication complexity is $O(k^2 n + nk \log q)$ and the round complexity is $O(\log k + \log n)$.

6.4. Approximate Truncation

We conclude this section by providing a protocol to perform approximate truncations. The algorithm takes as input polynomial shares of a secret a and a parameter k and returns as output shares of b such that $|b - a/2^k| \leq n + 1$. The protocol appears in Figure 2.

TRUNC Protocol

Common Input: A parameter k and a modulus $q > 2^{\rho + k + \log n + 4}$.
Private Input (for player P_j): A polynomial share $a_j \in \mathcal{Z}_q$ of the secret a.

Player P_j does as follows:

1. Obtain additive shares of a over the integers. (This is done by first running the polynomial to additive share conversion in $\mathcal{Z}_q$ and then by applying the SQ-to-SI protocol on the resulting shares). Let a'_j the additive share of a held by player P_j.

2. Locally compute $b_j = \texttt{trunc}\left(\frac{a'_j}{2^k}\right)$.

3. Obtain polynomial shares of b over $\mathcal{Z}_q$ (again using the conversion protocol described in previous sections).

FIGURE 2.   A protocol for distributed approximate truncation

---

It is very easy to see that the protocol is both correct and private for $|q| > \rho + k + \log n + 4$ (this is the same requirement we needed for the SQ-to-SI protocol). The bit complexity of the algorithm is $O(kn^2 \log n + k^2 n)$. The communication complexity is $O(kn)$ and the round complexity is $O(1)$.

7. Distributed Modular Reduction

In this section we present an efficient protocol to compute modular reductions, i.e. a distributed algorithm that taking on input shares of a and p returns as output shares of $a \bmod p$. Using such an algorithm it becomes immediately possible to (efficiently) perform distributed modular addition and multiplication. The proposed method uses an additional modulus q whose size is roughly twice that of p and which is publicly known by all players (note that this provides also an upper bound on the size of p).

We point out here that the modular reduction algorithm we are going to present it is actually an approximation one: it does not compute the actual $a \bmod p$ but a related value a' that is bounded by a small multiple of the modulus.

Before presenting the actual construction we highlight here the main ideas underlying it. We already defined $a \operatorname{rem} p = a - \lceil \frac{a}{p} \rfloor p$. Using this fact, the problem of computing $a \operatorname{rem} p$ reduces to compute shares of $\lceil \frac{a}{p} \rfloor p$. This last problem can be splitted in two: first we compute a distributed approximation of $1/p$ then, on top of this, we compute the shares of $\lceil \frac{a}{p} \rfloor p$. To compute the approximation of $1/p$ we employ the so-called Newton Iteration Method that we briefly recall in the next section. In Section 7.3 we will focus on how to compute a good approximation of $\lceil \frac{a}{p} \rfloor p$.

7.1. Newton Iteration Method

Newton's method provides a powerful way to approximate the roots of an equation. Let $f(x)$ be a differentiable function and let r and r_0 a root and a first approximation of this root respectively. Let us consider the point on the curve of the function $P \equiv (r_0, f(r_0))$. The slope of the tangent line in this point is clearly $f'(r_0)$. Moreover the tangent intersects the x-axis in a point having x-value r_1. It is easy to check that the value r_1 is a better approximation of r than r_0. From r_1 one can re-iterate the method to obtain a better approximation r_2 and so on. The equation of the tangent line in point P is given by

$$y - f(r_0) = f'(r_0)(x - r_0).$$

Thus for $y = 0$ we obtain the iteration formula

$$r_i = r_{i-1} - \frac{f(r_{i-1})}{f'(r_{i-1})}.$$

In our case we will employ Newton's method with the function $f(x) = \frac{1}{x} - \frac{p}{2^k}$. This leads to the iteration formula

$$x_{i+1} = x_i \left(2 - \frac{x_i p}{2^k} \right). \tag{2}$$

Recall that a sequence $\{z_k\}$ converges linearly to ω if for sufficiently large k, $|\{z_{k+1} - \omega\}| < c |\{z_k - \omega\}|$ where $0 < c < 1$ and it converges quadratically if for sufficiently large k, $|\{z_{k+1} - \omega\}| < c |\{z_k - \omega\}|^2$ for some constant c. It is easy to verify that the iteration formula 2 converges quadratically.

7.2. First Step: Computing Shares of an Approximation of $1/p$

Here we present a protocol to compute polynomial shares of an integer p' such that $p' 2^{-k-t} = 1/p + \epsilon$ (where $|\epsilon| < (n+1) 2^{-k-t+4}$ for some parameter t) starting from polynomial shares of $2^{k-1} < p < 2^k$. As already mentioned in the previous section we will adopt Newton's method using equation 2 as iteration formula. In particular we initialize it with the starting value $3/2$; this produces a starting error of $\left| \frac{2^k}{p} - \frac{3}{2} \right| < \frac{1}{2}$ and then we need about $\log t$ iterations to have a t-bit approximation x' of $2^k/p$. Then once this x' is computed we set $p' = x' 2^t$ which is an integer. The formal protocol is presented in Figure 3.

Remark 3. Note that in the formal protocol that appears in Figure 3 we initialize the iteration using $u_0 = 3 \cdot 2^{t-1}$ rather than with $u_0 = 3/2$. This has basically no consequences in practice because at the end of the algorithm we set $p' = u_{i+1} = x'$ (which is already of the correct form) rather than $p' = 2^t x'$.

Remark 4. The pseudo-code in Figure 3 contains a slight misuse of notation. Note that during the first execution of the cycle for (i.e. when $i = 0$) x_0 is a constant value known to all the participants. In this case, then, player P_j computes his share of $p \cdot x_0$ by simply multiplying by x_0 his share p_j (see Section 4.1). On the other hand all the other x_i's *are not* publicly known by the all the other players and thus resorting to the multiplication protocol becomes necessary.

Now we prove that the proposed protocol is secure.

Theorem 2. *Let ρ be a security parameter and $q > 2^{\rho + t + \mu + 6 + \log n}$, where $\mu = max(k, t)$ then for any $t > 5 + \log(n + 1)$ and any p satisfying $2^{k-1} < p < 2^k$ for some k, the protocol presented in Figure 3 securely computes shares of p' such that*

$$\left| \frac{2^k}{p} - \frac{p'}{2^t} \right| < \frac{n+1}{2^{t-4}}$$

where $0 < p' < 2^{t+2}$.
Thus $\frac{p'}{2^{t+k}}$ is an approximation of $\frac{1}{p}$ with (relative) error $\frac{n+1}{2^{t-4}}$.

Proof. We have to prove that p' is actually an approximation of $1/p$. Security trivially follows from the composability of the sub protocols used.

P-INVERT Protocol

Public Parameters: A value k such that $2^{k-1} < p < 2^k$ and an approximation parameter t.

A prime q such that $|q| > 2|p|$.

Private Input (for player P_j): A polynomial share $p_j \in Z_q$ of the secret p.

Player P_j does as follows:

1. Set $(x_0)_j = x_0 = 3 \cdot 2^{t-1}$.
2. **For** $i = 0$ to $\lceil \log(t - 3 - \log(n + 1)) \rceil - 1$ **do**
   - (a) Run the protocol $\text{MUL}((x_i)_j, p_j)$ to produce a polynomial sharing, over Z_q, of $x_i \cdot p$. Denote with z_j the local output of player P_j.
   - (b) Run the $\text{TRUNC}(z_j, k)$ protocol and let w_j the local output
   - (c) Run the protocol $\text{MUL}(w_j, (x_i)_j)$ to produce a polynomial sharing, over Z_q, of $w_j \cdot (x_i)$. Denote with W_j the local output of player P_j.
   - (d) Set $v_j = 2^{t+1}(x_i)_j - W_j$ rem q.
   - (e) Run the $\text{TRUNC}(v_j, t)$ protocol and set x_{i+1} as the local output.
3. The players run the protocol $\text{JRPZ}(Z_q)$ to produce polynomial shares of zero. Denote with σ_j the share obtained by P_j.
4. Set the output to $p'_j = (x_{i+1})_j + \sigma_j$ rem q.

FIGURE 3.   A protocol for distributed computation of an approximation of $1/p$

First note that x_0 and x_1 are both positive. Moreover one can write

$$
\begin{aligned}
x_{i+1} &= \tfrac{x_i}{2^t}\left(2^{t+1} - \tfrac{x_i p}{2^k}\right) \\
&= \tfrac{x_i}{2^t}\left(2^{t+1} - \tfrac{x_{i-1}p}{2^{k+t}}\left(2^{t+1} - \tfrac{x_{i-1}p}{2^k}\right)\right) \\
&= x_i\left(2 - 2\tfrac{x_{i-1}p}{2^{k+t}} + \left(\tfrac{x_{i-1}p}{2^{k+t}}\right)^2\right).
\end{aligned}
\tag{3}
$$

Thus

$$
x_{i+1} = x_i\left(1 + \left(\frac{x_{i-1}p}{2^{k+t}} - 1\right)^2\right)
$$

and then $x_i \geq 0$ for all i.

Now, because of the local truncations, one has that

$$
x_{i+1} \leq 2x_i - \frac{1}{2^t}\left(\frac{px_i}{2^k} - n - 1\right)x_i + n + 1
$$

and

$$
x_{i+1} \geq 2x_i - \frac{1}{2^t}\left(\frac{px_i}{2^k} + n + 1\right)x_i - n - 1.
$$

This means that

$$
\left|\frac{2^k}{p} - \frac{x_{i+1}}{2^t}\right| \leq \frac{2^k}{p} - \frac{2x_i}{2^t} + \frac{x_i}{2^{2t}}\left(\frac{px_i}{2^k} + n + 1\right) + \frac{n+1}{2^t}.
$$

That can be rewritten as

$$\frac{p}{2^k}\left(\left(\frac{2^k}{p}\right)^2 - 2\frac{x_i 2^k}{2^t p} + \frac{x_i^2}{2^{2t}}\right) + \frac{(n+1)}{2^t}\left(\frac{x_i}{2^t}+1\right).$$

Since $p/2^k < 1$ we have that the above relation is strictly smaller than

$$\left(\frac{2^k}{p} - \frac{x_i}{2^t}\right)^2 + \frac{n+1}{2^t}\left(\frac{x_i}{2^t}+1\right).$$

Now let us see how "big" every x_i can be. Observe that $x_i 2^{-t} < 4$ for all i's. This is because $x_{i+1} = \frac{x_i}{2^t}(2^{t+1} - \frac{x_i p}{2^k})$ and since $\frac{x_i}{2^t}, x_{i+1} \geq 0$ it has to be the case that $2^{t+1} - \frac{x_i p}{2^k} \geq 0$ which, in turn, implies that $x_i < 4 \cdot 2^t$. Thus we can conclude that

$$\left|\frac{2^k}{p} - \frac{x_{i+1}}{2^t}\right| \leq \left(\frac{2^k}{p} - \frac{x_i}{2^t}\right)^2 + \frac{n+1}{2^{t-3}}. \tag{4}$$

Now we define $\epsilon_0 = \frac{2^k}{p} - \frac{x_0}{2^t}$ and

$$\epsilon_i = \epsilon_{i-1}^2 + \frac{n+1}{2^{t-3}}.$$

Notice that $\epsilon_0 < 1/2$, moreover by imposing that $n < 2^{t-5}-1$ one has that $\epsilon_1 < 1/2$ and $\epsilon_i = 2^{2^{-i}} + \frac{n+1}{2^{t-3}}$. Thus to obtain an $\epsilon_i = \frac{n+1}{2^{t-4}}$, $i = \lceil\log(t - 3 - \log(n + 1))\rceil$ iterations suffice.

Finally note that the bound on the size of q comes from the fact that we need resort to the SQ-to-SI algorithm to properly deal with the shares v_j and z_j □

The cost of the protocol is dominated by the cost of the distributed multiplication protocol which has to be repeated $2\lceil\log(t - 3 - \log(n + 1))\rceil \approx O(\log t)$ times. Thus the cost of the protocol, in terms of bit operations is roughly $O(\log t(k^2 n + kn^2 \log n)$ bit operations per player. Its communication complexity is $O(kn \log t)$ and its round complexity is $O(\log t)$.

Remark 5. The previous theorem holds for any $t > 5 + \log(n + 1)$ but in order for the ℓ most significant bits of $1/p$ and $p'/2^{t+k}$ to be the same, this parameter should be set bigger than $\ell + 5 + \log(n + 1)$

7.3. Second Step: the Modular Reduction Protocol

Here we describe the actual modular reduction protocol. Assume the players are given polynomial shares (over $\mathbb{Z}_q$) of three integers: the modulus p (in the range $[2^{k-1}..2^k]$), the approximation of $1/p$, p' (in the range $[0..2^{t+2}]$) and a value c in the range $[-2^w..2^w]$. The proposed protocol distributively computes shares of an integer d that is an approximation of c rem p. More precisely $d = c$ rem $p + ip$ with $|i| \leq (n + 1)(1 + 2^{w+4-k-t})$.

The basic idea of the algorithm is to compute d as $c - \lceil cp'2^{-k-t}\rceil p$. Note that in order to avoid wrapping arounds for the product cp' it is important that the public

modulus q is (at least) $w + t$ bits long.
The formal protocol appears in Figure 4.

Remark 6. Notice that the $\ell \approx n$ least significant bits of c do not influence the computation of the quotient. For this reason we could eliminate these bits from c using the truncation algorithm described in Figure 2. Denoting with c' the "truncated" c, one can compute the required d as $c - \lceil c'p'2^{-k-t+\ell} \rfloor p$, which has the advantage of requiring a public modulus q of smaller size. This solution however requires a slightly more complicate analysis (more parameters have to be considered). Thus, even though reducing the size of public modulus is of primary importance for practical applications, in our context it may be preferable to describe a slightly less efficient but simpler solution.

MOD-RED Protocol

Public Parameters: A value k such that $2^{k-1} < p < 2^k$, a value w such that $-2^w < c < 2^w$, and an approximation parameter t such that $0 < p' < 2^{t+2}$. A security parameter ρ.
A prime q such that $|q| > 2^{\rho+w+2\log(n+1)+6+t}$.
Private Input (for player P_j): A polynomial share $p_j \in \mathbb{Z}_q$ of the secret modulus p. A polynomial share $p'_j \in \mathbb{Z}_q$ of an approximation of $1/p$ and a polynomial share $c_j \in \mathbb{Z}_q$ of the value c.

Player P_j does as follows:

1. Run the protocol $\texttt{MUL}(c_j, p'_j)$ Denote with z'_j the local output of player P_j.
2. Run the $\texttt{TRUNC}(z'_j, k+t)$ protocol and let z_j the local output pf player P_j.
3. Run the protocol $\texttt{MUL}(z_j, p_j)$ and denote with W_j the local output of player P_j.
4. Set the output to $d_j = c_j - W_j$.

FIGURE 4.   A protocol for distributely compute shares of c rem p

---

Theorem 3. *Assume the players are given polynomial shares (over $\mathbb{Z}_q$) of three integers $2^{k-1} < p < 2^k$, $-2^w < c < 2^w$, and $0 < p' < 2^{t+2}$. The protocol in Figure 4 securely computes shares of an integer d such that $d = c$ rem $p + ip$ where $|i| < (n+1)(2^{w-k-t+4} + 1)$, given that $\log q > 2^{\rho+w+2\log(n+1)+6+t}$.*

Proof. Here we prove that the protocol is correct. Security follows from the composability of the sub-protocols used.
First note that due to the local truncations (step 2 of the algorithm) one has that

$$\frac{cp'}{2^{k+t}} - n - 1 \le z \le \frac{cp'}{2^{k+t}} + n + 1.$$

As seen in previous section, however, $p'2^{-(k+t)}$ is only an approximation of $1/p$. This means that we can rewrite the relation above as

$$c\left(\frac{1}{p} - \frac{n+1}{2^{k+t-4}}\right) - n - 1 \leq z \leq c\left(\frac{1}{p} + \frac{n+1}{2^{k+t-4}}\right) + n + 1$$

and in particular

$$\left\lceil\frac{c}{p}\right\rceil - (n+1)\frac{c}{2^{k+t-4}} - n - 1 \leq z \leq \left\lceil\frac{c}{p}\right\rceil + (n+1)\frac{c}{2^{k+t-4}} + n + 1.$$

Moreover, since $-2^w < c < 2^w$, the above relation becomes

$$\left\lceil\frac{c}{p}\right\rceil - (n+1)(2^{w-k-t+4} + 1) \leq z \leq \left\lceil\frac{c}{p}\right\rceil + (n+1)(2^{w-k-t+4} + 1).$$

Which means that $d = c - pz = (c \text{ rem } p) + ip$ with $|i| < (n+1)(2^{w-k-t+4} + 1)$. Finally note that the bound on the size of q comes from the fact that we need resort to the SQ-to-SI algorithm to use the TRUNC algorithm. $\qquad\square$

Again the cost of the protocol is dominated by the cost of the TRUNC algorithm and by that of the MULT protocol. Since, this time, these protocols are run just one time we have that the MOD-RED protocol costs $O(k^2 n + kn^2 \log n)$ bit operations per player. The communication complexity is $O(kn)$ and the round complexity is $O(1)$.

Remark 7 (Size of the parameters).
Once we have described the algorithms to perform reductions modulo a shared integer, we are ready to discuss some applications that require computation with respect to a shared modulus. In other words we can now show how to build new protocols on top of those just described.
In order to do this properly, we need to clarify how to set the parameters of the MOD-RED and P-INVERT algorithms to make such an on going computation possible. As before, assume the players are given polynomial shares (over $\mathcal{Z}_q$) of the integers $2^{k-1} < p < 2^k$ and $0 < p' < 2^{t+2}$. If we set

$$t = \lceil k + 10 + 2\log(3(n+1)) \rceil$$

$$v = k + \log(3(n+1)) + 1$$

and

$$\log q > \rho + 2k36 + 6\log(n+1),$$

then starting with polynomial shares (over $\mathcal{Z}_q$) of an integer $-2^{2v} < c < 2^{2v}$, the players can compute shares of an integer $-2^v < d < 2^v$, by means of the MOD-RED protocol.
Moreover this means that if the players are given on input polynomial shares of $-2^v < a, b < 2^v$, they can compute shares of an integer $-2^v < d' < 2^v$ obtained as $a \cdot b \text{ rem } p$. Thus such a d' can later be used as input for further distributed computation modulo the shared p.

8. Exponentiation with a Shared Exponent

In this section we describe some useful applications of the protocols described in the previous section.

Our first application is a distributed version of the Square and Multiply algorithm. In a nutshell the Square and Multiply algorithm allows to efficiently compute $a^b \bmod p$. In particular it requires at most 2ℓ modular multiplications, where ℓ is the number of bits in the binary representation of b. The method assumes that the exponent is represented in binary notation and exploits the fact that

$$a^b = a^{2^{\ell-1}b_{\ell-1}} \cdots a^{b_0}.$$

The algorithm is presented in Figure 5

Square and Multiply Algorithm

1. $z \leftarrow 1$
2. **For** $i = \ell - 1$ down to 0 **do**
3. $\quad z \leftarrow z^2 \bmod p$
4. $\quad$ if $b_i = 1$, then $z \leftarrow z \cdot a \bmod p$

FIGURE 5. The basic (non distributed) Square and Multiply algorithm to compute $c = a^b \bmod p$

Now assume that the players want to compute shares of $c = a^b \bmod p$ when a, b, p, p' are shared secrets and p' is the usual approximation of $1/p$. Thus we need to build a distributed version of the Square and Multiply method discussed above. More specifically this means that we need to be able to efficiently "distribute" the operations in steps 3 and 4 of the algorithm in Figure 5. Computing $z^2 \bmod p$ is rather straightforward: it requires an execution of the MUL protocol and then an execution of the distributed modular reduction protocol MOD-RED. Implementing step 4 requires some thinking.

The problem here is that we need to implement an **if** condition on a secret value. This means that the players should be able to determine the actual value of the bits b_i's *without* revealing any information about these bits. We realize this as follows. First note that $a^{b_i} = (a-1)b_i + 1$; then, with this formula in mind, the step 4 in the Square and Multiply algorithm can be rewritten as $z \leftarrow z \cdot ((a-1)b_i+1) \bmod p$ and it can be easily implemented by resorting, once again, to one execution of the MUL protocol followed by an execution of the MOD-RED protocol.

The full details of the algorithm appear in Figure 6.

As per the bit complexity of the protocol, its cost is roughly that of $3k$ executions of the multiplication protocol and $2k$ distributed modular reductions. This leads to a total cost of $O(k^3n+k^2n^2 \log n)$ bit operations per player. The total communication complexity is about $O(nk^2)$ and it requires $O(k)$ rounds of communication among the parties.

Distr-Sq-Mult Protocol

Public Parameters: A value k such that $2^{k-1} < p, b < 2^k$, a value v such that $-2^v < c < 2^v$, a security parameter ρ.
A prime q of size as described in remark 7.
Private Input (for player P_j): A polynomial share $p_j \in \mathcal{Z}_q$ of the secret modulus p. A polynomial share $p'_j \in \mathcal{Z}_q$ of an approximation of $1/p$, a polynomial share $a_j \in \mathcal{Z}_q$ of the value a and shares $(b_i)_j$ for the bits of b.

Player P_j does as follows:

1. Run the protocol MUL($a_j - 1$ rem q, $(b_k)_j$). Denote with $(z_k)_j$ the local output of player P_j.
2. Set $(c_k)_j = (z_k)_j + 1$ rem q.
3. **For** $i = k - 1$ **to** 1 **do**
   (a) Run the protocol MUL($a_j - 1$ rem q, $(b_i)_j$). Denote with $(z_i)_j$ the local output of player P_j.
   (b) Set $(d_i)_j = (z_i)_j + 1$ rem q.
   (c) Run the protocol MUL($(c_{i+1})_j, (c_{i+1})_j$).
       Denote with $(z_{i+i})_j$ the local output of player P_j.
   (d) Reduce z_{i+i} modulo p by invoking the protocol MOD-RED($(z_{i+i})_j, p_j, p'_j$).
       Denote with $(z_{i+i}$ rem $p)_j$ the local output of player P_j
   (e) Run the protocol MUL($(z_{i+i}$ rem $p)_j, (d_i)_j$) and let $(\alpha_i)_j$ be the local output of player P_j.
   (f) Run the protocol MOD-RED($(\alpha_i)_j, p_j, p'_j$) and set $(c_i)_j$ the local output for player P_j.

4. Output $c_j = (c_1)_j$.

FIGURE 6. A distributed version of the Square and Multiply algorithm to compute $c = a^b \bmod p$

8.1. Set Membership

In this section we discuss a simple protocol that uses the distributed modular reduction algorithm as a subroutine to solve the so-called Set Membership problem. Assume that a set of n players wants to establish whether a shared value a belongs to a set of (shared) integers $b_1, \ldots, b_m$. A simple strategy to solve this problem is to check if there is a b_j for which $a \equiv b_j \bmod p$ holds. To perform this check in a distributed way one may simply compute (for each b_j) the value $a - b_j \bmod p$, multiply it with a jointly generated random element and check if the obtained result is zero or not.

Unfortunately, however, this solution does not quite solve the problem in our setting. Indeed the modular reduction protocol we have can only compute an approximation of the actual $a - b_j \bmod p$ (i.e. a value that is off by some small multiple i of p from the actual solution).

However since i is less (in absolute value) than $3n$ we can distributely compute $A = \prod_{j=-3(n+1)}^{j \leq 3(n+1)} a - b_j - jp \bmod p$ and then check if A is zero or not. Note that this holds also modulo q if q is sufficiently bigger than p.

We are not done however. The ideas described so far allow to test if a shared a is equivalent to a shared b_j modulo a shared p in a secure way. As a consequence one may think of computing several A_i's as before (one for each b_i to be tested) and then check which one is zero. This solution however would release some additional information (in addition to the set membership) and in particular to which b_j, a is equal to (one could, for example, learn that a is equal to the, say, third element in the set). To overcome this problem we can further multiply the A_i's with each other and test if the resulting product is zero or not.

The complete protocol for the set membership problem is presented in Figure 7 and assumes that a and the b_i's are all bounded in absolute value by 2^v.

SET-MEM Protocol

Public Parameters: A value k such that $2^{k-1} < p < 2^k$, a value v such that $-2^v < a, b_i < 2^v$, for $i = 1, \dots, m$. A security parameter ρ.
A prime q of size as described in remark 7.

Private Input (for player P_j): A polynomial share $p_j \in \mathcal{Z}_q$ of the secret modulus p. A polynomial share $p'_j \in \mathcal{Z}_q$ of an approximation of $1/p$. A polynomial share $a_j \in \mathcal{Z}_q$ of the value a and shares $(b_i)_j$ of the values b_i's.

Player P_j does as follows:

1. **For** $i = 1$ to m **do**
   Run the protocol MOD-RED$(a_j - (b_i)_j$ rem $q, p_j, p'_j)$.
   Denote with $(c_i)_j$ the local output of player P_j.
2. **For** $i = 1$ to m **do**
   (a) Set $A_{(-3(n+1),i)} = (c_i)_j - 3(n+1)p_j$ rem q.
   (b) **For** $\ell = -3(n+1) + 1$ to $3(n+1)$ **do**
      Run the protocol MUL$(A_{(\ell-1,i)}, (c_i)_j + \ell p_j$ rem $q)$.
      Denote with $(A_{(\ell,i)})_j$ the local output of player P_j.
3. Set $(B_1)_j = (A_{(3(n+1),1)})_j$.
4. **For** $i = 2$ to m **do**
   Run the protocol MUL$((B_{i-1})_j, (A_{(3(n+1),i)})_j$.
   Denote with $(B_i)_j$ the local output of player P_j.
5. Run the protocol JRP-INV to generate shares r_j of a random invertible element r.
6. Run the protocol MUL$((B_m)_j, r_j)$ and denote with z_j the share obtained by player P_j.
7. Publish z_j and using the values disclosed by the other players interpolate z. Output YES if $z \equiv 0$ rem q and NO otherwise.

FIGURE 7.   A distributed protocol to test if a belongs to the set $b_1, \dots, b_m$

The security of the protocol easily follows from the secure composability of the sub-protocols used. Furthermore no information about the shared inputs is disclosed when z is reconstructed because z is either zero or a completely random value.

The protocol requires $O(mn(nk^2 + kn^2 \log n))$ bit operations per player and $O(k + n)$ rounds of communication. The communication complexity is bounded by $O(mn^2k)$.

9. Generating Shared Random Primes

In this section we show how to generate a shared prime and a shared *safe* prime [3]. Our approach consists, essentially, in showing how to use the protocols presented so far to implement a distributed version of the Miller-Rabin algorithm [42, 47] on a candidate random secret, jointly chosen by the players.

We proceed step by step: first we present and discuss the basic (i.e. non distributed) Miller-Rabin test, then we show how to efficiently generate a shared candidate prime and finally we present a distributed version of the Miller-Rabin method.

9.1. The Basic Miller-Rabin Algorithm

We begin with a brief description of the Miller-Rabin algorithm to test if a given integer p is a prime (the pseudo-code appears on Figure 8). The Miller-Rabin test is a probabilistic algorithm that takes on input a candidate prime p and returns as output "yes" if it "thinks" that p is prime and "no" otherwise. If the algorithm answers "no", then this answer is always correct. On the other hand if the algorithm's output is "yes" this answer is correct only with probability $1/4$ (see [47] for a proof of this fact). This means that if we run the test ω times – on some candidate integer p – and the test always outputs "yes", then p is actually prime with probability $1 - (1/4)^\omega$.

The algorithm is based on the following basic idea. Fermat's Little Theorem states that if p is a prime and $a \in \mathbb{Z}_p^*$, then $a^{p-1} \equiv 1 \bmod p$. Thus one may think of using this fact the other way round as a possible way to test if a given number is prime. In particular one can choose a random a and test if $a^{p-1} \equiv 1 \bmod p$ holds. Unfortunately this strategy does not work, because there are composites p (known as *Carmichael numbers*) for which $a^{p-1} \equiv 1 \bmod p$ for all $a \in \mathbb{Z}_p^*$. The Miller-Rabin test overcomes this difficulty by choosing several random a's in $\mathbb{Z}_p^*$ for which a^{p-1} is computed via repeated squarings. After each exponentiation the algorithm checks if the obtained power of a is a non trivial square root of 1 (i.e. a root of 1 that is not congruent to $\pm 1 \bmod p$). If this is the case, then p has to be a composite. The quality of the test depends on a theorem that Rabin proved in [47]. The reader is referred to that paper for further details.

---

[3] Recall that a prime p is said to be safe if it is of the form $p = 2p' + 1$ where p' is a prime number itself. Safe primes are very useful objects in cryptography.

Miller-Rabin Primality Test

1. Let $p - 1 = 2^\ell m$ (m odd).
2. Choose a random integer a such that $1 \leq a \leq p - 1$.
3. Compute $b = a^m \bmod p$.
4. **if** $b \equiv 1 \bmod p$, **then**
   Answer *yes* and quit.
5. **For** $i = 0$ to $\ell - 1$ **do**
   if $b \equiv -1 \bmod p$, **then**
   Answer *yes* and quit.
   else $b = b^2 \bmod p$.
6. Answer *no*.

FIGURE 8.   The basic (non distributed) Miller-Rabin primality test for an odd integer p

---

9.2. Generation of a Shared Candidate Prime

In this section we will discuss a very elegant method, originally proposed by Boneh and Franklin [7], to efficiently generate a shared candidate prime of some size k. Every participant, but the first one, chooses a random $(k - \log n - 1)$-bit integer p_i such that $p_i \equiv 0 \bmod 4$. The first player, on the other hand, chooses a random $(k - \log n - 1)$-bit integer p_1' such that $p_1' \equiv 3 \bmod 4$ and sets $p_1 = 2^{k-1} + p_1'$.
In this way the players have an additive sharing (over the integers) of the candidate $p = \sum_i p_i$, which is clearly a k bit integer.
We point out that the original Boneh-Franklin technique does not require $p \equiv 3 \bmod 4$ as we are doing here. However, as we will see in Section 9.3, this restriction allows for a more efficient variant of the Miller-Rabin test.
Once the candidate p is shared the players engage in a secure distributed protocol to determine if p is divisible by any prime less than some (publicly known) bound B. This trial division protocol can be easily implemented as described in Figure 9

Remark 8. Observe that the method described in Figure 9 does not work correctly if e is smaller than n. This is because in this case $\mathcal{Z}_e$ is too small and there are not enough points to do a polynomial secret sharing among n players.
For such small e's one must resort to an extension field $\mathbb{F}_e$ that contains at least $n + 1$ points. See [7] for more details about this.

Since there are (approximately) $B/\log B$ primes in the interval $\{1, \ldots, B\}$ the proposed protocols costs ($\frac{B}{\log B}(n^2 \log B + n(\log B)^2)$) in terms of bit operations. Furthermore it requires $O(1)$ rounds and its bit complexity is $O(Bk)$.

9.3. Distributed Miller-Rabin Primality Test

Now we are ready to describe a distributed version of the Miller-Rabin algorithm. First notice that if p is of the form $p \equiv 3 \bmod 4$ it can be written as $p = 2\omega + 1$ where ω is odd (all the primes of this form are known as *Blum primes*). For such

---

Trial Division Protocol

Public Parameters: A bound B on the small prime divisors to test.
Private Input (for player P_j): An additive share p_j, over the integers, of a secret p.
Player P_j does as follows:
For each prime e smaller than B **do**

1. Re-share p_j rem e using polynomial sharing over $\mathcal{Z}_e$.
2. Sum all the received shares to get a share p'_j of p rem e over $\mathcal{Z}_e$.
3. Run the protocol JRP-INV (over $\mathcal{Z}_e$) to generate shares r_j of a random invertible element r.
4. Run the protocol MUL(r_j, p'_j) (over $\mathcal{Z}_e$) and denote with z_j the local output of player P_j.
5. Publish z_j and using the value disclosed by the other players interpolate z. If $z \equiv 0$ rem e, then e divides p.

FIGURE 9.   A simple protocol to check if a shared p is divisible by all small primes less than some bound B

integers the Miller-Rabin test reduces to choosing a random base a and checking if $a^{(p-1)/2} \equiv \pm 1 \bmod p$.

A technical problem arises from the fact that, since the players *don't know* the value of p, they cannot choose a uniformly and at random in $\mathcal{Z}_p$. To overcome this difficulty we allow the players to choose a in a large enough interval (say $\{0,1\}^{2k}$ where $p < 2^k$). The intuition underlying this solution is that if the interval where a is sampled is sufficiently larger than p, then $a \bmod p$ has a distribution that is statistically close to uniform.

The detailed protocol appears in Figure 10.

The cost of the Distributed Miller-Rabin test is dominated by the cost of the ADD-To-BIN protocol and the cost of τ executions of the Distr-Sq-Mult protocol. This leads to $O(kn^3 \log n\gamma + n^2k\gamma^2 + n^2k^2 \log n + \tau(nk^3 + n^2k^2 \log n))$ bit operations per player, where γ is the size of the small prime used in the ADD-To-BIN conversion protocol.

Its communication complexity is $O(k^2 n\tau)$ and it requires $O(k + \log n)$ rounds.

9.4. Generation of Shared Random Safe Primes

We conclude this part by presenting a method to distributely generate a shared (random) safe prime. It should be pointed out here, that not very much has been proved about the density of these primes. In particular we don't even know if there are infinitely many safe primes. However it is widely conjectured that safe primes are sufficiently "dense" and this conjecture is supported by empirical evidence.

In order to generate such primes one can use the following protocol (which is a distributed variant of the single-player, safe-prime generation procedure, proposed by Cramer and Shoup in [14]).

Distributed Miller-Rabin Protocol

Public Parameters: A parameter k such that $p < 2^k$.
An approximation parameter τ. The usual prime q.
Private Input (for player P_j): An additive share p_j, over the integers, of a candidate prime p (obtained as described in Section 9.2).
Player P_j does as follows:

1. **If** $j \geq 2$ set $b_j = p_j/2$
2. **else** set $b_j = (p_j - 1)/2$.
3. Run the `ADD-To-BIN` protocol to obtain shares of the bits of b. Denote with $((b_1)_j, \ldots, (b_k)_j)$ the local output for player P_j.
4. Convert the additive shares of p into polynomial shares over $\mathcal{Z}_q$ (using the methods described in Section 6).
   Denote with $\hat{p}_j$ the local output for player P_j.
5. Run the `P-INVERT` protocol to produce shares of an approximation of $1/p$ and denote with p'_j the local share held by player P_j.
6. **Repeat** τ times (in parallel).
   (a) Choose r_j uniformly and at random in $\{0,1\}^{2k}$ (this implicitly defines $r = \sum_i r_i$ over the integers).
   (b) Convert the additive shares (of r) into polynomial shares over $\mathcal{Z}_q$ and let $\hat{r}_j$ the local output of player P_j.
   (c) Run the `MOD-RED` protocol on local input $\hat{r}_j$, $\hat{p}_j$ and p'_j. We denote with a_j the local output produced by the protocol.
   (d) Run the protocol `Distr-Sq-Mult` on input $(a_j, ((b_1)_j, \ldots, (b_k)_j), \hat{p}_j, p'_j)$. Let z_j be the local output for player P_j.
   (e) Run the protocol `SET-MEM` on input $(z_j, \{-1, 1\}, p_j, p'_j)$. If it outputs `NO`, output `NO`.
7. Output `YES`.

FIGURE 10.   A distributed version of the Miller-Rabin primality test.

First the players choose a random candidate p' as described in Section 9.2. Then player P_1 sets $p_1 = 2p'_1 + 1$ as his additive share for the candidate safe prime p. The remaining players set $p_j = 2p_j$.

The players run the `Trial Division` protocol on both p and p'. If this step fails they start over with a new candidate.

If, on the contrary, the trial division test is successfully passed, the players run the Distributed Miller-Rabin protocol on input p' with approximation parameter τ. Then, if the test indicates that p' is prime the players run the Distributed Miller-Rabin protocol on input p with approximation parameter τ. If the test succeeds the players accept p as a safe prime.

10. Efficient Generation of Shared RSA Keys

Using the algorithms described so far, we can easily generate (in a distributed way) a composite integer obtained as the product of two (standard or of some special form) primes. In other words we can use the protocols described in the previous sections to efficiently generate a shared RSA modulus for which none of the players knows the factorization.

In many situations, however, the parties are required to efficiently generate not only the modulus but also shares of the private exponent d. Of course one can still combine the previously described methods to obtain a protocol for this task as well. In the following, however, we decided to discuss a completely differ-ent approach to solve the problem. Specifically we describe a simple and efficient algorithm to compute polynomial shares of the private exponent d, starting from a public exponent e and shares of $\phi(N)$. More generally this algorithm can be used to compute the inverse of a public value modulo a shared integer (assuming, of course, that the greatest common divisor of the two integers is 1). In this sense it can be seen as a "dual" algorithm with respect to that discussed in Section 3 by Bar-Ilan and Beaver [4].

It must be noticed that an algorithm for the same problem was already proposed by Boneh and Franklin in [7]. The protocol we are going to present, however, improves on some of the features of the Boneh Franklin solution (see [11] for a discussion about this).

Remark 9. Note that in presenting the inversion protocol we go back to the stan-dard notations for modular arithmetic (see Section 5). In particular we go back to the symbol 'mod' to denote the operator for modular reduction.

11. Computing Inverses over a Shared Modulus

11.1. The Basic Idea

We start by presenting a very simple protocol which, although doesn't quite solve our problem, is rather useful for illustrating the ideas underlying the complete solution.

For this protocol, we assume that the players hold additive shares (over the integers) of some multiple of the secret modulus ϕ. This means that each player P_i has a share α_i such that $\sum_i \alpha_i = \lambda\phi$, where λ is some random integer, much larger than ϕ (say, of order $O(N^2)$). The protocol goes as follows. Each player P_i chooses a "randomizing integer" $r_i \in_R [0..N^3]$, and publishes the value $\gamma_i = \alpha_i + r_i e$. Using this public data all the players compute $\gamma = \sum_i \gamma_i$. Clearly,

$$\gamma = \sum_i \gamma_i = \sum_i \alpha_i + r_i e = \lambda\phi + Re$$

(where $R = \sum_i r_i$). If one assumes that $GCD(\gamma, e) = 1$, then there exist a, b such that $a\gamma + be = 1$ and thus $d = aR + b = e^{-1} \mod \phi$. At this point additive shares of d can be easily obtained by having player P_1 set $d_1 = ar_1 + b$, and the other players set $d_i = ar_i$. Obviously $d = \sum_i d_i$.

Note that the only information leaked by the protocol is the integer $\gamma = \lambda\phi + Re$. However it is possible to prove (and we do that for the general protocol in the next section) that the distribution of γ is (almost) independent of ϕ. More precisely, it can be shown that, when λ and R follow the probability distribution described above, then the distributions $\{\gamma = \lambda\phi + Re\}$ and $\{\gamma' = \lambda N + Re\}$ are statistically close.

The above protocol, however, is not secure when it is used more than once with the same λ and different e's. Indeed, for each input e, the protocol leaks the value $\lambda\phi \bmod e$, and so after sufficiently many runs with different e's we can then recover the integer $\lambda\phi$ via the Chinese Remainder Theorem (see for example [40] for details about this theorem). To overcome this, it is necessary to use a "fresh" λ for each input e. In the next section we show how to do this, and at the same time get a t-out-of-n threshold solution (but still in the "honest but curious" model).

Note that, for the case of RSA key generation, having a protocol that is secure only if used once can be perfectly fine. After all, once N is computed, only one inverse modulo $\phi(N)$ has to be computed to obtain shares of the private exponent. For the sake of completeness, however, we prefer to present here the full protocol (i.e. the one secure even when used more than once with the same secret $\phi(N)$).

11.2. The Full Protocol

The protocol in this section achieves a t-out-of-n sharing. However the most important difference between this solution and the one given in the previous section is that all the secrets are shared via polynomials over the integers (rather than sums), and the multiple λ is chosen afresh with each new execution. The rest of the protocol is similar to the basic case. The protocol is described in detail in Figure 11. On a high-level description, it goes as follows:

- Each player starts by holding as input a share of the secret modulus ϕ (multiplied by a factor of $L = n!$ for technical reasons, as discussed in 3.4), via a t-degree polynomial $f(z)$ with free term $L\phi$.
- In the first round of the protocol, the players jointly generate two random t-degree polynomials $g(z)$ and $h(z)$ with free terms $L\lambda$ and LR, respectively, and a random $2t$-degree polynomial $\rho(z)$ with free term 0.
- In the second round they reconstruct the $2t$-degree polynomial $F(z)=f(z)g(z) + e \cdot h(z) + \rho(z)$ and recover its free term $\gamma = F(0) = L^2\lambda\phi + LRe$.
- Finally, they use the GCD algorithm to (publicly) compute a, b such that $a\gamma + be = 1$ and set $d = aLR + b = e^{-1} \bmod \phi$. To conclude the protocol, each player P_i computes its share of d by setting $d_i = ah(i) + b$.

Theorem 4. *If all the players carry out the prescribed protocol and $n > 2t$, then the protocol in Figure 11 is a secure Modular Inversion Protocol according to the Definition given in Section 2.2.*

Inversion Protocol

Private inputs: Sharing of $L\phi$ using a t-degree polynomial over the integers.
   Player P_i has private input $f_i = f(i)$, where $f(z) = L\phi + a_1 z + \ldots + a_t z^t$,
   and $\forall j, a_j \in [-L^2 N .. L^2 N]$.

Public input: prime number $e > n$, an approximate bound N on ϕ.

[Round 1] Each player P_i does the following:

1. Choose $\lambda_i \in_R [0..N^2]$ and $b_{i,1}, \ldots, b_{i,t} \in_R [-L^2 N^3 .. L^2 N^3]$.
   Choose $r_i \in_R [0..N^3]$ and $c_{i,1}, \ldots, c_{i,t} \in_R [-L^2 N^4 .. L^2 N^4]$.
   Choose $\rho_{i,1}, \ldots, \rho_{i,2t} \in_R [-L^2 N^5 .. L^2 N^5]$.
2. Set $g_i(z) = L\lambda_i + b_{i,1} z + \ldots + b_{i,t} z^t$, $h_i(z) = Lr_i + c_{i,1} z + \ldots + c_{i,t} z^t$, and
   $\rho_i(z) = 0 + \rho_{i,1} z + \ldots + \rho_{i,2t} z^{2t}$.
3. Send to each player P_j the values $g_i(j), h_i(j), \rho_i(j)$, computed over the
   integers.

[Round 2] Each player P_j does the following:

1. Set $g_j = \sum_{i=1}^{n} g_i(j)$, $h_j = \sum_{i=1}^{n} h_i(j)$, and $\rho_j = \sum_{i=1}^{n} \rho_i(j)$.
   (These are its shares of the polynomials $g(z) = \sum_i g_i(z)$, $h(z) = \sum_i h_i(z)$, and $\rho(z) = \sum_i \rho_i(z)$.)
2. Broadcast the value $F_j = f_j g_j + e h_j + \rho_j$

[Output] Each player P_i does the following:

1. From the broadcast values interpolate the $2t$-degree polynomial $F(z) = f(z)g(z) + e \cdot h(z) + \rho(z)$.
2. Using the GCD algorithm, find a, b such that $aF(0) + be = 1$. If no such
   a, b exist, go to Round 1.
3. The inverse of e is $d = ah(0) + b$. Privately output the share of the inverse,
   $d_i = ah(i) + b$.

FIGURE 11.   A protocol to compute inverses over a
shared modulus

Sketch of Proof   In this proof we assume that $N - \phi = O(\sqrt{N})$ (which is true
for the case we are interested in, where N is an RSA modulus and $\phi = \phi(N)$). In
the more general case where we can bound $N - \phi$ with $O(N)$, the bounds in the
proof have to be adjusted slightly.

INITIAL INPUTS.   First we show that the distribution of t shares of the secret ϕ
with polynomial $f(z)$ is statistically indistinguishable from the distribution of t
shares that result from sharing the value N via the polynomial $\hat{f}(z)$. Intuitively,
this allows us to show that t players have no information about the shared secret
$\phi(N)$.

   We prove this fact by showing that, with very high probability, there is a
sharing of N with a polynomial $\hat{f}$, having integer coefficients in the same range as
f, such that $\hat{f}(i) = f(i)$ for $i = 1, \ldots, t$. Let $h(z)$ be a t-degree polynomial such

that $h(0) = (\phi - N)L$ and $h(1) = \ldots = h(t) = 0$. Formally this means that,

$$h(z) = \sum_{i=0}^{t} h(i) \prod_{j\neq i, j=0,\ldots,t} \frac{z-j}{i-j} = L(\phi - N) \prod_{j=1,\ldots,t} \frac{z-j}{-j}$$

and the coefficient of z^i is

$$L(\phi - N) \sum_{B\subseteq\{1,\ldots,t\}, |B|=i} \frac{\prod_{j\in B}(-j)}{\prod_{j=1\ldots,t}(-j)}.$$

Since $L = n!$ the value above is an integer. Furthermore it can be bounded – in absolute value – by

$$\sum_{B\subseteq\{1,\ldots,t\}, |B|=i} L(\phi - N) \leq (\phi - N)L \binom{t}{i} \leq \frac{(\phi - N)Lt!}{i!(t-i)!} \leq (\phi - N)Lt! \leq 3L^2\sqrt{N}.$$

The desired polynomial is then $\hat{f}(z) = f(z) - h(z)$ and clearly $\hat{f}(0) = LN$. Moreover the coefficient of this polynomial are integers in the range $[-L^2N - 3L^2\sqrt{N}..L^2N + 3L^2\sqrt{N}]$, thus the probability that the coefficients are outside the legal range is

$$t\frac{6L^2\sqrt{N}}{2(L^2N + 3L^2\sqrt{N})} \leq O(\frac{t}{\sqrt{N}})$$

which is negligible.

CORRECTNESS. It is easy to see that the protocol is correct. As a matter of fact, since all players are honest, the interpolation at step 2 of the last round, will give as the unique polynomial $F(z)$ a polynomial with integer coefficients. Thus $F(0) = L^2\lambda\phi + LRe$ is an integer and we can compute its GCD with respect to e. If e does not divide ϕ, the probability that $GCD(e, F(0)) = 1$ is roughly $1/e$ (i.e. actually this is the probability that e divides λ).

Thus, for sufficiently big e, it is very unlikely that the protocol has to be repeated more than once. Once we obtain $aF(0) + be = 1$, it can be re-written as

$$a(L^2\lambda\phi + LRe) + be = 1.$$

That becomes $(aLR + b)e = 1 \bmod \phi$ when reduced mod ϕ. This means that we have $d = aLR + b = e^{-1} \bmod \phi$. Thus the t-degree polynomial $ag(z) + b$ interpolates to the correct value d and the shares d_i correctly lie on such polynomial. Notice that in order to interpolate $F(z)$ we need the shares of at least $2t + 1$ players.

SIMULATION OF THE INVERSION PROTOCOL. Without loss of generality assume that the simulator controls players $P_{t+1}, \ldots, P_n$. For these players it holds initial values $\hat{f}_i$ which comes from a sharing of N (instead of ϕ, as discussed before).

For Round 1 the simulator simply follows the same instructions as the protocol. This produces shared polynomials $\hat{h}(z)$, $\hat{g}(z)$ and $\hat{\rho}(z)$ and shared values $\hat{\lambda} = \hat{g}(0)$ and $\hat{R} = \hat{h}(0)$. Clearly $\hat{\lambda}$ and $\hat{R}$ follow the same distribution as λ, R.

Moreover notice that, using an argument very similar to the one used for the sharing of the initial values, it is possible to prove that the adversary has no information about $\hat{\lambda}$ and $\hat{R}$.

During Round 2 the simulator publishes the values $\hat{F}(i) = \hat{f}(i)\hat{g}(i) + e\hat{h}(i) + \hat{\rho}(i)$ (for $i = t+1, \ldots, n$). Because the polynomials $\rho(z)$ and $\hat{\rho}(z)$ have coefficients which are much larger than $f(z)g(z)$ and $\hat{f}(z)\hat{g}(z)$, both polynomials $F(z)$ and $\hat{F}(z)$ follow a distribution which is statistically close to $\rho(z)$, except for the free term.

Indeed the $2t$-degree polynomial $\hat{F}(z)$ interpolating those values has free term equal to $L^2\hat{\lambda}N + L\hat{R}e$ (while in the real execution it interpolates to $L^2\lambda\phi + LRe$.) This is the only difference between the simulated and the real execution.

It is then sufficient to prove that the distributions of these two values are statistically close. We do that with the following lemma. □

11.3. A Fundamental Lemma

Let $\lambda = \lambda_1 + \ldots + \lambda_n$ where each λ_i is an integer chosen uniformly at random in the interval $[0..N^2]$. Let us denote with $\sum_n[N^2]$ the probability distribution of λ (i.e. the sum of n independent random variables uniformly distributed in $[0..N^2]$). Similarly let R be distributed according to $\sum_n[N^3]$. Finally let N be a bound on ϕ (here too for simplicity we assume $N - \phi = O(\sqrt{N})$) and e a prime number, relatively prime with ϕ. We assume that e is at most $O(N)$.

Lemma 1. *Let $\lambda, \hat{\lambda}$ distributed according to $\sum_n[N^2]$. Let $R, \hat{R}$ distributed according to $\sum_n[N^3]$. Consider the random variables $X_\phi = \lambda\phi + Re$ and $X_N = \hat{\lambda}N + \hat{R}e$. Then X_ϕ and X_N are statistically indistinguishable, namely*

$$\sum_x |Prob[X_\phi = x] - Prob[X_N = x]| < N^{-c}$$

for some constant $c > 0$.

Remark 10. The proof of this lemma is quite technical and it is just sketched in these notes. The interested reader is referred to the full version of [11] for a complete and detailed proof.

Sketch of Proof   We begin the proof by proving the following fact.

Proposition 1. *Let x, y be two integers such that $GCD(x, y) = 1$ and A, B two integers such that $A < B$, $x, y < A$ and $B > Ax$. Then every integer in the closed interval $[xy - x - y + 1..Ax + By - xy + x + y - 1]$ can be written as $ax + by$ where $a \in [0..A]$ and $b \in [0..B]$.*

Proof. (Proposition 1.) It is a well known fact from the theory of integer programming that any integer larger than $xy - x - y$ can be written as $ax + by$ where a and b are non-negative integers (this is a special instance of the Frobenius problem, see [50] for example).

Clearly if $z = ax + by$ with $a \in [0..A], b \in [0..B]$, then $z \in [0..Ax + By]$. We will call an integer $z \in [0..Ax + By]$ *reachable* if can be written as $z = ax + by$ with $a \in [0..A]$ and $b \in [0..B]$.

Note that the interval $[0..Ax+By]$ is symmetric. I.e. if $z \in [0..(Ax+By)/2]$ is reachable, then $z' = Ax + By - z$ is also reachable. Thus to prove the Proposition it will be sufficient to prove that any $z \in [xy - x - y + 1..By]$ is reachable (since $By > (Ax + By)/2$).

Fix $z \in [xy - x - y + 1..By]$. Consider the equation with unknown a

$$z - ax = 0 \bmod y$$

since $GCD(x, y) = 1$ there exists an unique solution $a = zx^{-1} \bmod y$. Notice that $0 \le a < y < A$. Then $z - ax = by$ and $b \le B$ (since $z \le By$).

To prove that $b \ge 0$ let us consider

$$b = \frac{z - ax}{y} \ge x - 1 - \frac{x(a+1) - 1}{y} \ge x - 1 - \frac{xy - 1}{y} \ge \frac{1}{y} - 1.$$

Note that the quantity $(1/y) - 1$ is strictly greater than -1, thus, being b an integer, $b \ge 0$.
This completes the proof.   □

Consider now the sets

$$L = \{\lambda\phi + Re \mid \lambda \in [0..nN^2], \ R \in [0..nN^3]\}$$

and

$$\hat{L} = \{\hat{\lambda}N + \hat{R}e \mid \hat{\lambda} \in [0..nN^2], \ \hat{R} \in [0..nN^3]\}.$$

A consequence of Proposition 1 is that we can bound the intersection of L and $\hat{L}$ as the interval $[\delta..\Delta]$ where $\delta = Ne - e + 1$ and $\Delta = n(N^2\phi + N^3e) - \phi e + \phi + e - 1$.

It is very easy to see (by Chernoff's bounds) that the probability that X_ϕ or X_N fall outside the interval $[\delta..\Delta]$ is negligible since both bounds are very far away from the means of X_ϕ and X_N.

Let ϵ be a negligible quantity upper bounding all the following probabilities: $Prob[X_\phi < \delta], Prob[X_\phi > \Delta], Prob[X_N < \delta], Prob[X_N > \Delta]$. Then we have that

$$\sum_x |Prob[X_\phi = x] - Prob[X_N = x]| < 4\epsilon + \sum_{x=\delta}^{\Delta} |Prob[X_\phi = x] - Prob[X_N = x]|$$

so we can focus on the last term.

Let $x \in [\delta..\Delta]$. Given a pair λ, R such that $x = \lambda\phi + Re$ we present a mapping that produces $\hat{\lambda}, \hat{R}$ such that $x = \hat{\lambda}N + \hat{R}e$. That is

$$\lambda\phi - \hat{\lambda}N = (\hat{R} - R)e.$$

Since $GCD(N, e) = 1$, for any given λ there exists a unique $\hat{\lambda} \in [\lambda..\lambda + e - 1]$ such that $\lambda\phi - \hat{\lambda}N$ is a multiple of e. Once fixed this $\hat{\lambda}$ one can then solve for $\hat{R}$.

We are not done however. We need to prove that the probability weight of the pair $\hat{\lambda}, \hat{R}$ is very close to that of the pair λ, R. This is true because the points

$\lambda, \hat{\lambda}$ and $R, \hat{R}$ "close enough" relatively to the size of the interval they were chosen from. Indeed

$$\frac{|\lambda - \hat{\lambda}|}{nN^2} \le \frac{e}{nN^2} \le \frac{1}{nN},$$

also

$$|R - \hat{R}| = \frac{|\lambda\phi - \hat{\lambda}N|}{e} = \left| \frac{(\lambda - \hat{\lambda})\phi}{e} + \frac{\hat{\lambda}(\phi - N)}{e} \right|.$$

$$\le \left| \phi + \frac{nN^2\sqrt{N}}{e} \right| \le nN^2\sqrt{N}.$$

So

$$\frac{|R - \hat{R}|}{nN^3} \le \frac{1}{\sqrt{N}}$$

which again is negligible.   □

Remark 11 (Size of shares). Note that the shares d_i of $d = e^{-1}$ mod ϕ have order $O(N^5)$. However, the shares do *not* have to be this large. We chose these bounds to make the presentation and the proof simpler. It is possible to improve (a lot) on those bounds as discussed in [11]

Acknowledgments

Some of the results presented in these notes are from a paper [11] I co-authored with Rosario Gennaro and Shai Halevi. I would like to thank Rosario and Shai for this. Moreover I should thank Rosario for all the endless discussion we had, over the years, on subjects related to those described here. He also kindly explained me how to actually do the share-conversion I only sketched in Section 6.3.

In addition I wish to thank Joy Algesheimer for having answered my questions about her paper [1] and David Pointcheval for numerous discussions and many helpful comments about these notes. I would also like to thank Javier Herranz Sotoca, Germán Sáez and Pierre-Alain Fouque for having kindly read these notes and having carefully pointed out a number of typos and mistakes. Finally I thank Marc Heymann for having pointed out a mistake in Section 5.

References

[1] J. Algesheimer, J. Camenish and V. Shoup. Efficient Computation Modulo a Shared Secret with Applications to the Generation of Shared Safe Prime Products. In *Advances in Cryptology – Crypto '02*, LNCS vol. 2442, Springer, 2002, pages 417-432.

[2] G. Ateniese, J. Camenish, M. Joye and G. Tsudik. A practical and provably secure coalition resistant group signature scheme. In *Advances in Cryptology – Crypto '00*, LNCS vol. 1880, Springer, 2000, pages 255-270.

[3] N. Barić, and B. Pfitzmann. Collision-free accumulators and Fail-stop signature schemes without trees. In *Advances in Cryptology – Eurocrypt '97*, LNCS vol. 1233, Springer, 1997, pages 480-494.

[4] J. Bar-Ilan and D. Beaver. Non cryptographic fault tolerant computing in a constant number of rounds of iteration. In *Proceedings of the ACM Symposium on Principles of Distributed Computation*, pp.201–209, 1989.

[5] M. Ben-or, S. Goldwasser and A. Widgerson. Completeness Theorems for non-cryptographic fault tolerant distributed computation. In *Proc. of 20th Annual Symposium on Theory of Computing*, 1988.

[6] E. Berlekamp and L. Welch. Error correction of algebraic block codes. US Patent 4,633,470.

[7] D. Boneh and M. Franklin. Efficient Generation of Shared RSA Keys. In *Advances in Cryptology – Crypto '97*, LNCS vol. 1294, Springer, 1997, pages 425-439. Extended version available from `http://crypto.stanford.edu/~dabo/pubs.html`.

[8] R. Canetti. Security and Composition of Multy-Party Cryptographic Protocols. In *Journal of Cryptology* 13 (1) pages 143-202, 2000.

[9] R. Canetti, R. Gennaro, S. Jarecki, H. Krawczyk and T. Rabin. Adaptive Security for Threshold Cryptosystems. In *Advances in Cryptology – Crypto '99*, LNCS vol. 1666, Springer, 1999, pages 98-115.

[10] D. Catalano and R. Gennaro. New Efficient and Secure Protocols for Verifiable Signature Sharing and Other Applications. In *Advances in Cryptology – Crypto '98*, LNCS vol. 1462, Springer, 1998, pages 105-120.

[11] D. Catalano, R. Gennaro and S. Halevi. Computing Inverses over a Shared Secret Modulus. In *Proc. of EUROCRYPT 2000, LNCS vol. 1807 pages 190-206, 2000*. Full version available from `http://www.di.ens.fr/~catalano`.

[12] D. Chaum, C. Crepeau, and I. Damgård. Multiparty Unconditionally Secure Protocols. *20th ACM Symposium on the Theory of Computing*, pp.11–19, ACM Press, 1988.

[13] T. Cormen, C. Leiserson and R. Rivest. Introduction to Algorithms. MIT Press, Cambridge, 1992.

[14] R. Cramer and V. Shoup. Signature Schemes Based on the Strong RSA Assumption. In Proceedings of the *6th ACM Conference in Computer and Communication Security*, 1999.

[15] I. Damgård and M. Koprowski. Practical Threshold RSA Signatures without a trusted dealer. In *Advances in Cryptology–Eurocrypt '01*, Lecture Notes in Computer Science Vol. 2045, pp. 152–165, Springer-Verlag, 2001.

[16] A. De Santis, Y. Desmedt, Y. Frankel and M. Yung. How to share a function securely. In *Proc. of the 26th ACM Annual Symposium on the Theory of Computing*, pp.522–533, ACM Press, 1994.

[17] Y. Desmedt. Society and group oriented cryptography: A new concept. In Carl Pomerance, editor, *Advances in Cryptology–CRYPTO'87*, Lecture Notes in Computer Science Vol. 293, pp. 120–127, Springer-Verlag, 1988.

[18] Y. Desmedt. Threshold cryptography. *European Transactions on Telecommunications*, 5(4):449–457, July 1994.

[19] Y. Desmedt and Y. Frankel. Shared Generation of authenticators and signatures. *Advances in Cryptology–CRYPTO'91*, Lecture Notes in Computer Science Vol. 576, pp. 457–469, Springer-Verlag, 1992.

[20] T. ElGamal. A public-key cryptosystem and a signature scheme based on discrete logarithms. *IEEE Transactions on Information Theory*, IT-31(4):469–472, 1985.

[21] U. Feige, A. Fiat, A. Shamir. Zero Knowledge Proofs of Identity. In *Journal of Cryptology* 1 pages 77-94, 1988.

[22] A. Fiat, A. Shamir. How to prove yourself: Practical solutions to identification and signature problems. In *Advances in Cryptology–CRYPTO'86*, Lecture Notes in Computer Science Vol. 263, pp. 186–194, Springer-Verlag, 1986.

[23] P.A. Fouque and J. Stern. Fully Distributed Threshold RSA under Standard Assumptions. In *Advances in Cryptology–Asiacrypt '01*, Lecture Notes in Computer Science Vol. 2248, pp. 310–330, Springer-Verlag, 2001.

[24] Y. Frankel. A Practical protocol for large group oriented networks. In *Advances in Cryptology–Eurocrypt'89*, Lecture Notes in Computer Science Vol. 434, pp. 56–61, Springer-Verlag, 1990.

[25] Y. Frankel, P. Gemmell, P. Mackenzie, and M. Yung. Optimal Resilience Proactive Public-Key Cryptosystems. *38th IEEE Symposium on the Foundations of Computer Science*, pp.384–393, IEEE Computer Society Press, 1997.

[26] Y. Frankel, P. Gemmell, and M. Yung. Witness-based Cryptographic Program Checking and Robust Function Sharing. *28th ACM Symposium on the Theory of Computing*, pp.499–508, ACM Press, 1996.

[27] Y. Frankel, P. Mackenzie, and M. Yung. Robust Efficient Distributed RSA-Key Generation. In *STOC* 1998, pp.663–672.

[28] M. Franklin ans S. Haber. Joint Encryption and Message Efficient Secure Computation. *Journal of Cryptology*, Vol.9, pp. 217-232, 1996.

[29] E. Fujisaki and T. Okamoto. Statistical Zero-Knowledge Protocols to Prove Modular Polynomial Relations. In *Advances in Cryptology – Crypto '97*, LNCS vol. 1294, Springer, 1997, pages 16-30.

[30] P. Gemmell. An Introduction to Threshold Cryptography. *RSA Laboratories CryptoBytes*, Vol.2, No.3, Winter 1997.

[31] R. Gennaro, S. Halevi and T. Rabin. Secure Hash-and-Sign Signatures without the Random Oracle. In *Advances in Cryptology – Eurocrypt '99*, LNCS vol. 1592, Springer, 1999, pages 123-139.

[32] R. Gennaro, S. Jarecki, H. Krawczyk, and T. Rabin. Robust and efficient sharing of RSA functions. *Crypto'96*, pp.157–172, Lecture Notes in Computer Science vol.1109, Springer-Verlag, 1996.

[33] R. Gennaro, S. Jarecki, H. Krawczyk, and T. Rabin. Secure Distributed Key Generation for Discrete-Log Public-Key Cryptosystems. *Eurocrypt'99*, pp.295–310, Lecture Notes in Computer Science vol.1592, Springer-Verlag, 1999.

[34] R. Gennaro, M. Rabin and T. Rabin. Simplified VSS and fast-track multiparty computations with applications to threshold cryptography. In *Proc. 17th ACM Symposium on Principle of Distributed Computing*, 1998.

[35] N. Gilboa. Two party RSA key Generation. In *Advances in Cryptology – Crypto '99*, LNCS vol. 1666, Springer, 1999, pages 116-129.

[36] O. Goldreich, S. Micali, and A. Wigderson. Proofs that Yield Nothing but the Validity of the Assertion, and a Methodology of Cryptographic Protocol Design. *27th IEEE Symposium on the Foundations of Computer Science*, pp.174–187. IEEE Computer Society Press, 1986.

[37] O. Goldreich, S. Micali, and A. Wigderson. How to play any mental game. *19th ACM Symposium on Theory of Computing*, pp.218–229, ACM Press, 1987.

[38] S. Goldwasser, S. Micali, and C. Rackoff. The knowledge complexity of interactive proof-systems. *SIAM. J. Computing*, 18(1):186–208, February 1989.

[39] L. Guillou and J. Quisquater. A practical Zero Knowledge protocol fitted to secure microprocessor minimizing both transmission and memory. In *Advances in Cryptology – Eurocrypt '88*, LNCS vol. 330, Springer, 1988, pages 123-128.

[40] N. Koblitz. *A course in number theory and cryptography*, 2nd ed., Springer Verlag.

[41] S. Micali and P. Rogaway. Secure Computation. In *Advances in Cryptology – Crypto '91*, LNCS vol. 576, Springer, 1992, pages 392-404.

[42] G. L. Miller. Riemann's Hypothesis and tests for primality. In *Journal of Computers and System Sciences*, 13 (1976) 300-317.

[43] K. Ohta and T. Okamoto. A modification of the Fiat-Shamir scheme. In *Advances in Cryptology – Crypto '88*, LNCS vol. 403, Springer, 1990, pages 232-243.

[44] H. Ong and C. Schnorr. Fast Signature Generation with a Fiat-Shamir-like Scheme. In *Advances in Cryptology – Eurocrypt '90*, LNCS vol. 473, Springer, 1991, pages 432-440.

[45] T. Pedersen. A threshold cryptosystem without a trusted party. *Eurocrypt'91*, pp.522–526, Lecture Notes in Computer Science vol.547, Springer-Verlag, 1991.

[46] T. Pedersen. Non-interactive and information-theoretic secure verifiable secret sharing. *Crypto'91*, pp.129-140, Lecture Notes in Computer Science vol.576, Springer-Verlag, 1992.

[47] M. O. Rabin. Probabilist Algorithms for testing primality. *Journal of Number Theory*, 12 (1980), 128-138.

[48] T. Rabin. A Simplified Approach to Threshold and Proactive RSA. *Crypto'98*, pp.89–104, Lecture Notes in Computer Science vol.1462, Springer-Verlag, 1998.

[49] R. Rivest, A. Shamir and L. Adelman. A Method for Obtaining Digital Signature and Public Key Cryptosystems. *Comm. of ACM*, 21 (1978), pp. 120–126.

[50] A. Schrijver. *Theory of Linear and Integer Programming*. John Wiley & Sons. 1986.

[51] A. Shamir. How to share a secret . In *Communications of the ACM* 22 (11) pages 612–613, 1979.

[52] V. Shoup. Practical Threshold Signatures. In *Eurocrypt'00* pp.207–220, Lecture Notes in Computer Science vol.1807, Springer-Verlag, 2000.

[53] D. Stinson. Cryptography: Theory and Practice CRC Press 1995.

[54] M. Sudan. Efficient Checking of Polynomials and Proofs and the Hardness of Approximation Problems. Lecture Notes in Computer Science, vol.1001, Springer-Verlag, 1995.

[55] A. Yao. How to Generate and exchange Secrets. In *Proc. 18th IEEE Annual Symposium on Foundations of Computer Science*, pp.162–167 1986.

Multiparty Computation, an Introduction

Ronald Cramer and Ivan Damgård

1. Introduction

These lecture notes introduce the notion of secure multiparty computation. We introduce some concepts necessary to define what it means for a multiparty protocol to be secure, and survey some known general results that describe when secure multiparty computation is possible. We then look at some general techniques for building secure multiparty protocols, including protocols for commitment and verifiable secret sharing, and we show how these techniques together imply general secure multiparty computation.

Our goal with these notes is to convey an understanding of some basic ideas and concepts from this field, rather than to give a fully formal account of all proofs and details. We hope the notes will be accessible to most graduate students in computer science and mathematics with an interest in cryptography.

2. What is Multiparty Computation?

2.1. The MPC and VSS Problems

Secure *multi-party computation* (MPC) can be defined as the problem of n players to compute an agreed function of their inputs in a secure way, where security means guaranteeing the correctness of the output as well as the privacy of the players' inputs, even when some players cheat. Concretely, we assume we have inputs $x_1, \ldots, x_n$, where player i knows x_i, and we want to compute $f(x_1, \ldots, x_n) = (y_1, \ldots, y_n)$ such that player i is guaranteed to learn y_i, but can get nothing more than that.

As a toy example we may consider Yao's *millionaire's problem*: two millionaires meet in the street and want to find out who is richer. Can they do this without having to reveal how many millions they each own? The function computed in this case is a simple comparison between two integers. If the result is that the first millionaire is richer, then he knows that the other guy has fewer millions than him, but this should be all the information he learns about the other guy's fortune.

Another example is a voting scheme: here all players have an integer as input, designating the candidate they vote for, and the goal is to compute how many votes each candidate has received. We want to make sure that the correct result of the vote, but *only* this result, is made public. In these examples all players learn the same result, i.e, $y_1 = \cdots = y_n$, but it can also be useful to have different results for different players. Consider for example the case of a blind signature scheme, which is useful in electronic cash systems. We can think of this as a two-party secure computation where the signer enters his private signing key sk as input, the user enters a message m to be signed, and the function $f(sk, m) = (y_1, y_2)$, where y_1 is for the signer and is empty, and where y_2 is for the user and the signature on m. Again, security means exactly what we want: the user gets the signature and nothing else, while the signer learns nothing new.

It is clear that if we can compute *any* function securely, we have a very powerful tool. However, some protocol problems require even more general ways of thinking. A secure payment system, for instance, cannot naturally be formulated as secure computation of a single function: what we want here is to continuously keep track of how much money each player has available and avoid cases where for instance people spend more money than they have. Such a system should behave like a secure general-purpose computer: it can receive inputs from the players at several points in time and each time it will produce results for each player computed in a specified way from the current inputs and from previously stored values. Therefore, the definition we give later for security of protocols, will be for this more general type, namely a variant of the *Universally Composable* security definition of Canetti. Another remark is that although the general protocol constructions we give are phrased as solutions to the basic MPC problem, they can in fact also handle the more general type of problem.

A key tool for secure MPC, interesting in its own right, is *verifiable secret sharing* (VSS): a dealer distributes a secret value s among the players, where the dealer and/or some of the players may be cheating. It is guaranteed that if the dealer is honest, then the cheaters obtain no information about s, and all honest players are later able to reconstruct s, even against the actions of cheating players. Even if the dealer cheats, a unique such value s will be determined already at distribution time, and again this value is reconstructable even against the actions of the cheaters.

2.2. Adversaries and their Powers

It is common to model cheating by considering an *adversary* who may *corrupt* some subset of the players. For concreteness, one may think of the adversary as a hacker who attempts to break into the players' computers. When a player is corrupted, the adversary gets all the data held by this player, including complete information on all actions and messages the player has received in the protocol so far. This may seem to be rather generous to the adversary, for example one might claim that the adversary will not learn that much, if the protocol instructs players to delete sensitive information when it is no longer needed. However, first other players cannot check that such information really is deleted, and second even if a player has every intention of deleting for example a key that is outdated, it may be quite difficult to ensure that the information really is gone and cannot be retrieved if the adversary breaks into this player's computer. Hence the standard definition of corruption gives the entire history of a corrupted player to the adversary.

One can distinguish between passive and active corruption. *Passive* corruption means that the adversary obtains the complete information held by the corrupted players, but the players still execute the protocol correctly. *Active* corruption means that the adversary takes full control of the corrupted players.

It is (at least initially) unknown to the honest players which subset of players is corrupted. However, no protocol can be secure if *any* subset can be corrupted. For instance, we cannot even define security in a meaningful way if all players are corrupt. We therefore need a way to specify some limitation on the subsets the adversary can corrupt. For this, we define an *adversary structure* $\mathcal{A}$, which is simply a family of subsets of the players. And we define an $\mathcal{A}$-adversary to be an adversary that can only corrupt a subset of the players if that subset is in $\mathcal{A}$. The adversary structure could for instance consist of all subsets with cardinality less than some threshold value t. In order for this to make sense, we must require for any adversary structure that if $A \in \mathcal{A}$ and $B \subset A$, then $B \in \mathcal{A}$. The intuition is that if the adversary is powerful enough to corrupt subset A, then it is reasonable to assume that he can also corrupt any subset of A.

Both passive and active adversaries may be *static*, meaning that the set of corrupted players is chosen once and for all before the protocol starts, or *adaptive* meaning that the adversary can at any time during the protocol choose to corrupt a new player based on all the information he has at the time, as long as the total corrupted set is in $\mathcal{A}$.

2.3. Models of Communication

Two basic models of communication have been considered in the literature. In the *cryptographic* model, the adversary is assumed to have access to all messages sent, however, he cannot *modify* messages exchanged between honest players. This means that security can only be guaranteed in a cryptographic sense, i.e. assuming that the adversary cannot solve some computational problem. In the *information-theoretic* (abbreviated i.t., sometimes also called secure channels) model, it is assumed that the players can communicate over pairwise secure channels, in other words, the adversary gets no information at all about messages exchanged between honest players. Security can then be guaranteed even when the adversary has unbounded computing power.

For active adversaries, there is a further problem with broadcasting, namely if a protocol requires a player to broadcast a message to everyone, it does not suffice to just ask him to send the same message to all players. If he is corrupt, he may say different things to different players, and it may not be clear to the honest players if he did this or not (it is certainly not clear in the i.t. scenario). One therefore in general has to make a distinction between the case where a broadcast channel is given for free as a part of the model, or whether such a channel has to be simulated by a subprotocol. We return to this issue in more detail later.

We assume throughout that communication is *synchronous*, i.e., processors have clocks that are to some extent synchronized, and when a message is sent, it will arrive before some time bound. In more detail, we assume that a protocol proceeds in rounds: in each round, each player may send a message to each other player, and all messages are delivered before the next round begins. We assume that in each round, the adversary first sees all messages sent by honest players to corrupt players (or in the cryptographic scenario, all messages sent). If he is adaptive, he may decide to corrupt some honest players at this point. And only then does he have to decide which messages he will send on behalf of the corrupted players. This fact that the adversary gets to see what honest players say before having to act himself is sometimes referred to as a *rushing* adversary.

In an asynchronous model of communication where message delivery or bounds on transit time is not guaranteed, it is still possible to solve most of the problems we consider here. However, we stick to synchronous communication – for simplicity, but also because problems can only be solved in a strictly weaker sense using asynchronous communication. Note, for instance, that if messages are not necessarily delivered, we cannot demand that a protocol generates any output.

2.4. Definition of Security

2.4.1. How to not do it. Defining security of MPC protocols is not easy, because the problem is so general. A good definition must automatically lead to a definition, for instance, of secure electronic voting because this is a special case of MPC. The classical approach to such definitions is to write down a list of requirements: the inputs must be kept secret, the result must be correct, etc. However, apart from the fact that it may be hard enough technically to formalize such requirements, it can be very difficult to be sure that the list is complete. For instance, in electronic voting, we would clearly be unhappy about a solution that allowed a cheating voter to vote in a way that relates in a particular way to an honest player's vote. Suppose, for instance, that the vote is a yes/no vote. Then we do not want player P_1 to be able to behave such that his vote is always the opposite of honest player P_2's vote. Yet a protocol with such a defect may well satisfy the demand that all inputs of honest players are kept private, and that all submitted votes of the right form are indeed counted. Namely, it may be that a corrupt P_1 does not know how he votes, he just modifies P_2's vote in some clever way and submits it as his own. So maybe we should demand that all players in a multiparty computation *know* which input values they contribute? Probably yes, but can we then be sure that there are no more requirements we should make in order to capture security properly?

2.4.2. The Ideal vs. Real World Approach. To get around this seemingly endless series of problems, we will take a completely different approach: in addition to the *real world* where the actual protocol and attacks on it take place, we will define an *ideal world* which is basically a specification of what we would like the protocol to do. The idea is then to say that a protocol is good if what it produces cannot be distinguished from what we could get in the ideal scenario.

To be a little more precise, we will in the ideal world assume that we have access to an uncorruptible computer, a so called *Ideal Functionality F*. All players can privately send inputs to and receive outputs from F. F is programmed to execute a certain number of commands, and will, since it is uncorruptible, always execute them correctly according its (public) specification, without leaking any information other than the outputs it is supposed to send to the players. A bit more precisely, the interface of F is as follows: F has an input and an output port for every player. Furthermore, it has two special, so called *corrupt* input and output ports, used for communication with the adversary. In every round, F reads inputs from its input ports, and returns results on the output ports. The general rule is that whenever a player P_i is corrupted, F stops using the i'th input/output ports and the adversary then communicates on behalf of P_i over the corrupted input/output ports.

In the following, we will sometimes talk about a corrupted P_j communicating with F, to make the text easier to understand, but this should be taken to mean that the adversary communicates on behalf of P_j as we just described.

The goal of a protocol π is to create, without help from trusted parties, and in presence of some adversary, a situation "equivalent" to the case where we have F available. If this is the case, we say that π *securely realizes F*. For instance, the goal of computing a function securely can be specified by an ideal functionality that receives inputs from the players, evaluates the function and returns results to the players. But in fact, any cryptographic task, such as commitment schemes or payments systems can be naturally modelled by an ideal functionality.

In order to give a precise definition, we need to say exactly what we mean by the protocol being "equivalent" to F. Let us reason a little about this. A couple of things are immediately clear: when F is used, corrupting some player P_i means you see the inputs and outputs of that player – but you will learn nothing else. An active attack can change the inputs that P_i uses, but can influence the results computed in no other way – F always returns results to players that are correctly computed based on the inputs it received. So clearly, a protocol that securely realizes F must satisfy something similar.

But more is true: we want that protocol and functionality are equivalent, *no matter* in which context the protocol is used. And we have to realize that this context contains more than just the adversary. It also consists, for instance, of human users or computer systems that supply inputs to the protocol. Or if the protocol is used as a subroutine in a bigger system, that system is certainly part of the environment. So in general, we can think of the environment as an entity that chooses inputs that players will use in the protocol and receives the results they obtain. We will define equivalence to mean that *the entire environment* cannot tell any essential difference between using the protocol and using the ideal functionality.

Towards formalizing this, an important observation is that the adversary is not really an entity separate from the environment, he is actually an integrated part of it. Consider for instance the case where the protocol is used as a subroutine in a higher level protocol. In such a case, the honest players may choose their inputs as a result of what they experience in the higher level protocol. But this higher level protocol may also be attacked by the adversary, and clearly this may give him some influence on the inputs that are chosen. In other words, the choice of inputs at some point in time may be a result of earlier adversarial activity. A second observation relates to the results that honest players compute. Again, if we think of the situation where our protocol is used as a subroutine in a bigger construction, it is clear that the result an honest player obtains may be used in the bigger construction, and may affect his behavior later. As a result of this, the adversary may be able to deduce information about these results. In other words, adversarial activity now may be a function of results computed by the protocol earlier.

2.4.3. The Definition: Universal Composability. The definition we give here is a variant of the *universally composable* (UC) security definition given by Canetti in [8]. This definition builds on several earlier works (see e.g. [1, 24, 6]). The variant is due to Nielsen [25] and adapts the UC definition to the synchronous model of communication. We generalize it slightly here to cover both the i.t. and the cryptographic scenario.

We now go to the actual definition of the model:

The real world contains the environment Z and the players $P_1, \ldots, P_n$ all of whom are modelled as interactive Turing machines (ITM's). The players communicate on a synchronous network using open channels or perfectly secure pairwise communication as specified earlier. In line with the discussion above, the environment Z should be thought of as a conglomerate of everything that is external to the protocol execution. This includes the adversary, so therefore Z can do everything we described earlier for an adversary, i.e., it can corrupt players passively/actively and statically/adaptively, according to an adversary structure $\mathcal{A}$. This is called a $\mathcal{A}$-environment. The players follow their respective programs specified in protocol π, until they are corrupted and possibly taken over by Z. In addition to this, Z also communicates with the honest players, as follows: in every round Z sends a (possibly empty) input to every honest player, and at the end of every round each honest player computes a result that is then given to Z.

When the protocol is finished, Z outputs a single bit, the significance of which we will return to shortly. In addition to other inputs, all entities get as initial input a security parameter value k, which is used to control the security level of the execution, e.g., the size of keys to use in the cryptographic scenario. To fully formalize the description, more details need to be specified, such as the exact order in which the different ITM's are activated. Details on this can be found in the appendix.

The ideal world contains the same environment we have in the real world, but there are no players. Instead, we have an *ideal functionality F*, and a *simulator S*. As mentioned above, F cannot be corrupted, and it will be programmed to carry out whatever task we want to execute securely, such as computing a function. Recall that we described the interface of F: F has an input and an output port for every player in the real protocol, and *corrupt* input/output ports, for communication with the environment/adversary.

The whole idea is that the environment Z we looked at in the real world should be able to act in the same way in the ideal world. Now, Z has two kinds of activities. First, it is allowed to send inputs to the honest players and see their outputs. We handle this by relaying these data directly to the relevant input/output ports of F. Second, Z expects to be able to attack the protocol by corrupting players, seeing all data they send/receive and possibly control their actions. For this purpose, we have the simulator S. Towards Z, S attempts to provide all the data Z would see in a real attack, namely internal data of newly corrupted players and protocol messages that corrupted players receive. We want Z to work exactly like it does in the real world, so therefore S must go through the protocol in the right time ordering and in every round show data to Z that look like what it would see in the real world. S is *not* allowed to rewind Z. The only help S gets to complete this job is that it gets to use the corrupt input/output ports of F, i.e., towards F, it gets to provide inputs and see outputs on behalf of corrupted players. Concretely, as soon as Z issues a request to corrupt player P_i, both S and F are notified about this. Then the following happens: S is given all input/outputs exchanged on the i'th input/output ports of F until now. F then stops using input/output port number i. Instead it expects S to provide inputs "on behalf of P_i" on the corrupt input port and sends output meant for P_i to S on the corrupt output port. One way of stating this is: we give to S exactly the data that the protocol *is supposed to release* to corrupt players, and based on this, it should be possible to simulate towards Z all the rest that corrupted players would see in a real protocol execution.

It is quite obvious that whatever functionality we could possibly wish for, could be securely realized simply by programming F appropriately. *However*, do not forget that the ideal world does not exist in real life, it only provides a specification of a functionality we would like to have. The point is that we can have confidence that any reasonable security requirement we could come up with will be automatically satisfied in the ideal world, precisely because everything is done by an uncorruptible party – and so, if we can design a protocol that is in a strong sense equivalent to the ideal functionality, we know that usage of the protocol will guarantee the same security properties – even those we did not explicitly specify beforehand!

We can now start talking about what it means that a given protocol π securely realizes ideal functionality F. Note that the activities of Z have the same form in real as in ideal world. So Z will output one bit in both cases. This bit is a random variable, whose distribution in the real world may depend on the programs of π, Z and also on the security parameter k and Z's input z. We call this variable $REAL_{\pi,Z}(k,z)$. Its distribution is taken over the random choices of all ITM's that take part. Similarly, in the ideal world, the bit output by Z is a random variable called $IDEAL_{F,S,Z}(k,z)$. We then have:

Definition 1. *We say that π $\mathcal{A}$-securely realizes F, if there exists a polynomial time simulator S such that for any $\mathcal{A}$-environment Z and any input z, we have that*

$$|Pr(REAL_{\pi,Adv}(k,z) = 0) - Pr(IDEAL_{F,S,Adv}(k,z) = 0)|$$

is negligible in k.

Here, negligible in k means, as usual, that the entity in question is smaller than $1/f(k)$ for any polynomial $f()$ and all sufficiently large k.

Some remarks on how to interpret this definition: The output bit of Z can be thought of as its guess at which world it is in. So the definition basically demands that there is a simulator S using not too much computing power such that for every environment in which the protocol is used, the protocol can be replaced by the ideal functionality without the environment noticing this. So in this sense, the definition says that using the protocol is "equivalent" to using the ideal functionality.

For instance, the definition implies that the protocol does not release more information to corrupt players than it is "allowed to": in the ideal world, the simulator S gets results for corrupted players directly from F, and based on only this, S can produce a view of the protocol that looks exactly like what corrupt players would see in the real world. The definition also implies that honest players get correct results: this is automatically ensured in the ideal world, and any mismatch in the real world could be detected by Z so that the definition could not be satisfied.

There are several possible variants of this definition. The one we gave requires so-called *statistical security*, but can be made stronger by requiring that the two involved probabilities are equal for all k, and not just close. This is called *perfect security*. In both cases we consider all (potentially unbounded) adversaries and environments. This fits with the i.t. scenario. For the cryptographic scenario, we need to restrict adversaries and environments to polynomial time, and we will only be able to prove protocols relative to some complexity assumption – we then speak of *computational security*.

2.4.4. Composition of Protocols. The most useful feature of universally composable security as defined here is exactly the composability: Let us define a *G-hybrid model*, as follows: G is assumed to be an ideal functionality, just like we described above. A protocol π in the G-hybrid model is a real-world protocol that is also allowed to make calls to G through the usual interface, that is, honest player P_i may privately specify inputs to G by sending data directly to the i'th input port, and G returns results to P_i on the i'th output port. If the environment corrupts a player, it uses the corrupt input/output ports of G to exchange data on behalf of the corrupted player. The model allows the protocol to run several independent instances of G, and there is no assumption on the timing of different calls, in particular, they may take place simultaneously.

Of course, π may itself be a secure realization of some ideal functionality F, or put another way: π describes how to implement F securely, assuming functionality G is available. This is defined formally in the same way as in Definition 1, but with two changes: first, we replace in the definition the real world with the G-hybrid model. And second, the ideal world is modified: the simulator must create a setting that to the environment looks like a protocol execution in the G-hybrid model, even though no G is available. So therefore all messages Z wants to send to G will go to the simulator S, and S must then create responses "from G".

Now suppose we have a protocol ρ that securely realizes G in the real world. We let π^ρ denote the real-world protocol that is obtained by replacing each call in π to G by a call to ρ. Note that this may cause several instances of ρ to be running concurrently. We make no assumption on any synchronization between these instances. Then we have the following, which is proved in the appendix:

Theorem 1. *If protocol π in the G-hybrid model securely realizes F, and protocol ρ in the real world securely realizes G, then protocol π^ρ securely realizes F in the real world.*

As we shall see, this result is incredibly useful when constructing and proving protocols: when building π, we can assume that ideal functionality G is "magically" available, and not worry about how to implement it. When we build ρ, we only have to worry about realizing G, and not about how the protocol will be used later.

3. Results on MPC

We now list some important known results on MPC. A remark on terminology: the security definition works with an environment Z, that includes the adversary as an integrated part that may potentially influence everything the environment does. It is therefore really a matter of taste whether one wants to speak of Z as "the environment" or "the adversary". In the following, we will use both terms, but the formal interpretation will always be the entity Z as defined above. Furthermore, when we speak below of "securely computing" a function, this formally means securely realizing a functionality F_{MPC} that is defined in more detail later.

3.1. Results for Threshold Adversaries

The classical results for the information-theoretic model due to Ben-Or, Goldwasser and Wigderson [4] and Chaum, Crépeau and Damgård [10] state that every function can be securely computed with perfect security in presence of an adaptive, passive (adaptive, active) adversary, if and only if the adversary corrupts less than $n/2$ ($n/3$) players. The fastest known protocols can be found in Gennaro, Rabin and Rabin[19].

When a broadcast channel is available, then every function can be securely computed with statistical security in presence of an adaptive, active adversary if and only if the adversary corrupts less than $n/2$ players. This was first shown by

Rabin and Ben-Or[29]. The most efficient known protocols in this scenario are by Cramer, Damgård, Dziembowski, Hirt and Rabin [12].

The most general results for the cryptographic model are by Goldreich, Micali and Wigderson [20] who showed that, assuming trapdoor one-way permutations exist, any function can be securely computed with computational security in presence of a static, active adversary corrupting less than $n/2$ players and by Canetti et al. who show [7] that security against adaptive adversaries in the cryptographic model can also be obtained, although at the cost of a significant loss of efficiency. Under specific number theoretic assumptions, Damgård and Nielsen have shown that adaptive security can be obtained without essential loss of efficiency, compared to the best known statically secure solutions [17].

3.2. Results for General Adversaries

Hirt and Maurer [21] introduced the scenario where the adversary is restricted to corrupting any set in a general adversary structure.

In the field of secret sharing we have a well-known generalization from threshold schemes to secret sharing over general access structures. Hirt and Maurer's generalization does the same for multiparty computation. One may think of the sets in their adversary structure as corresponding in secret sharing terminology to those subsets that cannot reconstruct the secret.

Let $Q2$ (and $Q3$) be the conditions on a structure that no two (no three) of the sets in the structure cover the full player set. The result of [21] can be stated as follows: In the information-theoretic scenario, every function can be securely computed with perfect security in presence of an adaptive, passive (adaptive, active) $\mathcal{A}$-adversary if and only if $\mathcal{A}$ is $Q2$ ($Q3$). This is for the case where no broadcast channel is available. The threshold results of [4], [10], [20] are special cases, where the adversary structure contains all sets of size less than $n/2$ or $n/3$.

This general model leads to strictly stronger results. Consider, for instance, the following infinite family of examples: Suppose our player set is divided into two groups X and Y of m players each ($n = 2m$) where the players are on friendly terms within each group but tend to distrust players in the other group. Hence, a coalition of active cheaters might consist of almost all players from X or from Y, whereas a mixed coalition with players from both groups is likely to be quite small. Concretely, suppose we assume that a group of active cheaters can consist of at most $9m/10$ players from only X or only Y, *or* it can consist of less than $m/5$ players coming from both X and Y. This defines an adversary structure satisfying $Q3$, and so multiparty computations are possible in this scenario. Nevertheless, no threshold solution exists, since the largest coalitions of corrupt players have size more than $n/3$[1]. The intuitive reason why threshold protocols fail here is that they will by definition have to attempt protecting against *any* coalition of size $9m/10$ – an impossible task. On the other hand this is overkill because not every coalition

---

[1] It can be shown that no weighted threshold solution exists either for this scenario, i.e., a solution using threshold secret sharing, but where some players are given several shares.

of this size actually occurs, and therefore multiparty computation is still possible using more general tools.

The protocols of [21] rely on quite specialized techniques. Cramer, Damgård and Maurer [13] show that any linear secret sharing scheme can be used to build MPC protocols. A linear secret sharing scheme is one in which each share is fixed linear function (over some finite field) of the secret and some random field elements chosen by the dealer. Since all the most efficient general techniques for secret sharing are linear, this gives the fastest known protocols for general adversary structures. They also show that the $Q2$ condition is necessary and sufficient for MPC in the cryptographic scenario.

4. MPC Protocols

In this section we will sketch how to show some of the general results we listed above. More precisely, we will look at ways to securely realize the following functionality, where we assume a threshold adversary that can corrupt at most t players, and the function to be computed is a function $f : (\{0,1\}^*)^n \to (\{0,1\}^*)^n$.

Some notation: when we say that a functionality receives a message of form $(P_i : mes)$, this means that if P_i is honest at this point, mes was received on the i'th input port, and if P_i has been corrupted, $P_i : mes$ was received on the corrupt input port, i.e., it was sent by environment or simulator as a message on behalf of a corrupted player.

Functionality F_{MPC}

The behavior of the functionality depends on two integer parameters $Input Delay$, $Compute Delay$, that are explained in more detail below.

1. Initially, set $x_i = \perp$ (the empty string) for $i = 1, \ldots, n$.
2. In the first round, collect all messages received of form $(P_i : Input, v)$, and let I be the set of P_i's occurring as senders. If I includes all honest players, set $x_i = v$, for each $P_i \in I$ and send "Inputs received" on the corrupt output port. If I does not include the set of honest players, send all internal data to the corrupt output port and stop.

   If in a round before round number $Input Delay$, $(P_i : change, v')$ for corrupt player P_i is received, set $x_i = v'$ (note that we may have $v' = \perp$.)
3. If any non-empty message is received from an honest player after Step 2, send all internal data to the corrupt output port and stop. Wait $Compute Delay$ rounds, then set $(y_1, \ldots, y_n) = f(x_1, \ldots, x_n)$, send y_i to P_i (on the i'th output port if P_i is honest, and otherwise on the corrupt output port).

Two remarks on this functionality: The intended way to use the functionality is that all honest players should send their inputs in the first round (along with those corrupt players that want to contribute input), and after this point no honest player should send input. The functionality is defined such that security is only required if it is used as intended. If anything else happens, all internal data are revealed (to environment or simulator) and it becomes trivial to simulate. The reason for the peculiar way to define the input step is to model that honest players must know from the start what input they contribute, but a corrupt player need not be bound to its input until after $InputDelay$ rounds, and may for instance start the protocol honestly and then stop. The functionality waits for $ComputeDelay$ rounds before it sends the results out. This is to model the fact that the protocol implementing the actual computation takes some number of rounds to finish.

To build a concrete protocol for this problem, we assume that a fixed finite field K is given, and that the function we want to compute is specified as an arithmetic circuit over K. That is, all input values are elements in K and the desired computation is specified as a number of additions and multiplications in K of the input values (or intermediate results). This is without loss of generality: Any function that is feasible to compute at all can be specified as a polynomial size Boolean circuit using, for instance, and, or and not-operations. But any such circuit can be simulated by operations in K: Boolean values true or false can be encoded as 1 resp. 0. Then the negation of bit b is $1 - b$, the and of bits b, b' is $b \cdot b'$ and the or becomes $1 - (1 - b)(1 - b')$.

The only necessary restriction on K is that $|K| > n$, but we will assume for concreteness and simplicity that $K = Z_p$ for some prime $p > n$.

Our main tool to build the protocol will be *Secret Sharing*, in particular Shamir's scheme, which is based on polynomials over K. A value $s \in K$ is shared by choosing a random polynomial $f_s()$ of degree at most t such that $f_s(0) = s$. And then sending privately to player P_j the value $f_s(j)$. The well known facts about this methods are that any set of t or fewer shares contain no information on s, whereas it can be reconstructed from any $t + 1$ or more shares. Both of these facts are proved using *Lagrange interpolation*:

If $h(X)$ is a polynomial of degree at most l and if C is a subset of K with $|C| = l + 1$, then

$$h(X) = \sum_{i \in C} h(i)\delta_i(X),$$

where $\delta_i(X)$ is the degree l polynomial such that, for all $i, j \in C$, $\delta_i(j) = 0$ if $i \neq j$ and $\delta_i(j) = 1$ if $i = j$. In other words,

$$\delta_i(X) = \prod_{j \in C, j \neq i} \frac{X - j}{i - j}.$$

We briefly recall why this holds. The right hand side $\sum_{i \in C} h(i)\delta_i(X)$ is clearly a polynomial of degree at most l that on input i evaluates to $h(i)$ for $i = 1, \ldots, n$. Therefore, if it were not equal to $h(X)$, the difference of the two polynomials would be a non-zero polynomial whose number of zeroes exceeds its degree – a contradiction.

Another consequence of Lagrange interpolation is that if $h(X)$ is a polynomial of degree at most $n - 1$, then there exist easily computable values $r_1, \ldots, r_n$, such that

$$h(0) = \sum_{i=1}^{n} r_i h(i).$$

Namely, $r_i = \delta_i(0)$. We call $(r_1, \ldots, r_n)$ a *recombination vector*.

We are going to need the following simple fact about recombination vectors:

Lemma 1. *Let* $(r_1, \ldots, r_n)$ *be any recombination vector, and let* I *be any subset of* $\{1, 2, \ldots, n\}$ *of size less than* $n/2$. *Then there always exists an* $i \notin I$ *with* $r_i \neq 0$.

Proof. Suppose we share values a, b resulting shares $a_1, \ldots, a_n, b_1, \ldots, b_n$, using polynomials f, g of degree $\leq t$, where t is maximal such that $t < n/2$. Then $a_1 b_1, a_2 b_2, \ldots, a_n b_n$ is a sharing of ab based on fg which is of degree at most $2t \leq n - 1$. If the Lemma was false, there would exist a set I of size at most t which could use r and their shares in a, b to compute ab, but this contradicts the fact that any t or fewer shares contain no information on a, b. □

Since the function we are to compute is specified as an arithmetic circuit over K, our task is, loosely speaking to compute a number of additions and multiplications in K of the input values (or intermediate results), while revealing nothing except for the final result(s).

Exercise. A useful first step to build MPC protocols is to design a secret sharing scheme with the property that a secret can be shared among the players such that corruptible set has any information, whereas any non-corruptible set can reconstruct the secret. Shamir's scheme shows how to do this for a threshold adversary structure, i.e., where the corruptible sets are those of size t or less. In this exercise we will build a scheme for the non-threshold example we saw earlier. Here we have $2m$ players divided in subsets X, Y with m players in each, and the corruptible sets are those with at most $9m/10$ players from only X or only Y, and sets of less than $m/5$ players with players from both X and Y (we assume m is divisible by 10, for simplicity).

- Suppose we shared secrets using Shamir's scheme, with $t = 9m/10$, or with $t = m/5 - 1$. What would be wrong with these two solutions in the given context?
- Design a scheme that does work in the given context. Hint: in addition to the secret s, create a random element $u \in K$, and come up with a way to share it such that only subsets with players from *both* X and Y can compute u. Also use Shamir's scheme with both $t = 9m/10$ and $t = m/5 - 1$.

4.1. The Passive Case

This section covers the i.t. scenario with a passive adversary. We assume a threshold adversary that can corrupt up to t players, where $t < n/2$. The protocol starts by

Input Sharing: Each player P_i holding input $x_i \in K$ secret shares x_i using Shamir's secret sharing scheme: he chooses at random a polynomial f of degree $\leq t$ and sends a share to each player, i.e., he sends $f(j)$ to P_j, for $j = 1, \ldots, n$.

We then work our way gate by gate through the given arithmetic circuit over K, maintaining the following

Invariant: All input values and all outputs from gates processed so far are secret shared, i.e. each such value $a \in K$ is shared into shares $a_1, \ldots, a_n$, where P_i holds a_i. Remark: if a depends on an input from an honest player, this must be a *random* set of shares with the only constraint that it determines a. From the start, no gates are processed, and only the inputs are shared.

To determine which gate to process next, we simply take an arbitrary gate for which both of its input have been shared already.

Once a gate producing one of the final output values y has been processed, y can be reconstructed in the obvious way by broadcasting the shares $y_1, \ldots, y_n$, or if y is a value that should go to only player P_j, the shares are sent privately to P_j.

It is therefore sufficient to show how addition and multiplication gates are handled. Assume the input values to a gate are a and b, determined by shares $a_1, \ldots, a_n$ and $b_1, \ldots, b_n$, respectively.

Addition: For $i = 1, \ldots, n$, P_i computes $a_i + b_i$. The shares $a_1 + b_1, \ldots, a_n + b_n$ determine $a + b$ as required by the invariant.

Multiplication: For $i = 1, \ldots, n$, P_i computes $a_i \cdot b_i = \tilde{c}_i$.

Resharing step: P_i secret shares $\tilde{c}_i$, resulting in shares $c_{i1}, \ldots, c_{in}$, and sends c_{ij} to player P_j.

Recombination step: For $j = 1, \ldots, n$, player P_j computes $c_j = \sum_{i=1}^{n} r_i c_{ij}$, where $(r_1, \ldots, r_n)$ is the recombination vector. The shares $c_1, \ldots, c_n$ determine $c = ab$ as required by the invariant.

Note that we can handle addition and multiplication by a constant c by using a default sharing of c generated from, say, the constant polynomial $f(x) = c$. We are going to assume that every output from the circuit comes out of a multiplication gate. This is without loss of generality since we can always introduce a multiplication by 1 on the output without changing the result. This is not strictly necessary, but makes life easier in the proof of security below.

4.1.1. Proof of Security for the Passive Case. In this section, we will argue the following result:

Theorem 2. *The protocol described in the previous section realizes F_{MPC} in the i.t. scenario with perfect security against an unbounded, adaptive and passive environment corrupting at most $t < n/2$ players, and with InputDelay $= 1$ and ComputeDelay equal to the depth of the circuit used to implement the function computed.*

For simplicity, we show here a proof of security assuming that each player P_i gets as input a single value $x_i \in K$, and is to receive a single value $y_i \in K$. This generalizes trivially to the case where inputs and outputs can be several values in K.

Recall that to prove security, our task is to build a simulator S which interacts with the environment Z and the ideal functionality.

Since corruptions are passive, we may assume that Z specifies messages for corrupt players to send by following the protocol, by definition of the model these messages are given to S, and S must generate messages on behalf of honest players and show these to Z.

As a result of this, the algorithm of S is as follows, where throughout, A denotes the currently corrupted set, specified as a set of indices chosen from $\{1, 2, \ldots, n\}$:

1. Whenever Z requests to corrupt a new player P_i, S will as a result see the inputs (if any) specified so far for P_i by Z and results received from F_{MPC} (and will from now on learn future inputs and outputs). Now, S will use this information to reconstruct a complete view of P_i taking part in the protocol up the point of corruption, and will show this view to Z. The view must, of course, be consistent with what Z has seen so far. We describe this reconstruction procedure in more detail below. Finally, we set $A := A \cup \{i\}$. Note that these corruptions may take place at any point during the simulation below of input sharing, computation and output generation.

2. In the first round, S will learn, by definition of F_{MPC}, whether Z has used the functionality correctly, i.e., whether it has specified inputs for all honest players or not. If not, all inputs are revealed, and it becomes trivial to simulate. So we continue, assuming inputs were specified as expected. S specifies arbitrary input values for corrupt players and send them to F_{MPC} (this is no problem, we will learn the correct values soon).

    In the next round, S does the following for each player P_i: if $i \in A$, record the shares Z has generated on behalf of corrupt players, and reconstruct x_i (which is easy by the assumption that Z follows the protocol). Send $(P_i : change, x_i)$ to F_{MPC}.

    If $i \notin A$, choose t random independent elements in K send these to Z and record them for later use. These elements play the role of the shares of x_i held by corrupt players.

3. S must now simulate towards Adv the computation and reconstruction of the outputs. To simulate the computation, S goes through the circuit with the same order of gates as in the real protocol.

    For each addition gate, where we add intermediate results a, b, each corrupt P_i holds shares a_i, b_i (which are known to S). S now simply records the fact that P_i now should add the shares to get $c_i = a_i + b_i$, and also records c_i as the share of $a + b$ known by P_i.

For each multiplication gate, where we multiply intermediate results a, b, each corrupt P_i holds shares a_i, b_i of a, b. S sets $\tilde{c}_i = a_i b_i$, watches perform Z a normal secret sharing of $\tilde{c}_i$ and record for later use the shares c_{ij} generated. For each honest P_i, S chooses random values $\{c_{ij} | P_j \in A\}$ to simulate the resharing done by honest players and sends the values to Adv. Finally, S records the fact that each corrupt P_j now computes $c_j = \sum_{i=1}^{n} r_i c_{ij}$, and also records c_j as the share of ab known by P_j.

4. To simulate the computation of the final result, S uses the fact that it knows from F_{MPC} the result y_i for each corrupt P_i. From the simulation of the circuit in the previous step, S has created a value s_j for each $P_j \in A$, and this value plays the role as P_j's share of y_i.

S now computes a polynomial $f_{y_i}()$ of degree at most t such that $f_{y_i}(0) = y_i$ and $f_{y_i}(j) = s_j$ for all $P_j \in A$. Then S sets $s_j = f_{y_i}(j)$ for all $P_j \notin A$, and sends these values to Adv, pretending that these are the shares in y_i sent to P_i by honest players.

5. Finally, we describe how S can reconstruct the view of a player P_i taking part in the protocol up to a given point, such that this is consistent with the data generated by S so far. This can be thought of as a list of polynomials chosen by P_i in order to secret share various values and a list of shares received from other players. We describe how to do the reconstruction when the entire computation has already taken place. This is without loss of generality: if P_i is corrupted earlier, we just truncate the reconstruction procedure in the natural way.

Input sharing: We now know x_i, the input of P_i, and S has already specified random shares r_j for $P_j \in A$. Now choose a random polynomial $f_{x_i}()$ of degree at most t subject to $f_{x_i}(0) = x_i, f_{x_i}(j) = r_j$. List $f_{x_i}()$ as the polynomial used by P_i to share x_i. As for inputs shared by another player P_k, do as follows: if $P_k \in A$, a polynomial $f_{x_k}()$ for x_k has already been chosen, so just list $f_{x_k}(i)$ as the share received by P_i. If $P_k \notin A$, choose a random value as the share in x_k received by P_i.

Additions: We may assume that we already listed a_i, b_i as P_i's shares in the summands, so we just list $a_i + b_i$ as his share in the sum.

Multiplications: The following method will work for all multiplication operations except those leading to output values of already corrupted players, which are handled in the next item. We may assume that we already listed a_i, b_i as P_i's share in the factors, so we compute $\tilde{c}_i = a_i b_i$. We now reconstruct P_i's sharing of $\tilde{c}_i$ in exactly the same way as we reconstructed his sharing of x_i above. We also follow the method from input sharing to reconstruct the shares P_i receives of $\tilde{c}_j$'s of other players. Finally we can compute c_i, P_i's share in the product following the normal interpolation algorithm from the protocol.

Output generation: As for y_i, the output of P_i, this is now known from F_{MPC}, and the shares in y_i held by corrupt players have been fixed earlier, so we follow the same method for simulating the shares P_i receives in output reconstruction stage that we already described above.

For an output value y_j of an already corrupted player P_j, we have the problem that we already showed to the adversary what was supposed to be P_i's share s_i in y_j. Recall we assumed that any output y_j comes from a multiplication gate. So we have to specify the values involved in P_i's handling of this multiplication such that they will be consistent with s_i, but also consistent with the view of P_i we generated so far. This is done as follows: Let the multiplication in the circuit leading to y_j be $y_j = ab$, let a_i, b_i be the shares in a, b we already specified for P_i, and let $\tilde{c}_i = a_i b_i$. The multiplication protocol involves sharing $\tilde{c}_i$, and this has already taken place, in the sense that S has sent random values c_{ij} to players in A pretending they came from P_i. So we now choose a random polynomial $f_{\tilde{c}_i}()$ of degree at most t such that $f_{\tilde{c}_i}(0) = \tilde{c}_i, f_{\tilde{c}_i}(j) = c_{ij}, j \in A$, list this as the polynomial chosen by P_i for the multiplication. Finally, P_i receives in the real protocol shares c_{ji}, for every j, and is supposed to compute his share in the product as $s_i = \sum_j r_j c_{ji}$. Of the c_{ji}'s, we have already fixed the ones coming from corrupt players, $\{c_{ji} | j \in A\}$ and $c_{ii} = f_{\tilde{c}_i}(i)$, altogether at most t values (P_i has just been corrupted, so there could be at most $t - 1$ corruptions earlier). We now choose the remaining values c_{ji} as random independent values, subject only to $s_i = \sum_j r_j c_{ji}$. So actually, we select a random solution to a linear equation. By Lemma 1, there always exists a solution.

This concludes the description of S. To show that S works as required, we begin by fixing, in both the real and ideal world, arbitrary values for the input and random tape of Z. This means that the only source of randomness is the random choices of the players in the real world and those of S in the ideal world. We claim that, for every set of fixed values, Z sees exactly the same distribution when interacting with the ideal as with the real world, if we use S in the ideal world as described above. This of course implies that the protocol realizes F_{MPC} with perfect security since Z will then output 1 with the same probability in the two cases.

What Z can observe is the outputs generated by the players, plus it sees the view of the corrupt players as they execute the protocol. It will clearly be sufficient to prove the following

Claim: In every round j, for $j = 0$ up to the final round, the view of Z has the same distribution in ideal as in real world, given the fixed input and random tape for Z.

We argue this by induction on j. The basis $j = 0$ is trivial as nothing has happened in the protocol before the first round. So assume we have completed round j having produced some correctly distributed view for Z so far. We need to argue that given this, what S shows to Z in round $j+1$ is correctly distributed.

Assume first that $j+1$ is not the final round. Then the only messages Z will see from honest players are sharings of values they hold. This is simulated perfectly: both in simulation and in real protocol, the adversary sees $\leq t$ independent random values in K as a result of every such sharing. Indeed, it is straightforward to show, using interpolation, that any vector of $\leq t$ shares of a random threshold-t Shamir sharing consists of independent random values. The only other source of information for Z is what it will see as a result of corrupting a player P_i in round $j+1$. Since round $j+1$ is not the final round, the view reconstruction procedure will only execute the input sharing, addition and multiplication steps. By definition of the model, we start with the correct value of x_i, and also with correctly distributed shares of inputs of other players. It is then straightforward to see that the rest of the values in the view follow in a correct way from the starting values.

Then assume that round $j+1$ is the final round. This means that Z will see results for all players. In the ideal world, these results are computed according to the given function by F_{MPC} from the inputs specified by Z. But in the real world, one can check by straightforward inspection of the protocol that all players will compute the same function of the inputs specified by Z. In addition, Z will see the corrupted players' view of the output reconstruction. Note that by induction hypothesis, the shares in a final result y_i held by corrupted players just before the output reconstruction stage has the same distribution in simulation as in real life. If y_i goes to an honest player, nothing further is revealed. If y_i goes to a corrupt player, observe that in the real protocol, the polynomial that determines y_i is random of degree at most t with the only constraint that it determines y_i and is consistent with the shares held by corrupt players – since by Lemma 1, at least one random polynomial chosen by an honest player is added into the polynomial determining y_i. It is now clear that the procedure used by S to construct a corresponding polynomial leads to the same distribution. Finally, one can check by inspection and arguments similar to the above, that also the output generation step of the procedure for reconstructing the view of a newly corrupted player P_i chooses data with the correct distribution, again conditioned on inputs and random tapes we fixed for Z and everything Z has seen earlier.

4.1.2. Optimality of Corruption Bound. What if $t \geq n/2$? We will argue that then there are functions that cannot be computed securely.

Towards a contradiction, suppose there is a protocol Π, with *perfect privacy* and *perfect correctness* for two players P_1, P_2 to securely evaluate the logical AND of their respective private input bits b_1, b_2, i.e., $b_1 \wedge b_2$.

Assume that the players communicate using a perfect *error-free communication channel*. One of the players may be corrupted by an *infinitely powerful, passive* adversary.

Without loss of generality, we may assume the protocol is of the following form.

1. Each player P_i has a private input bit b_i. Before the protocol starts, they select private random strings $\rho_i \in \{0,1\}^*$ of appropriate length.

   Their actions in the forthcoming protocol are now uniquely determined by these initial choices.

2. P_1 sends the first message m_{11}, followed by P_2's message m_{21}.

   This continues until P_2 has sent sufficient information for P_1 to compute $r = b_1 \wedge b_2$. Finally, P_1 sends r (and some halting symbol) to P_2.

   The *transcript* of the conversation is

$$\mathcal{T} = (m_{11}, m_{21}, \ldots, m_{1t}, m_{2t}, r).$$

For $i = 1, 2$, the *view* of P_i is

$$\text{view}_i = (b_i, \rho_i, \mathcal{T}).$$

Perfect correctness means here that the protocols always halts (in a number of rounds t that may perhaps depend on the inputs and the random coins) and that always the correct result is computed.

Perfect privacy means that given their respective views, each of the players learns nothing more about the other player's input b' than what can be inferred from the own input b and from the resulting function output $r = b \wedge b'$.

Note that these conditions imply that if one of the players has input bit equal to 1, then he learns the other player's input bit with certainty, whereas if his input bit equals 0, he has no information about the other player's input bit.

We now argue that there is a strategy for a corrupted P_1 to always correctly determine the input bit b_2 of P_2, even if his input b_1 equals 0, thereby contradicting privacy.

Let P_1 have input bit $b_1 = 0$, and let the players execute the protocol, resulting in some particular transcript $\mathcal{T}$.

If P_2 has input bit $b_2 = 0$, he doesn't learn anything about b_1 by privacy. Hence, the transcript is also consistent with $b_1 = 1$.

But if $b_2 = 1$, then by correctness, the transcript cannot also be consistent with $b_1 = 1$: in that case its final message r is not equal to the AND of the input bits.

This gives rise to the following strategy for P_1.

1. P_1 sets $b_1 = 0$.
2. P_1 and P_2 execute the assumed protocol Π. This results in a fixed transcript $\mathcal{T}$.
3. P_1 verifies whether the transcript $\mathcal{T} = (m_{11}, m_{21}, \ldots, m_{1t}, m_{2t}, r)$ is also consistent with $b_1 = 1$.

   The consistency check can be performed as follows. P_1 checks whether there exists a random string σ_1 such that the same transcript $\mathcal{T}$ results, given that P_1 starts with $b_1 = 1$ and σ_1.

   P_1 can do this with an exhaustive search over all σ_1 and "simulating" P_2 by having him "send" the same messages as in the execution.

More precisely, he first checks whether $(b_1 = 1, \sigma_1)$ leads to m_{11}. If so, he "receives" P_2's message m_{21}, and checks whether his own next message would equal m_{22}, and so forth, until perhaps exactly the same transcript T results.

This process may take a long time, but that doesn't hurt since we have assumed an all powerful adversary.

4. If so, he decides that $b_2 = 0$. Otherwise he decides that $b_2 = 1$.

Similar arguments can be given if we relax the assumptions on privacy and correctness.

The assumptions about the players' computational resources and the communication channel are essential.

It can be shown that any of the following conditions is sufficient for the existence of a secure two-party protocol for the AND function (as well as OR).

1. Existence of trapdoor one-way permutations.
2. Both players are memory bounded.
3. The communication channel is noisy.

In principle, this leads to secure two-party protocols for any function. For more information, see for instance [14].

4.2. The Active Case

In this section, we show how to modify the protocol secure against a passive adversary to make it secure also against active cheating. We will postulate in the following that we have a certain ideal functionality F_{Com} available. This functionality can then be implemented both in the i.t. and the cryptographic scenario. We consider such implementations later.

We note already now, however, that in the cryptographic scenario, F_{Com} can be implemented if $t < n/2$ (or in general, the adversary is $Q2$) and we make an appropriate computational assumption. In the i.t. scenario we need to require $t < n/3$ in case of protocols with zero error and no broadcast given. If we assume a broadcast channel and allow a non-zero error, then $t < n/2$ will be sufficient. All these bounds are tight.

Before we start, a word on broadcast: with passive corruption, broadcast is by definition not a problem, we simply ask a player to send the same message to everyone. But with active adversaries where no broadcast is given for free, a corrupt player may say different things to different players, and so broadcast is not immediate. Fortunately, in this case, we will always have that $t < n/3$ for the i.t. scenario and $t < n/2$ for the cryptographic scenario, as mentioned. And in these cases there are in fact protocols for solving this so called Byzantine agreement problem efficiently. So we can assume that broadcast is given as an ideal functionality. In the following, when we say that a player broadcasts a message, this means that we call this functionality. Although real broadcast protocols take several rounds to finish, we will assume here for simplicity that broadcast happens in one round.

4.2.1. Model for Homomorphic Commitments and Auxiliary Protocols. We will assume that each player P_i can commit to a value $a \in K$. This will later be implemented by distributing and/or broadcasting some information to other players. We model it here by assuming that we have an ideal functionality F_{Com}. To commit, one simply sends a to F_{Com}, who will then keep it until P_i asks to have it revealed. Formally, we assume F_{Com} is equipped with the two commands Commit and Open described below (more will be defined later).

Some general remarks on the definition of F_{Com}: since the implementation of any of the commands may require all (honest) players to take part actively, we require that all honest players in a given round send the same command to F_{Com} in order for the command to be executed. In some cases, such as a commitment we can of course not require that all players send exactly the same information since only the committing players knows the value to be committed to. So in such a case, we require that the committer sends the command and his secret input, while the others just send the command. If F_{Com} is not used as intended, e.g., the honest players do not agree on the command to execute, F_{Com} will send all it's private data to all players and stop working. As with F_{MPC}, this is just a way to specify that no security is required if the functionality is not used as intended.

Notation: $CurrentRound$ always denotes the index of the current round. Some commands take some number of rounds to finish. This number for command Xxx is called $XxxDelay$.

Commit: This command is executed if in some round player P_i sends $(commit, i, cid, a)$ and in addition all honest players send $(commit, i, cid, ?)$. In this case F_{Com} records the triple (i, cid, a). Here, cid is just an identifier, and a is the value committed to. We require that all honest players agree to the fact that a commitment should be made because an implementation will require the active participation of all honest players. If P_i is corrupted and in a round before $CurrentRound + CommitDelay$ sends $(commit, i, cid, a')$, then (i, cid, a) is replaced by (i, cid, a'). A corrupt player may choose to have a be $\perp$ and not a value in K. This is taken to mean that the player refuses to commit.

In round $CurrentRound + CommitDelay$, if $i, cid, a, a \in K$ is stored, send $(commit, i, success)$ to all players. If $a = \perp$ send $(Commit, i, fail)$.
Open: This command is executed if in some round all honest players send $(open, i, cid)$. In addition P_i should send x, where x may be $accept$ or $refuse$, and where $x = accept$ if P_i is honest. In this case F_{Com} looks up the triple (i, cid, a), and if $x = accept$, it sends in the next round $(open, cid, a)$ to all players, else it sends $(open, cid, fail)$.

As a minor variation, we also consider *private* opening of a commitment. This command is executed if in some round all honest players send $(open, i, cid, j)$. The only difference in its execution is that F_{Com} sends its output to player P_j only, rather than to all players. The effect is of course that only P_j learns the committed value.

The symbol $[\cdot]_i$ denotes a variable in which F_{Com} keeps a committed value received from player P_i. Thus when we write $[a]_i$, this means that player P_i has committed to a. It is clear from the above that all players know at any point which committed values have been defined. Of course, such a value is not known to the players (except the committer), but nevertheless, they can ask F_{Com} to manipulate committed values, namely to add committed values, multiply them by public constants, or transfer a committed value to another player (the final operation is called a Commitment Transfer Protocol (CTP)):

CommitAdd: This command is executed if all honest players send (*commitadd*, $cid1, cid2, cid3$) (in the same round), and if triples $(i, cid1, a), (i, cid2, b)$ have been stored previously. Then F_{Com} stores the triple $(i, cid3, a + b)$.

ConstantMult: This command is executed if all honest players send (*constantmult*, $cid1, cid2, u$) (in the same round) where $u \in K$, and if a triple $(i, cid1, a)$ has been stored previously. Then F_{Com} stores the triple $(i, cid2, u \cdot a)$.

CTP: This command is executed if all honest players send ($ctp, i, cid1, j, cid2$) (in the same round), and if a triple $(i, cid1, a)$ has been stored earlier. If P_i is corrupt, he may send ($cid1, refuse$) in some round before $CurrentRound + CTPDelay$. If this happens, then F_{Com} sends ($cid1, cid2, fail$) to all players. Otherwise, F_{Com} stores ($j, cid2, a$), sends a to P_j, and ($cid1, cid2, success$) to everyone.

In our abbreviated language, writing $[a]_i + [b]_i = [a + b]_i$ means that the CommitAdd command is executed, creating $[a + b]_i$, and $u \cdot [a]_i = [ua]_i$ refers to executing the ConstantMult command. The CTP command can be thought of as creating $[a]_j$ from $[a]_i$. Note that we only require that the addition can be applied to two commitments made by *the same* player. Note also that there is no delay involved in the CommitAdd and ConstantMult commands, so an implementation cannot use any interaction between players.

A last basic command we assume is that F_{Com} can be asked to confirm that three commitments $[a]_i, [b]_i, [c]_i$ satisfy that $ab = c$. This is known as a Commitment Multiplication Protocol (CMP).

CMP: This command is executed if all honest players send ($cmp, cid1, cid2, cid3$) (in the same round), and if triples $(i, cid1, a), (i, cid2, b), (i, cid3, c)$ have been stored earlier. If P_i is corrupt, he may send ($cid1, cid2, cid3, refuse$) in some round before $CurrentRound + CMPDelay$. If this happens, or if $ab \neq c$, then in round $CurrentRound + CMPDelay$, F_{Com} sends ($cid1, cid2, cid3, fail$) to all players. Otherwise, F_{Com} sends ($cid1, cid2, cid3, success$) to everyone.

The final command we need from F_{Com} is called a Commitment Sharing Protocol (CSP). It starts from $[a]_i$ and produces a set of commitments to shares of a: $[a_1]_1, \ldots, [a_n]_n$, where $(a_1, \ldots, a_n)$ is a correct threshold-t Shamir-sharing of a, generated by P_i. More formally:

CSP: This command is executed if all honest players send $(csp, cid0, cid1, \ldots,$ $cidn)$ (in the same round), and if a triple $(i, cid0, a)$ has been stored earlier. If P_i is honest, he should also send (coefficients of) a polynomial $f_a()$ of degree at most t, such that $f_a(0) = a$. If P_i is corrupt, he may send a correct polynomial in some round before number $CurrentRound + CSPDelay$, or he may send $(cid0, cid1, \ldots, cidn, refuse)$. When we reach round number $CurrentRound + CSPDelay$, if a correct polynomial has been received, store triples $(j, cidj, f_a(j))$ for $j = 1..n$, and send $(cid0, cid1, \ldots, cidn, success)$ to everyone, else send $cid0, cid1, \ldots, cidn, fail)$.

The CTP, CMP, and CSP commands are special: although they can be implemented "from scratch" like the other commands, they can also be implemented using the commands we already defined. For CTP, we have the following.

Generic CTP Protocol

1. Given a commitment $[a]_i$, P_i sends privately to P_j his total view of the protocol execution in which $[a]_i$ was created [2]. If this information is in any way inconsistent, P_j broadcasts a complaint, and we go to Step 4.

   Otherwise (if P_i was honest) P_j is a situation equivalent to having made $[a]_i$ himself.

2. P_j commits himself to a, resulting in $[a]_j$.

3. We use the ConstantMult command to get $[-a]_j$ and the CommitAdd command to get $[a]_i + [-a]_j$ Note that, assuming that the information P_j got in step 1 was correct, this makes sense since then the situation is equivalent to the case where P_j had been the committer when $[a]_i$ was created. Then $[a]_i + [-a]_j$ is opened, and we of course expect this to succeed with output 0. If this happens, the protocol ends. Otherwise do Step 4.

4. If we arrive at this step, it is clear that at least one of P_i, P_j are corrupt, so P_i must then open $[a]_i$ in public, and we either end with fail (if the opening fails) or a becomes public. We then continue with a default commitment to a assigned to P_j.

For CMP, we describe this protocol for a prover and a single verifier. To convince all the players, the protocol is simply repeated independently (for instance in parallel), each other player P_j taking his turn as the verifier. In the end, all verifying players broadcast their decision, and the prover is accepted by everyone if there are more than t accepting verifiers. This guarantees that at least one honest verifier has accepted the proof.

Generic CMP Protocol

1. Inputs are commitments $[a]_i, [b]_i, [c]_i$ where P_i claims that $ab = c$. P_i chooses a random β and makes commitments $[\beta]_i, [\beta b]_i$.

2. P_j generates a random challenge $r \in K$, and sends it to P_i.

---

[2] As is standard, the view of a protocol consists of all inputs and random coins used, plus all messages received during the protocol execution.

3. P_i opens the commitments $r[a]_i + [\beta]_i$ to reveal a value r_1. P_i opens the commitment $r_1[b]_i - [\beta b]_i - r[c]_i$ to reveal 0.
4. If any of these opening fail, P_j rejects the proof, else he accepts it.

It is easy to show that if P_i remains honest, then all values opened are random (or fixed to 0) and so reveal no extra information to the adversary. If P_i is corrupt, then it is also straightforward to show that if, after committing in step 2, P_i can answer correctly two different challenges, then $ab = c$. Thus the error probability is at most $1/|K|$.

Finally, for CSP, assuming $[a]_i$ has been defined, P_i chooses a random polynomial f_a of degree at most t such that $f_a(0) = a$. He makes commitments to the coefficients of f: $[v_1]_i, \ldots, [v_t]_i$ (the degree-0 coefficient of f_a is a and has already been committed). Let $(a_1, \ldots, a_n) = (f_a(1), \ldots, f_a(n))$ be the shares resulting from sharing a using the polynomial f_a. Then the a_i's are a linear function of the committed values, and commitments to the shares $([a_1]_i, \ldots, [a_n]_i)$ can be created by calling the CommitAdd and ContstantMult commands, e.g.,

$$[a_j]_i = [a]_i + [v_1]_i \cdot j + [v_2]_i \cdot j^2 + \cdots + [v_t]_i \cdot j^t$$

Finally, we call CTP to create $[a_j]_j$ from $[a_j]_i$, for $j = 1, \ldots, n$.

Committing to a and then performing CSP is equivalent to what is known as *verifiably secret sharing a* (VSS): the value a is uniquely defined when the CSP is executed, and it is guaranteed that the honest players can reconstruct it: the commitments to shares prevent corrupted players from contributing false shares when the secret is reconstructed. All we need is that at least $t+1$ good shares are in fact revealed.

4.2.2. An MPC Protocol for Active Adversaries. The protocol starts by asking each player to verifiably secret-share each of his input values as described above: he commits to the value and then performs CSP. If this fails, the player is disqualified and we take default values for his inputs.

We then work our way through the given arithmetic circuit, maintaining as invariant that all inputs and intermediate results computed so far are verifiably secret shared as described above, i.e. each such value a is shared by committed shares $[a_1]_1, \ldots, [a_n]_n$ where *all* these shares are correct, also those held by corrupted players. Moreover, if a depends on an input from an honest player, this must be a *random* set of shares determining a. From the start, only the input values are classified as having been computed.

Once an output value y has been computed, it can be reconstructed in the obvious way by opening the commitments to the shares $y_1, \ldots, y_n$. This will succeed, as the honest players will contribute enough correct shares, and a corrupted player can only choose between contributing a correct share, or have the opening fail.

It is therefore sufficient to show how addition and multiplication gates are handled. Assume the input values to a gate are a and b, determined by committed shares $[a_1]_1, \ldots, [a_n]_n$ and $[b_1]_1, \ldots, [b_n]_n$.

Addition: For $i = 1..n$, P_i computes $a_i + b_i$ and CommitAdd is called to create $[a_i + b_i]_i$. By linearity of the secret sharing, $[a_1 + b_1]_1, \ldots, [a_n + b_n]_n$ determine $a + b$ as required by the invariant.

Multiplication: For $i = 1..n$, P_i computes $a_i \cdot b_i = \tilde{c}_i$, commits to it, and performs CMP on inputs $[a_i]_i, [b_i]_i, [\tilde{c}_i]_i$.

  Resharing step: P_i performs CSP on $[\tilde{c}_i]_i$, resulting in commitments $[c_{i1}]_1, \ldots$, $[c_{in}]_n$.

  We describe below how to recover if any of this fails.

  Recombination step: For $j = 1..n$, player P_j computes $c_j = \sum_{i=1}^{n} r_i c_{ij}$, where $(r_1, \ldots, r_n)$ is the recombination vector. Also all players compute (non-interactively) $[c_j]_j = \sum_{i=1}^{n} r_i [c_{ij}]_j = [\sum_{i=1}^{n} r_i c_{ij}]_j$. By definition of the recombination vector and linearity of commitments, the commitments $[c_1]_1, \ldots, [c_n]_n$ determine $c = ab$ as required by the invariant.

It remains to be described what should be done if a player P_i fails in the multiplication and resharing step above. In general, the simplest way to handle such failures is to go back to the start of the computation, open the input values of the players that have just been disqualified, and restart the computation, simulating openly the disqualified players. This allows the adversary to slow down the protocol by a factor at most linear in n. This solution works in all cases. However, in the i.t. case when $t < n/3$, we can do better: after multiplying shares locally, we have points on a polynomial of degree $2t$, which in this case is less than the number of honest players, $n - t$. In other words, reconstruction of a polynomial of degree $2t$ can be done by the honest players on their own. So the recombination step can always be carried out, we just tailor the recombination vector to the set of players that actually completed the multiplication step correctly.

4.3. Realization of F_{Com}: Information Theoretic Scenario

We assume throughout this subsection that we are in the i.t. scenario and that $t < n/3$.

  We first look at the commitment scheme: The idea that immediately comes to mind in order to have a player D commit to a is to ask him to secret share a. At least this will hide a from the adversary if D is honest, and will immediately ensure the homomorphic properties we need, namely to add commitments, each player just adds his shares, and to multiply by a constant, all shares are multiplied by the constant.

  However, if D is corrupt, he can distribute false shares, and can then easily "open" a commitment in several ways, as detailed in the exercise below.

Exercise A player P sends a value a_i to each player P_i (also to himself). P is supposed to choose these such that $a_i = f(i)$ for all i, for some polynomial $f()$ of degree at most t where $t < n/3$ is maximal number of corrupted players. At some later time, P is supposed to reveal the polynomial $f()$ he used, and each P_i reveals a_i. The polynomial is accepted if values of at most t players disagree with

$f()$ (we cannot demand fewer disagreements, since we may get t of them even if P was honest).

1. We assume here (for simplicity) that $n = 3t + 1$. Suppose the adversary corrupts P. Show how to choose two different polynomials $f(), f'()$ of degree at most t and values $\tilde{a}_i$ for P to send, such that P can later reveal and have accepted both $f()$ and $f'()$.
2. Suppose for a moment that we would settle for computational security, and that P must send to P_i, not only a_i, but also his digital signature s_i on a_i. We assume that we can force P to send a valid signature even if he is corrupt. We can now demand that to be accepted, a polynomial must be consistent with *all* revealed and properly signed shares. Show that now, the adversary cannot have two different polynomials accepted, even if up to $t \leq n/3$ players may be corrupted before the polynomial is to be revealed. Hint: First argue that the adversary must corrupt P before the a_i, s_i are sent out (this is rather trivial). Then, assume $f_1()$ is later successfully revealed and let C_1 be the set that is corrupted when f_1 is revealed. Assume the adversary could also choose to let P reveal $f_2()$, in which case C_2 is the corrupted set. Note that since the adversary is adaptive, you cannot assume that $C_1 = C_2$. But you can still use the players outside C_1, C_2 to argue that $f_1() = f_2()$.
3. (Optional) Does the security proved above still hold if $t > n/3$? why or why not?

To prevent the problems outline above, we must find a mechanism to ensure that the shares of all uncorrupted players after committing consistently determine a polynomial f of degree at most t, without harming privacy of course.

Before we do so, it is important to note that n shares out of which at most t are corrupted still uniquely determine the committed value a, even if we don't know which t of them are.

Concretely, define the shares

$$\mathbf{s}_f = (f(1), \ldots, f(n)),$$

and let $\mathbf{e} \in K^n$ be an arbitrary "error vector" subject to

$$w_H(\mathbf{e}) \leq t,$$

where w_H denotes the Hamming-weight of a vector (i.e., the number of its non-zero coordinates), and define

$$\tilde{\mathbf{s}} = \mathbf{s} + \mathbf{e}.$$

Then a is uniquely defined by $\tilde{\mathbf{s}}$.

In fact, more is true, since the entire polynomial f is. This is easy to see from Lagrange Interpolation and the fact that $t < n/3$.

Namely, suppose that $\tilde{\mathbf{s}}$ can also be "explained" as originating from some other polynomial g of degree at most t together with some other error vector $\mathbf{u}$

with Hamming-weight at most t. In other words, suppose that

$$\mathbf{s}_f + \mathbf{e} = \mathbf{s}_g + \mathbf{u}.$$

Since $w_H(\mathbf{e}), w_H(\mathbf{u}) \leq t$ and $t < n/3$, there are at $\geq n - 2t > t$ positions in which the coordinates of both are simultaneously zero. Thus, for more than t values of i we have

$$f(i) = g(i).$$

Since both polynomials have degree at most t, this means that

$$f = g.$$

Assuming that we have established the mechanism for ensuring correct sharings as discussed above, there is a simple open protocol for this commitment scheme.

Open Protocol (*Version I*):

1. Each player P_i simply reveals his share s_i to all other players P_j.
2. Each of them individually recovers the committed value a that is uniquely defined by them. This can be done by exhaustive search, or by the efficient method described below.

Note that broadcast is not required here.

We now show one particular method to *efficiently* recover the committed value. In fact, we'll recover the entire polynomial f. [3]

Write

$$\tilde{\mathbf{s}} = (\tilde{s}_1, \ldots, \tilde{s}_n).$$

The method "interpolates" the points $(i, \tilde{s}_i)$ by a bi-variate polynomial Q of a special form (which from a computational view comes down to solving a system of linear equations), and "extracts" the polynomial f from Q in a very simple way.

Concretely, let $Q(X,Y) \in K[X,Y]$, $Q \neq 0$ be any polynomial such that, for $i = 1 \ldots n$,

$$Q(i, \tilde{s}_i) = 0,$$

and such that

$$Q(X,Y) = f_0(X) - f_1(X) \cdot Y,$$

for some $f_0(X) \in K[X]$ of degree at most $2t$ and some $f_1(X) \in K[X]$ of degree at most t.

Then we have that

$$f(X) = \frac{f_0(X)}{f_1(X)}.$$

Clearly, the conditions on Q can be described in terms of a linear system of equations with Q's coefficients as the unknowns.

To recover f, we simply select an arbitrary solution to this system, which is a computationally efficient task, define the polynomial Q by the coefficients

---

[3] What we show is actually the Berlekamp-Welch decoder for Reed-Solomon error-correcting codes.

thus found, extract f_0, f_1 from it by appropriately ordering its terms, and finally perform the division of the two, which is again a computationally efficient task.

We now show correctness of this algorithm. First, we verify that this system is solvable. For this purpose, we may assume that we are given the polynomial f and the positions A in which an error is made (thus, A is a subset of the corrupted players). Define

$$k(X) = \prod_{i \in A} (X - i).$$

Note that its degree is at most t. Then

$$Q(X, Y) = k(X) \cdot f(X) - k(X) \cdot Y$$

satisfies the requirements for Q, as is verified by simple substitution.

It is now only left to show that whenever some polynomial Q satisfies these requirements, then indeed $f(X) = f_0(X)/f_1(X)$.

To this end, define

$$Q'(X) = Q(X, f(X)) \in K[X],$$

and note that its degree is at most $2t$.

If $i \notin A$, then $(i, s_i) = (i, \tilde{s}_i)$. Thus, for such i,

$$Q'(i) = Q(i, f(i)) = Q(i, s_i) = Q(i, \tilde{s}_i) = 0.$$

Since $t < n/3$,

$$n - |A| \geq n - t > 2t.$$

We conclude that the number of zeroes of $Q(X)$ exceeds its degree, and that it must be the zero polynomial. Therefore,

$$f_0 - f_1 \cdot f = 0,$$

which establishes the claim (note that $f_1 \neq 0$ since $Q \neq 0$).

Below we describe an alternative open protocol that is less efficient in that it uses the broadcast primitive. The advantage, however, is that it avoids the above "error correction algorithm" which depends so much on the fact that Shamir's scheme is the underlying secret sharing scheme. In fact, it can be easily adapted to a much wider class of commitment schemes, namely those based on general linear secret sharing schemes.

Open Protocol (*Version II*):

1. D broadcasts the polynomial f.
   Furthermore, each player P_i broadcasts his share.
2. Each player decides for himself by the following rule.
   If all, except for possibly $\leq t$, shares are consistent with the broadcast polynomial and its degree is indeed at most t, the opening is accepted. The opened value is $a = f(0)$.
   Else, the opening is rejected.

This works for essentially the same reasons as used before.

Note that both open protocols allow for *private opening* of a commitment to a designated player P_j. This means that only P_j learns the committed value a. This is achieved by simply requiring that all information is privately sent to P_j, and it works because of the privacy of the commit protocol (as shown later) and because the open protocol only depends on local decisions made by the players.

We now describe the *commit protocol*. Let $F(X, Y) \in K[X, Y]$ be a symmetric polynomial of degree at most t in both variables, i.e.,

$$F(X, Y) = \sum_{k,l=0}^{t} c_{kl} X^k Y^l,$$

and

$$F(X, Y) = F(Y, X),$$

which is of course equivalent to $c_{kl} = c_{lk}$ for all $1 \leq k, l \leq t$.
We define

$$f(X) = F(X, 0),$$

$$f(0) = a,$$

and, for $i = 1..n$,

$$f(i) = s_i.$$

Note that

$$\deg f \leq t.$$

We call f the *real sharing polynomial*, a the *committed value*, and s_i a *real share*.
We also define, for $i, j = 1 \ldots n$,

$$f_i(X) = F(X, i),$$

and

$$f_i(j) = s_{ij}.$$

Note that

$$\deg f_i \leq t.$$

We call f_i a *verification polynomial*, and s_{ij} a *verification share*.
By symmetry we have

$$s_i = f(i) = F(i, 0) = F(0, i) = f_i(0).$$

$$s_{ij} = f_i(j) = F(j, i) = F(i, j) = f_j(i) = s_{ji}.$$

Commit Protocol:

1. To commit to $a \in K$, D chooses a random, symmetric bivariate polynomial $F(X, Y)$ of degree at most t in both variables, such that

$$F(0,0) = a.$$

   D sends the verification polynomial f_i (i.e., its $t+1$ coefficients) privately to P_i for each i.

   P_i sets $s_i = f_i(0)$, his real share.

2. For all $i > j$, P_i sends the verification share s_{ij} privately to P_j.

3. It must hold that

$$s_{ij} = s_{ji}.$$

   If P_j finds that

$$s_{ij} \neq s_{ji},$$

   he broadcasts a *complaint*.

   In response to each such complaint (if any), D must broadcast the correct value s_{ij}.

   If P_j finds that the broadcast value differs from s_{ji}, he knows that D is *corrupt* and broadcasts an *accusation* against D, and *halts*.

   A similar rule applies to P_i if he finds that the broadcast value differs from s_{ij}.

4. For all players P_j who accused D in the previous step (if any), D must now broadcast the correct verification polynomial f_j.

5. Each player P_i that is "still in the game" verifies each of the broadcast verification polynomials f_j (if any) against his own verification polynomial f_i, by checking that, for each of those, $s_{ij} = s_{ji}$.

   If there is any inequality, P_i knows that D is corrupt, and broadcasts an *accusation* against D and halts.

6. If there are $\leq t$ accusations in total, D is *accepted*.

   In this case, each player P_j who accused D in Step 5, replaces the verification polynomial received in Step 1 by the polynomial f_i broadcast in Step 4, and defines $s_j = f_j(0)$ as his real share.

   All others stick to their real shares as defined from the verification polynomials received in in Step 1.

7. If there are $> t$ accusations in total, the dealer is deemed *corrupt*.

We sketch a proof that this commitment scheme works. For simplicity we assume that the adversary is static.

Honest D Case: It is immediate, by inspection of the protocol, that honest players never accuse an honest D. Therefore, there are at most t accusations and the commit protocol is always *accepted*.

In particular, each honest player P_i accepts $s_i = f(i)$ as defined in step 1 as his real share. This means that in the open protocol $a = f(0)$ is accepted as the committed value.

For *privacy*, i.e., the adversary does not learn the committed value a, note first that steps 2–4 of the commit protocol are designed such that the adversary learns nothing he was not already told in step 1.

Indeed, the only information that becomes available to the adversary afterwards, is what is broadcast by the dealer. This is either a verification share s_{ij} where P_i is corrupt or P_j is corrupt, or a verification polynomial f_i of a corrupt player P_i. All of this is already implied by the information the adversary received in step 1.

Therefore, it is sufficient to argue that the information in step 1 does not reveal a to the adversary.

Denote by A the set of corrupted players, with $|A| \leq t$. It is sufficient to show that for each guess a' at a, there is the same number of appropriate polynomials $F'(X, Y)$ consistent with the information received by the adversary in step 1.

By appropriate we mean that $F'(X, Y)$ should be symmetric, of degree at most t in both variables, and for all $i \in A$ we must have $f'_i(X) = f_i(X)$.

Consider the polynomial

$$h(X) = \prod_{i \in A} (\frac{-1}{i} \cdot X + 1) \in K[X]$$

Note that its degree is at most t, $h(0) = 1$ and $h(i) = 0$ for all $i \in A$.

Now define

$$Z(X, Y) = h(X) \cdot h(Y) \in K[X, Y].$$

Note that $Z(X, Y)$ is symmetric and of degree at most t in both variables, and that it has the further property that $Z(0, 0) = 1$ and $z_i(X) = Z(X, i) = 0$ for all $i \in A$.

If D in reality used the polynomial $F(X, Y)$, then for all possible a', the information held by the adversary is clearly also consistent with the polynomial

$$F'(X, Y) = F(X, Y) + (a' - a) \cdot Z(X, Y).$$

Indeed, it is symmetric, of degree at most t in both variables, and, for $i \in A$,

$$f'_i(X) = f_i(X) + (a - a') \cdot z_i(X) = f_i(X),$$

and

$$f'(0) = F'(0, 0) = F(0, 0) + (a' - a) \cdot Z(0, 0) = a + (a - a') = a'.$$

This construction immediately gives a one-to-one correspondence between the consistent polynomials for committed value a and those for a'. Thus all values are equally likely from the point of view of the adversary.

Corrupt D Case: Let B denote the set of honest players, and let s_i, $i \in B$, be the real shares as defined *at the end* of the protocol. In other words, $s_i = f_i(0)$, where f_i is the verification polynomial as defined at the end of the protocol.

We have to show that if the protocol was accepted, then there exists a polynomial $g(X) \in K[X]$ such that its degree is at most t and $g(i) = s_i$ for all $i \in B$.

It is important to realize that we have to argue this from the acceptance assumption alone; we cannot make any apriori assumptions on how a corrupt D computes the various pieces of information.

Write C for the set of honest players that did not accuse D at any point. Note that

$$|C| \geq n - \#\text{Accusations} - \#\text{Corruptions} \geq n - 2t > t.$$

Furthermore, there is consistency between the players in C on the one hand, and the players in B on the other hand. Namely, for all $P_i \in C$, $P_j \in B$, it follows from the acceptance assumption that

$$f_i(j) = f_j(i),$$

where the verification polynomials are defined as at end of the protocol.

Indeed, let $P_i \in C$ be arbitrary and let $P_j \in B$ be an arbitrary honest player who did not accuse the dealer before step 5. Then their verification polynomials f_i, f_j as defined at the end are the ones given in step 1. If it were so that $f_i(j) \neq f_j(i)$, then at least one of the two would have accused D in step 3.

On the other hand, if P_j is a player who accused D in step 3, and if the *broadcast* polynomial f_j is not consistent with P_i's verification polynomial, P_i would have accused D in step 5.

Let $r_i, i \in C$, be the coefficients of the recombination vector for C. Define

$$g(X) = \sum_{i \in C} r_i \cdot f_i(X).$$

Note that its degree is at most t.

We now only have to verify that for all $j \in B$, we have $s_j = g(j)$.

Indeed, we have that

$$g(j) = \sum_{i \in C} r_i \cdot f_i(j) = \sum_{i \in C} r_i \cdot f_j(i) = f_j(0) = s_j.$$

The first equality follows by definition of $g(X)$, the second by the observed consistency, the third by Lagrange interpolation and the fact that $|C| > t$ and that the degree of g is at most t, and the final equality follows by definition of the real shares at the end of the protocol.

This concludes the analysis of the commit protocol. Note that both the commit and the open protocol consume a constant number of rounds of communication.

So this commitment scheme works with no probability of error, if $t < n/3$. If instead we have $t < n/2$, the commit protocol can be easily adapted so that the proof that all honest players have consistent shares still goes through; basically, the process of accusations with subsequent broadcast of verification polynomials as in step 5 will be repeated until there are no new accusations (hence the commit protocol may no longer be constant round).

However, the proof that the opening always succeeds fails. The problem is that since honest players cannot prove that the shares they claim to have received are genuine, we have to accept up to $n/2$ complaints in the opening phase, and this will allow a corrupt D to open a commitment any way he wants. Clearly, if D could digitally sign his shares, then we would not have to accept any complaints and we would be in business again. Of course, digital signatures require computational assumptions, which we do not want to make in this scenario. However, there are ways to make unconditionally secure authentication schemes which ensure the same functionality (except with negligibly small error probability, see [12]).

Finally, this commitment scheme generalizes nicely to a scenario in which the underlying secret sharing scheme is not Shamir's but in fact a general linear secret sharing scheme (see later for more details on this).

We now show a *Commitment Multiplication Protocol (CMP)* that works without error if $t < n/3$.

CMP:

1. Inputs are commitments $[a]_i, [b]_i, [c]_i$ where P_i claims that $ab = c$.
   First P_i performs CSP on commitments $[a]_i, [b]_i$ to get committed shares $[a_1]_1, \ldots, [a_n]_n$ and $[b_1]_1, \ldots, [b_n]_n$.
2. P_i computes the polynomial $g_c = f_a \cdot f_b$, where f_a (f_b) is the polynomial used for sharing a (b) in the previous step.
   He commits to the coefficients of g_c.
   Note that there is no need to commit to the degree 0 coefficient, since this should be c, which is already committed to.
3. Define $c_i = g_c(i)$.
   From the commitments made so far and $[c]_i$, the players can compute (by linear operations) commitments $[c_1]_i, \ldots, [c_n]_i$, where of course P_i claims that $a_j b_j = c_j$, for $1 \leq j \leq n$.
4. For $j = 1, \ldots, n$, commitment $[c_j]_i$ is opened privately to P_j, i.e. the shares needed to open it are sent to P_j (instead of being broadcast).
5. If the value revealed this way is not $a_j b_j$, P_j broadcasts a complaint and opens (his own) commitments $[a_j]_j, [b_j]_j$. In response, P_i must open $[c_j]_i$ and is disqualified if $a_j b_j \neq c_j$.

We argue the correctness of this protocol.

Clearly, no matter how a possible adversary behaves, there is a polynomial g_c of degree at most $2t$ such that $c = g_c(0)$ and each $c_j = g_c(j)$.

Consider the polynomial $f_a \cdot f_b$, which is of degree at most $2t$ as well.

Suppose that $c \neq ab$. Thus $g_c \neq f_a \cdot f_b$. By Lagrange Interpolation, it follows that for at most $2t$ values of j we have $g_c(j) = f_a(j) \cdot f_b(j)$, or equivalently, $c_j = a_j b_j$.

Thus at least $n - 2t$ players P_j have $c_j \neq a_j b_j$, which is at least one more than the maximum number t of corrupted players (since $t < n/3$).

Therefore, at least one honest player will complain, and the prover is exposed in the last step of the protocol.

CSP:

Although CSP can be bootstrapped in a generic fashion from homomorphic commitment and CTP using the Generic CSP Protocol given earlier, we now argue that in the information theoretic scenario with $t < n/3$, there is a much simpler and more efficient solution: a slightly more refined analysis shows that the commit protocol we presently earlier is essentially already a CSP!

Consider an execution of the commit protocol, assuming D is honest. It is immediate that, for each player P_i (honest or corrupt!), there exists a commitment $[s_i]_i$ to his share s_i in the value a that D is committed to via $[a]_D$. The polynomial underlying $[s_i]_i$ is of course the verification polynomial $f_i(X)$ and each honest player P_j obtains $f_i(j)$ as $f_j(i)$.

Therefore, if each honest player holds on to his verification polynomial for later use, *each player* P_i is committed to his share s_i in the value a via $[s_i]_i$.

Apart from handling the corrupt D case, the only thing to be settled is that, by definition, CSP takes as input a commitment $[a]_D$. This, however, can easily be "imported" into the protocol: D knows the polynomial f that underlies $[a]_D$, and the players know their shares in a. We simply modify the commit protocol by requiring that D chooses this particular f as the real sharing polynomial. Also, upon receiving his verification polynomial in the first step of the commit protocol, each player checks that his real share is equal to the share in a he already had as part of the input. If this is not so, he broadcasts an accusation. If there are at most t accusations, the commit protocol continues as before. Else, it is aborted, and D is deemed corrupt. It is easy to see that this works; if D is honest it clearly does, and if D is corrupt and uses a different real sharing polynomial, then, by similar arguments as used before, there are more than t accusations from honest players.

As for the case of a possibly corrupt D, the discussion above shows that it is sufficient to prove the following. If the commit protocol is accepted, then there exists a unique symmetric bi-variate polynomial $G(X,Y) \in K[X,Y]$, with the degrees in X as well as Y at most t, such that for an honest player P_i, $f_i(X) = G(X,i)$ is the verification polynomial held by him *at the end of the protocol*. In other words, if the protocol is accepted, then, regardless whether the dealer is honest or not, the information held by the honest players is "consistent with an honest D."

We have to justify the claim above from the acceptance assumption only; we cannot make any a priori assumptions about how a possibly corrupt D computes the various pieces of information.

Let C denote the subset of the honest players B that do not accuse D at any point. As we have seen, acceptance implies $|C| \geq t + 1$ as well as "consistency," i.e., for all $i \in C$ and for all $j \in B$, $f_i(j) = f_j(i)$. Without loss of generality, we now assume that $|C| = t + 1$.

Let $\delta_i(X) \in K[X]$ denote the polynomial of degree t such that for all $i, j \in C$,

$$\delta_i(j) = 1 \text{ if } i = j \text{ and } \delta_i(j) = 0 \text{ if } i \neq j,$$

or, equivalently,

$$\delta_i(X) = \prod_{j \in C, j \neq i} \frac{X - i}{i - j}.$$

Recall that the Lagrange Interpolation Theorem may be phrased as follows. If $h(X) \in K[X]$ has degree at most t, then $h(X) = \sum_{i \in C} h(i)\delta_i(X)$.

Consider the polynomial

$$G(X, Y) = \sum_{i \in C} f_i(X)\delta_i(Y) \in K[X, Y].$$

This is clearly the unique polynomial in $K[X, Y]$ whose degree in Y is at most t and for which $G(X, i) = f_i(X)$ for all $i \in C$. This follows from Lagrange Interpolation applied over $K(X)$, i.e, the fraction field of $K[X]$, rather than over K. Note also that its degree in X is at most t.

We now verify that $G(X, Y)$ is symmetric:

$$G(X, Y) = \sum_{i \in C} f_i(X)\delta_i(Y) = \sum_{i \in C} \left(\sum_{j \in C} f_i(j)\delta_j(X) \right) \delta_i(Y)$$

$$= \sum_{i \in C} f_i(i)\delta_i(X)\delta_i(Y) + \sum_{i, j \in C, i \neq j} f_i(j)(\delta_i(X)\delta_j(Y) + \delta_j(X)\delta_i(Y)),$$

where the last equality follows from consistency.

Finally, for all $j \in B$, we have that

$$f_j(X) = \sum_{i \in C} f_j(i)\delta_i(X) = \sum_{i \in C} f_i(j)\delta_i(X)$$

$$= \sum_{i \in C} G(j, i)\delta_i(X) = \sum_{i \in C} G(i, j)\delta_i(X) = G(X, j),$$

as desired.

4.4. Formal Proof for the F_{Com} Realization

We have not given a full formal proof that the F_{Com} realization we presented really implements F_{Com} securely according to the definition. For this, one needs to present a simulator and prove that it acts as it should according to the definition. We will not do this in detail here, but we will give the main ideas one needs to build such a simulator – basically, one needs the following two observations:

- If player P_i is honest and commits to some value x_i, then since the commitment is based on secret sharing, this only results in the adversary seeing an unqualified set of shares, insufficient to determine x_i (we argued that anything else the adversary sees follows from these shares). The set of shares is easy to simulate even if x_i is not known, e.g., by secret sharing an arbitrary value and extracting shares for the currently corrupted players. This simulation is perfect because our analysis above shows that an unqualified set of shares have the same distribution regardless of the value of the secret.

If the (adaptive) adversary corrupts P_i later, it expects to see all values related to the commitment. But then the simulator can corrupt P_i in the ideal process and learn the value x_i that was committed to. It can then easily make a full set of shares that are consistent with x_i and show to the adversary. This can be done by solving a set of linear equations, since each share is a linear function of x_i and randomness chosen by the committer.

- If P_i is corrupt already when it is supposed to commit to x_i, the adversary decides all messages that P_i should send, and the simulator sees all these messages. As we discussed, either the commitment is rejected by the honest players and P_i is disqualified, or the messages sent by P_i determine uniquely a value x'_i. So then the simulator can in the ideal process send x'_i on behalf of P_i.

5. The Cryptographic Scenario

We have now seen how to solve the MPC problem in the i.t. scenario. Handling the cryptographic case can be done in various ways, each of which can be thought of as different ways of adapting the information theoretic solution to the cryptographic scenario.

5.1. Using Encryption to Implement the Channels

A very natural way to adapt the information theoretic solution is the following: since the i.t. protocol works assuming perfect channels connecting every pair of players, we could simply run the information theoretically secure protocol, but implement the channels using encryption, say by encrypting each message under the public key of the receiver. Intuitively, if the adversary is bounded and cannot break the encryption, he is in a situation no better than in the i.t. scenario, and security should follow from security of the information theoretic protocol.

This approach can be formalized by thinking of the i.t. scenario as being the cryptographic scenario extended with an ideal functionality that provides the perfect channels, i.e., it will accept from any player a message intended for another player, and will give the message to the receiver without releasing any information to the adversary, other than the length of the message. If a given method for encryption can be shown to securely realize this functionality, the result we wanted follows directly from the composition theorem.

For a static adversary, standard semantically secure encryption provides a secure realization of this communication functionality, whereas for an adaptive adversary, one needs a strong property known as non-committing encryption [9]. The reason is as follows: suppose player P_i has not yet been corrupted. Then the adversary of course does not know his input values, but it has seen encryptions of them. The simulator doesn't know the inputs either, so it must make fake encryptions with some arbitrary content to simulate the actions of P_i. This is all fine for the time being, but if the adversary corrupts P_i later, then the simulator gets an input for P_i, and must produce a good simulation of P_i's entire history to show to the adversary, and this must be consistent with this input and what the adversary already knows. Now the simulator is stuck: it cannot open its simulated encryptions the right way. Non-committing encryption solves exactly this problem by allowing the simulator to create "fake" encryptions that can later be convincingly claimed to contain any desired value.

Both semantically secure encryption and non-committing encryption can be implemented based on any family of trapdoor one-way permutations, so this shows that these general complexity assumptions are sufficient for general cryptographic MPC. More efficient encryption schemes exist based on specific assumptions such as hardness of factoring. However, known implementations of non-committing encryption are significantly slower, typically by a factor of k where k is the security parameter.

5.2. Cryptographic Implementations of Higher-Level Functionalities

Another approach is to use the fact that the general actively secure solution is really a general high-level protocol that makes use of the F_{Com} functionality to reach its goal.

Therefore, a potentially more efficient solution can be obtained if one can make a cryptographically secure implementation of F_{Com}, as well as the communication functionality.

If the adversary is static, we can use, e.g., the commitments from [11] based on q-one-way homomorphisms, which exists, e.g. if RSA is hard to invert or if the decisional Diffie-Hellman problem in some prime order group is hard. We then require that the field over which we compute is $GF(q)$. A simple example is if we have primes p, q, where $q|p-1$ and g, h, y are elements in Z_p^* of order q chosen as public key by player P_i. Then $[a]_i$ is of form $(g^r, y^a h^r)$, i.e. a Diffie-Hellman (El Gamal) encryption of y^a under public key g, h. In [11], protocols are shown for proving efficiently in zero-knowledge that you know the contents of a commitment, and that two commitments contains the same value, even if they were done with respect to different public keys. It is trivial to derive a CTP from this: P_i privately reveals the contents and random bits for $[a]_i$ to P_j (by sending them encrypted under P_j's public key). If this is not correct, P_j complains, otherwise he makes $[a]_j$ and proves it contains the same value as $[a]_i$. Finally, [11] also show a CMP protocol. We note that, in order to be able to do a simulation-based proof of security of this F_{Com} implementation, each player must give zero-knowledge, proof of knowledge of his secret key initially, as well as prove that he knows the contents of each commitment he makes.

If the adversary is adaptive, the above technique will not work, for the same reasons as explained in the previous subsection. It may seem natural to then go to commitments and encryption with full adaptive security, but this means we need to use non-committing encryption and so we will loose efficiency. However, under specific number theoretic assumptions, it is possible to build adaptively secure protocols using a completely different approach based on homomorphic public key encryption, without loosing efficiency compared to the static security case[17].

6. Protocols Secure for General Adversary Structures

It is relatively straightforward to use the techniques we have seen to construct protocols secure against general adversaries, i.e., where the adversary's corruption capabilities are not described only by a threshold t on the number of players that can be corrupt, but by a general adversary structure, as defined earlier.

What we have seen so far can be thought of as a way to build secure MPC protocols from Shamir's secret sharing scheme. The idea is now to replace Shamir's scheme by something more general, but otherwise use essentially the same high-level protocol.

To see how such a more general scheme could work, observe that the evaluation of shares in Shamir's scheme can be described in an alternative way. If the polynomial used is $f(x) = s + a_1 x + \cdots + a_t x^t$, we can think of the coefficients $(s, a_1, \ldots, a_t)$ as being arranged in a column vector $\mathbf{a}$. Evaluating f in points $1, 2, .., n$ is now equivalent to multiplying the vector by a Van der Monde matrix M, with rows of form $(i^0, i^1, \ldots, i^t)$. We may think of the scheme as being defined by this fixed matrix, and by the rule that each player is assigned 1 row of the matrix, and gets as his share the coordinate of $M\mathbf{a}$ corresponding to his row.

It is now immediate to think of generalizations of this: to other matrices than Van der Monde, and to cases where players can have more than one row assigned to them. This leads to general linear secret sharing schemes, also known as Monotone Span Programs (MSP). The term "linear" is motivated by the fact any such scheme has the same property as Shamir's scheme, that sharing two secrets s, s' and adding corresponding shares of s and s', we obtain shares of $s + s'$. The protocol constructions we have seen have primarily used this linearity property, so this is why it makes sense to try to plug in MSP's instead of Shamir's scheme. There are, however, several technical difficulties to sort out along the way, primarily because the method we used to do secure multiplication only generalizes to MSP's with a certain special property, so called multiplicative MSP's. Not all MSP's are multiplicative, but it turns that any MSP can be used to construct a new one that is indeed multiplicative.

Furthermore, it turns out that for *any* adversary structure, there exists an MSP-based secret sharing scheme for which the unqualified sets are exactly those in the adversary structure. Therefore, these ideas lead to MPC protocols for any adversary structure where MPC is possible at all.

For details on how to use MPS's to do MPC, see [13].

Appendix A. Formal Details of the General Security Model for Protocols

In this section we propose a notion of universally composable security of synchronous protocols.

A.1. The Real-Life Execution

A real-life protocol π consists of n parties $P_1, \ldots, P_n$, all PPT interactive Turing machines (ITMs). The execution of a protocol takes place in the presence of an environment $\mathcal{Z}$, also a PPT ITM, which supplies inputs to and receives outputs from the parties. Following Definition 4 from [8] $\mathcal{Z}$ also models the adversary of the protocol, and so schedules the activation of the parties, corrupts parties adaptively and controls corrupted parties. We assume that the parties are connected by open authenticated channels.

To simplify notation we assume that in each round r each party P_i sends a message $m_{i,j,r}$ to each party P_j, including itself. The message $m_{i,i,r}$ can be thought of as the state of P_i after round r. To further simplify the notation we assume that in each round $\mathcal{Z}$ inputs a value $x_{i,r}$ to P_i and receives an output $y_{i,r}$. A protocol not following this convention can easily be patched by introducing some dummy value $\epsilon =$ **not a value**. Using this convention we can write the r'th activation of P_i as $(m_{i,1,r}, \ldots, m_{i,n,r}, y_{i,r}) = P_i(k, m_{1,i,r-1}, \ldots, m_{n,i,r-1}, x_{i,r}; r_i)$, where k is the security parameter and r_i is the random bits used by P_i. We assume that the parties cannot reliably erase their state. To model this we give r_i to $\mathcal{Z}$ when P_i is corrupted. Since $\mathcal{Z}$ knows all the inputs of P_i this will allow $\mathcal{Z}$ to reconstruct the entire execution history of P_i. In detail the real-life execution proceeds as follows.

Init: The input to an execution is the security parameter k, the random bits $r_1, \ldots, r_n \in \{0,1\}^*$ used by the parties and an auxiliary input $z \in \{0,1\}^*$ for $\mathcal{Z}$.

Initialize the round counter $r = 0$ and initialize the set of corrupted parties $C = \emptyset$. In the following let $H = \{1, \ldots, n\} \setminus C$.

Let $m_{i,j,0} = \epsilon$ for $i, j \in [n]$.

Input k and z to $\mathcal{Z}$ and activate $\mathcal{Z}$.

Environment activation: When $\mathcal{Z}$ is activated it outputs one of the following commands: (**activate** $i, x_{i,r}, \{m_{j,i,r-1}\}_{j \in C}$) for $i \in H$ or (**corrupt** i) for $i \in H$ or (**end round**) or (**guess** b) for $b \in \{0,1\}$.

We require that no two (**activate** $i, \ldots$) commands for the same i are issued without being separated by an (**end round**) command and we require that between two (**end round**) commands an (**activate** $i, \ldots$) command was issued for $i \in H$, where H denotes the value of H when the second of the (**end round**) commands were issued.

When a (**guess** b) command is given the execution stops. The other commands are handled as described below. After the command is handled the environment is activated again.

Party activation: Values $\{m_{j,i,r-1}\}_{j \in H}$ were defined in the previous round; Add these to $\{m_{j,i,r-1}\}_{j \in C}$ from the environment and compute

$$(m_{i,1,r}, \ldots, m_{i,n,r}, y_{i,r}) = P_i(k, m_{1,i,r-1}, \ldots, m_{n,i,r-1}, x_{i,r}; r_i) \ .$$

Then give $\{m_{i,j,r}\}_{j \in [n] \setminus \{i\}}$ to $\mathcal{Z}$.

Corrupt: Give r_i to $\mathcal{Z}$. Set $C = C \cup \{i\}$.

End round: Give the values $\{y_{i,r}\}_{i \in H}$ defined in **Party activation** to $\mathcal{Z}$ and set $r = r + 1$.

The result of the execution is the bit b output by $\mathcal{Z}$. We are going to denote this bit by $\text{REAL}_{\pi, \mathcal{Z}}(k, r_1, \ldots, r_n, z)$. This defines a random variable $\text{REAL}_{\pi, \mathcal{Z}}(k, z)$, where we take the r_i to be uniformly random, and in turn defines a Boolean distribution ensemble $\text{REAL}_{\pi, \mathcal{Z}} = \{\text{REAL}_{\pi, \mathcal{Z}}(k, z)\}_{k \in \mathbf{N}, z \in \{0,1\}^*}$.

A.2. The Ideal Process

To define the security of a protocol an ideal functionality $\mathcal{F}$ is specified. The ideal functionality is a PPT ITM with n input tapes and n output tapes which we think of as being connected to n parties. The ideal functionality defines the desired input-output behaviour of the protocol and defines the desired secrecy by keeping the inputs secret. In the execution of an ideal functionality in an environment $\mathcal{Z}$, the inputs to P_i from $\mathcal{Z}$ is simply handed to $\mathcal{F}$ and the outputs from $\mathcal{F}$ to P_i is handed to $\mathcal{Z}$. To be able to specify protocols which leak some information about the inputs of the parties $\mathcal{F}$ has a special tape. To model protocols which are allowed to leak some specified information about the inputs of the parties the functionality simply outputs this information on the special tape. An example could be the following functionality modelling secure communication: It is connected to two parties S and R. If R inputs some value $m \in \{0,1\}^*$, then $|m|$ is output on the special tape and m is output to R.

The ideal functionality also has the special input tape on which it receives two kinds of messages. When a party P_i is corrupted it receives the input (`corrupt` i) in response to which it might produce some output which is written on the special output tape. This behaviour can be used when modelling protocols which are allowed to leak a particular information when a given party is corrupted. It can also receive the input (`activate` v) on the special tape in response to which it writes a value on the output tape for each party. The rules of the ideal process guarantees that $\mathcal{F}$ will have received exactly one input for each honest party between consecutive (`activate` v) commands. The value v can be thought of as the inputs to $\mathcal{F}$ from the corrupted parties, but can be interpreted by $\mathcal{F}$ arbitrarily, i.e., according to its specification.

We then say that a protocol π securely realizes an ideal functionality $\mathcal{F}$ if the protocol has the same input-output behaviour as the functionality (this captures correctness) and all the communication of the protocol can be simulated given only the inputs and the outputs of the corrupted parties and the values on the special tape of $\mathcal{F}$ (this captures secrecy of the honest parties' inputs). When $\mathcal{F}$ is executed in some environment $\mathcal{Z}$ the environment knows the inputs and the outputs of all parties, so $\mathcal{Z}$ cannot be responsible of simulating. We therefore introduce a so-called interface or simulator $\mathcal{S}$ which is responsible for the simulation. The interface is put between the environment $\mathcal{Z}$ and the ideal-process. The job of $\mathcal{S}$ is then to simulate a real-life execution by giving the environment correctly looking responses to the commands it issues. In doing this the interface sees the outputs from $\mathcal{F}$ on the special output tape (to model leaked information) and can specify the value v to $\mathcal{F}$ on the special input tape (to specify inputs of the corrupted parties or e.g. non-deterministic behaviour, all depending on how $\mathcal{F}$ is defined to interpret v). We note that $\mathcal{S}$ does not see the messages sent between $\mathcal{F}$ and $\mathcal{Z}$ for honest parties (which is exactly the purpose of introducing $\mathcal{S}$). In detail the ideal process proceeds as follows.

Init: The input to an ideal process is the security parameter k, the random bits $r_{\mathcal{F}}$ and $r_{\mathcal{S}}$ used by $\mathcal{F}$ and $\mathcal{S}$ and an auxiliary input $z \in \{0,1\}^*$ for $\mathcal{Z}$.
    Initialize the round counter $r = 0$ and initialize the set of corrupted parties $C = \emptyset$.
    Provide $\mathcal{S}$ with $r_{\mathcal{S}}$, provide $\mathcal{F}$ with $r_{\mathcal{F}}$ and give k and z to $\mathcal{Z}$ and activate $\mathcal{Z}$.

Environment activation: $\mathcal{Z}$ is defined exactly as in the real-word, but now commands are handled by $\mathcal{S}$, as described below.

Party activation: The values $\{m_{j,i,r-1}\}_{i \in C}$ are input to $\mathcal{S}$ and the value $x_{i,r}$ is input to $\mathcal{F}$ on the input tape for P_i and $\mathcal{F}$ is run and outputs some value $v_{\mathcal{F}}$ on the special tape. This value is given to $\mathcal{S}$ which is then required to compute some values $\{m_{i,j,r}\}_{j \in [n] \setminus \{i\}}$ and return these to $\mathcal{Z}$.

Corrupt: When $\mathcal{Z}$ corrupts P_i, $\mathcal{S}$ is given the values $x_{i,0}, y_{i,0}, x_{i,1}, \ldots$ exchanged between $\mathcal{Z}$ and $\mathcal{F}$ for P_i. Furthermore (corrupt i) is input to $\mathcal{F}$ in response to which $\mathcal{F}$ returns some value $v_{\mathcal{F}}$ which is also given to $\mathcal{S}$. Then $\mathcal{S}$ is required to compute some value r_i and return it to $\mathcal{Z}$. Set $C = C \cup \{i\}$.

End round: When a (end round) command is issued $\mathcal{S}$ is activated and produces a value v. Then (activate v) is input to $\mathcal{F}$ which produces outputs $\{y_{i,r}\}_{i \in [n]}$. The values $\{y_{i,r}\}_{i \in C}$ are then handed to $\mathcal{S}$ and the values $\{y_{i,r}\}_{i \in H}$ are handed to $\mathcal{Z}$. Set $r = r + 1$.

The result of the ideal-process is the bit b output by $\mathcal{Z}$. We are going to denote this bit by $\text{IDEAL}_{\mathcal{F},\mathcal{S},\mathcal{Z}}(k, r_{\mathcal{F}}, r_{\mathcal{S}}, z)$. This defines a random variable $\text{IDEAL}_{\mathcal{F},\mathcal{S},\mathcal{Z}}(k, z)$ and in turn defines a Boolean distribution ensemble $\text{IDEAL}_{\mathcal{F},\mathcal{S},\mathcal{Z}} = \{\text{IDEAL}_{\mathcal{F},\mathcal{S},\mathcal{Z}}(k, z)\}_{k \in \mathbf{N}, z \in \{0,1\}^*}$.

Notice that the interaction of $\mathcal{Z}$ with the real-world and the ideal process has the same pattern. The goal of the interface is then to produce the values that it hands to $\mathcal{Z}$ in such a way that $\mathcal{Z}$ cannot distinguish whether it is observing the real-life execution or a simulation of it in the ideal process. Therefore the bit b output by $\mathcal{Z}$ can be thought of as a guess on which of the two it is observing. This gives rise to the following definition.

Definition 2. *We say that π t-securely realizes $\mathcal{F}$ if there exists an interface $\mathcal{S}$ such that for all environments $\mathcal{Z}$ corrupting at most t parties it holds that* $\text{IDEAL}_{\mathcal{F},\mathcal{S},\mathcal{Z}} \overset{c}{\approx} \text{REAL}_{\pi,\mathcal{Z}}$.

Here, the notation $\overset{c}{\approx}$ means that the two distributions involved are computationally indistinguishable.

A.3. The Hybrid Models

We now describe the $\mathcal{G}$-hybrid model for a synchronous ideal functionality $\mathcal{G}$. Basically the $\mathcal{G}$-hybrid model is the real-life model where in addition the parties have access to an ideal functionality $\mathcal{G}$ to aid them in the computation. In each round r party P_i will receive an output $t_{i,r-1}$ from $\mathcal{G}$ from the previous round and will produce and input $s_{i,r}$ for $\mathcal{G}$ for round r. This means that the r'th activation of P_i now is given by $(m_{i,1,r}, \ldots, m_{i,n,r}, y_{i,r}, s_{i,r}) = P_i(k, m_{1,i,r-1}, \ldots, m_{n,i,r-1}, x_{i,r}, t_{i,r-1}; r_i)$. In the hybrid model, still $\mathcal{Z}$ models the adversary. Therefore, the output from $\mathcal{G}$ on its special tape, which models public information, is given to $\mathcal{Z}$, and the inputs to $\mathcal{G}$ on its special input tape, which can be thought of as modelling the inputs from corrupted parties, is provided by $\mathcal{Z}$. In detail the hybrid execution proceeds as follows.

Init: The input to an execution is the security parameter k, the random bits $r_1, \ldots, r_n \in \{0,1\}^*$ used by the parties, the random bits $r_\mathcal{G}$ for $\mathcal{G}$ and an auxiliary input $z \in \{0,1\}^*$ for $\mathcal{Z}$.

Initialize the round counter $r = 0$ and initialize the set of corrupted parties $C = \emptyset$.

Let $m_{i,j,0} = \epsilon$ for $i, j \in [n]$ and let $t_{i,-1} = \epsilon$.

Provide $\mathcal{G}$ with $r_\mathcal{G}$ and input k and z to $\mathcal{Z}$ and activate $\mathcal{Z}$.

Environment activation: $\mathcal{Z}$ is defined exactly as in the real-word except that the (**end round**) command has the syntax (**end round** v) for some value v and that $\mathcal{Z}$ receives some extra values in response to the commands as described below.

Party activation: Values $\{m_{j,i,r-1}\}_{j \in H}$ and $t_{i,r-1}$ were defined in the previous round. Add these to $\{m_{j,i,r-1}\}_{j \in C}$ from the environment and compute

$$(m_{i,1,r}, \ldots, m_{i,n,r}, y_{i,r}, s_{i,r}) = P_i(k, m_{1,i,r-1}, \ldots, m_{n,i,r-1}, x_{i,r}, t_{i,r-1}; r_i) .$$

Then the value $s_{i,r}$ is input to $\mathcal{G}$ on the input tape for P_i and $\mathcal{G}$ is run and produces some value $v_\mathcal{G}$ on the special tape. Then $v_\mathcal{G}$ is given to $\mathcal{Z}$ along with $\{m_{i,j,r}\}_{j \in [n] \setminus \{i\}}$.

Corrupt: Give r_i to $\mathcal{Z}$ along with the values $s_{i,0}, t_{i,0}, s_{i,1} \ldots$ exchanged between P_i and $\mathcal{G}$, see below in **End round**. Furthermore (**corrupt** i) is input to $\mathcal{G}$ in response to which $\mathcal{G}$ returns some value $v_\mathcal{G}$ which is also given to $\mathcal{Z}$. Set $C = C \cup \{i\}$.

End round: Give the values $\{y_{i,r}\}_{i \in H}$ defined in **Party activation** to $\mathcal{Z}$. Furthermore, input (**activate** v) to $\mathcal{G}$ and receive the output $\{t_{i,r}\}_{i \in [n]}$. The values $\{t_{i,r}\}_{i \in C}$ are then handed to $\mathcal{Z}$ and the values $\{t_{i,r}\}_{i \in H}$ are used as input for the honest parties in the next round. Set $r = r + 1$.

The result of the hybrid execution is the bit b output by $\mathcal{Z}$. We will denote this bit by $\mathrm{HYB}^\mathcal{G}_{\pi, \mathcal{Z}}(k, r_1, \ldots, r_n, r_\mathcal{G}, z)$. This defines a random variable $\mathrm{HYB}^\mathcal{G}_{\pi, \mathcal{Z}}(k, z)$ and in turn defines a Boolean distribution ensemble $\mathrm{HYB}^\mathcal{G}_{\pi, \mathcal{Z}}$.

As for an interface $\mathcal{S}$ simulating a real-life execution of a protocol π in the ideal process for ideal functionality $\mathcal{F}$ we can define the notion of a hybrid interface

$\mathcal{T}$ simulating a hybrid execution of a hybrid protocol $\pi[\mathcal{G}]$ in the ideal process for ideal functionality $\mathcal{F}$. This is defined equivalently. The only difference is that an ideal interface $\mathcal{T}$ has to return more values to $\mathcal{Z}$ to be successful. For completeness we give the ideal process with a hybrid simulator in detail.

Init: The input to an ideal process is the security parameter k, the random bits $r_{\mathcal{F}}$ and $r_{\mathcal{T}}$ used by $\mathcal{F}$ and $\mathcal{T}$ and an auxiliary input $z \in \{0,1\}^*$ for $\mathcal{Z}$.

Initialize the round counter $r = 0$ and initialize the set of corrupted parties $C = \emptyset$.

Provide $\mathcal{T}$ with $r_{\mathcal{T}}$, provide $\mathcal{F}$ with $r_{\mathcal{F}}$ and give k and z to $\mathcal{Z}$ and activate $\mathcal{Z}$.

Environment activation: $\mathcal{Z}$ is defined exactly as in the hybrid world, but now the commands are handled by $\mathcal{T}$, as described below.

Party activation: The values $\{m_{j,i,r-1}\}_{i \in C}$ are input to $\mathcal{T}$ and the value $x_{i,r}$ is input to $\mathcal{F}$ on the input tape for P_i and $\mathcal{F}$ is run and outputs some value $v_{\mathcal{F}}$ on the special tape. This value is given to $\mathcal{T}$ which is then required to compute some values $\{m_{i,j,r}\}_{j \in [n] \setminus \{i\}}$ and a value value $v_{\mathcal{G}}$ and return these to $\mathcal{Z}$.

Corrupt: When $\mathcal{Z}$ corrupts a party $\mathcal{T}$ is given the values $x_{i,0}, y_{i,0}, x_{i,1}, \ldots$ exchanged between $\mathcal{Z}$ and $\mathcal{F}$ for P_i. Furthermore (**corrupt** i) is input to $\mathcal{F}$ in response to which $\mathcal{F}$ returns some value $v_{\mathcal{F}}$ which is also given to $\mathcal{T}$. Then $\mathcal{T}$ is required to compute some value r_i, some value $s_{i,0}, t_{i,0}, s_{i,1}, \ldots$ and some value $v_{\mathcal{G}}$ and return it to $\mathcal{Z}$. Set $C = C \cup \{i\}$.

End round: When a (**end round** v) command is issued $\mathcal{T}$ is activated with input (**end round** v) and produces a value v'. Then (**activate** v') is input to $\mathcal{F}$ which produces outputs $\{y_{i,r}\}_{i \in [n]}$. The values $\{y_{i,r}\}_{i \in C}$ are then handed to $\mathcal{T}$ which produces an output $\{t_{i,r}\}_{i \in C}$ and the values $\{t_{i,r}\}_{i \in C}$ and $\{y_{i,r}\}_{i \in H}$ are handed to $\mathcal{Z}$. Set $r = r + 1$.

Notice that the interaction of $\mathcal{Z}$ with the hybrid model and the ideal process has the same pattern. The goal of the interface $\mathcal{T}$ is then to produce the values that it hands to $\mathcal{Z}$ in such a way that $\mathcal{Z}$ cannot distinguish whether it is observing the hybrid execution or a simulation of it in the ideal process.

Definition 3. *We say that π t-securely realizes $\mathcal{F}$ in the $\mathcal{G}$-hybrid model if there exists an hybrid interface $\mathcal{T}$ such that all environments $\mathcal{Z}$ corrupting at most t parties it holds that $IDEAL_{\mathcal{F},\mathcal{T},\mathcal{Z}} \stackrel{c}{\approx} HYB^{\mathcal{G}}_{\pi,\mathcal{Z}}$.*

A.4. Composing Protocols

Assume that we are given two protocols $\gamma = (P_1^{\gamma}, \ldots, P_n^{\gamma})$ for the real-life model and $\pi[\cdot] = (P_1^{\pi}[\cdot], \ldots, P_n^{\pi}[\cdot])$ for a hybrid model. We describe how to compose such protocol to obtain a real-life protocol $\pi[\gamma] = (P_1^{\pi}[P_1^{\gamma}], \ldots, P_n^{\pi}[P_n^{\gamma}])$, which is intended to be the two protocols run in lock-step while replacing the ideal functionality access of $\pi[\cdot]$ by calls to γ. The messages send by the parties $P_i = P_i^{\pi}[P_i^{\gamma}]$ will consist of a message from each of the two protocols. For this purpose we fix some bijective encoding $(\cdot,\cdot) : \{0,1\}^* \times \{0,1\}^* \rightarrow \{0,1\}^*$ which can be computed and inverted efficiently.

The activation $(m_{i,1,r}, \ldots, m_{i,n,r}, y_{i,r}) = P_i(k, m_{1,i,r-1}, \ldots, m_{n,i,r-1}, x_{i,r}; r_i)$ is computed as follows. If while running $P_i^\pi[\cdot]$ and P_i^γ these machines request a random bit, give them a fresh random bit from r_i. For notational convenience we let r_i^π and r_i^γ denote the bits used by $P_i^\pi[\cdot]$ respectively P_i^γ. For $j \in [n] \setminus \{i\}$ let $(m_{i,j,r-1}^\pi, m_{i,j,r-1}^\gamma) = m_{i,j,r-1}$ and let $((m_{i,i,r-1}^\pi, m_{i,i,r-1}^\gamma), t_{i,r-1}) = m_{i,i,r-1}$. Then compute $(m_{1,i,r}, \ldots, m_{n,i,r}, y_{i,r}, s_{i,r}) = P_i^\pi(k, m_{1,i,r-1}^\pi, \ldots, m_{n,i,r-1}^\pi, x_{i,r}^\pi, t_{i,r-1}; r_i^\pi)$ and then compute $(m_{1,i,r}, \ldots, m_{n,i,r}, t_{i,r}) = P_i^\gamma(k, m_{1,i,r-1}^\gamma, \ldots, m_{n,i,r-1}^\gamma, s_{i,r}; r_i^\gamma)$. Then for $j \in [n] \setminus \{i\}$ let $m_{i,j,r} = (m_{i,j,r}^\pi, m_{i,j,r}^\gamma)$ and let $m_{i,i,r} = ((m_{i,i,r}^\pi, m_{i,i,r}^\gamma), t_{i,r})$.

The following composition theorem follows directly from Lemma 2 in the below section.

Theorem 3. *Assume γ t-securely realizes $\mathcal{G}$ and that $\pi[\cdot]$ t-securely realizes $\mathcal{F}$ in the $\mathcal{G}$-hybrid model. Then $\pi[\gamma]$ t-securely realizes $\mathcal{F}$.*

A.5. Composing Interfaces

We now describe how to compose two interfaces. Assume that we are given a real-life interface $\mathcal{S}$ and a hybrid model interface $\mathcal{T}[\cdot]$. We now describe how to construct a new real-life interface $\mathcal{T}[\mathcal{S}]$. The idea behind the composition operation is as follows. Assume that $\mathcal{T}[\cdot]$ simulates a protocol $\pi[\mathcal{G}]$ while having access to the ideal functionality $\mathcal{F}$, and assume that $\mathcal{S}$ simulates a protocol π while having access to $\mathcal{G}$. We then want $\mathcal{U} = \mathcal{T}[\mathcal{S}]$ to simulate the protocol $\pi[\gamma]$ while having access to $\mathcal{F}$. This is done as follows. First of all $\mathcal{U}$ runs $\mathcal{T}[\cdot]$ using $\mathcal{U}$'s access to $\mathcal{F}$. This provides $\mathcal{U}$ with a simulated version of $\pi[\mathcal{G}]$ consistent with $\mathcal{F}$, which in particular provides it with a simulated access to $\mathcal{G}$. Using the simulated access to $\mathcal{G}$ it then runs $\mathcal{S}$ and gets a simulated version of γ consistent with $\mathcal{G}$ from the simulated $\pi[\mathcal{G}]$ consistent with $\mathcal{F}$. It then merges the values of the simulated version of $\pi[\mathcal{G}]$ and the simulated γ as defined by the composition operation on protocols and obtains a simulated version of $\pi[\gamma]$ consistent with $\mathcal{F}$. The notation used to describe the composition operation will reflect the above idea. The composed interface works as follows.

Init: $\mathcal{U}$ receives k and random bits r. When $\mathcal{S}$ or $\mathcal{T}[\cdot]$ request a random bit $\mathcal{U}$ gives them a random bit from r.

Party activation: $\mathcal{U}$ receives $\{m_{i,j,r-1}\}_{i \in C}$ from $\mathcal{Z}$ and $v_\mathcal{F}$ from $\mathcal{F}$ and must provide outputs $\{m_{i,j,r}\}_{j \in [n] \setminus \{i\}}$. This is done as follows.

1. For $i \in C$ compute $(m_{i,j,r-1}^\pi, m_{i,j,r-1}^\gamma) = m_{i,j,r-1}$.
2. Input $\{m_{i,j,r-1}^\pi\}_{i \in C}$ and $v_\mathcal{F}$ to $\mathcal{T}[\cdot]$ which generates values $\{m_{i,j,r}^\gamma\}_{j \in [n] \setminus \{i\}}$ and $v_\mathcal{G}$.
3. Input $\{m_{i,j,r-1}^\gamma\}_{i \in C}$ and $v_\mathcal{G}$ to $\mathcal{S}$ which generates values $\{m_{i,j,r}^\gamma\}_{j \in [n] \setminus \{i\}}$.
4. Output $\{m_{i,j,r}\}_{j \in [n] \setminus \{i\}}$, where $m_{i,j,r} = (m_{i,j,r}^\pi, m_{i,j,r}^\gamma)$.

Corrupt: $\mathcal{U}$ receives $x_{i,0}, y_{i,1}, x_{i,1}, \ldots$ and $v_\mathcal{F}$ and must provide an output r_i. This is done as follows.

1. Input $x_{i,0}, y_{i,1}, x_{i,1}, \ldots$ and $v_{\mathcal{F}}$ to $\mathcal{T}[\cdot]$ which generates values r_i^π and $s_{i,0}, t_{i,1}, s_{i,1}, \ldots$ and $v_{\mathcal{G}}$.
2. Input $s_{i,0}, t_{i,1}, s_{i,1}, \ldots$ and $v_{\mathcal{G}}$ to $\mathcal{S}$ which generates a value r_i^γ.
3. Outputs $r_i = [r_i^\pi, r_i^\gamma]$.

End round: $\mathcal{U}$ is given (**end round**) and must produce an output for $\mathcal{F}$ in response to which it receives $\{y_{i,r}\}_{i \in C}$. To run $\mathcal{S}$ and $\mathcal{T}[\cdot]$ as they expect this is done as follows.

1. Activate $\mathcal{S}$ on input (**end round**) and receive as output a value v.
2. Activate $\mathcal{T}[\cdot]$ on input (**end round** v) and receive as output a value v'.
3. Outputs v' and receive $\{y_{i,r}\}_{i \in C}$.
4. Hand $\{y_{i,r}\}_{i \in C}$ to $\mathcal{T}[\cdot]$ and get the output $\{t_{i,r}\}_{i \in C}$.
5. Then input $\{t_{i,r}\}_{i \in C}$ to $\mathcal{S}$.

Using the proof techniques from [8] it is straight forward to construct a proof for the following lemma. The proof contains no new ideas and have been excluded for that reason and to save space.

Lemma 2. *Assume that for all environments $\mathcal{Z}$ corrupting at most t parties, it holds that $IDEAL_{\mathcal{G},\mathcal{S},\mathcal{Z}} \overset{c}{\approx} REAL_{\gamma,\mathcal{Z}}$, and assume that for all hybrid environments $\mathcal{Z}$ corrupting at most t parties it holds that $IDEAL_{\mathcal{F},\mathcal{T},\mathcal{Z}} \overset{c}{\approx} HYB_{\pi,\mathcal{Z}}^{\mathcal{G}}$, then for all environments $\mathcal{Z}$ corrupting at most t parties it holds that $IDEAL_{\mathcal{F},\mathcal{T}[\mathcal{S}],\mathcal{Z}} \overset{c}{\approx} REAL_{\pi[\gamma],\mathcal{Z}}$.*

As mentioned, this lemma is essentially the composition theorem listed in the main text of this note. It trivially generalizes from the threshold adversaries assumed here to general adversary structures.

References

[1] D. Beaver: *Foundations of Secure Interactive Computing*, Proc. of Crypto 91.

[2] L. Babai, A. Gál, J. Kollár, L. Rónyai, T. Szabó, A. Wigderson: *Extremal Bipartite Graphs and Superpolynomial Lowerbounds for Monotone Span Programs*, Proc. ACM STOC '96, pp. 603–611.

[3] J. Benaloh, J. Leichter: *Generalized Secret Sharing and Monotone Functions*, Proc. of Crypto '88, Springer Verlag LNCS series, pp. 25–35.

[4] M. Ben-Or, S. Goldwasser, A. Wigderson: *Completeness theorems for Non-Cryptographic Fault-Tolerant Distributed Computation*, Proc. ACM STOC '88, pp. 1–10.

[5] E. F. Brickell: *Some Ideal Secret Sharing Schemes*, J. Combin. Maths. & Combin. Comp. 9 (1989), pp. 105–113.

[6] R. Canetti: *Studies in Secure Multiparty Computation and Applications*, Ph. D. thesis, Weizmann Institute of Science, 1995. (Better version available from Theory of Cryptography Library).

[7] R.Canetti, U.Fiege, O.Goldreich and M.Naor: *Adaptively Secure Computation*, Proceedings of STOC 1996.

[8] R.Canetti: *Universally Composable Security*, The Eprint archive, www.iacr.org.

[9] R. Canetti, U. Feige, O. Goldreich, M. Naor: *Adaptively Secure Multi-Party Computation*, Proc. ACM STOC '96, pp. 639–648.

[10] D. Chaum, C. Crépeau, I. Damgård: *Multi-Party Unconditionally Secure Protocols*, Proc. of ACM STOC '88, pp. 11–19.

[11] R. Cramer, I. Damgård: *Zero Knowledge for Finite Field Arithmetic or: Can Zero Knowledge be for Free?*, Proc. of CRYPTO'98, Springer Verlag LNCS series.

[12] R. Cramer, I. Damgård, S. Dziembowski, M: Hirt and T. Rabin: *Efficient Multiparty Computations With Dishonest Minority*, Proceedings of EuroCrypt 99, Springer Verlag LNCS series.

[13] R. Cramer, I. Damgård and U. Maurer: *Multiparty Computations from Any Linear Secret Sharing Scheme*. In: Proc. EUROCRYPT '00.

[14] R. Cramer. Introduction to Secure Computation. Latest version: January 2001. Available from http://www.brics.dk/~cramer

[15] C. Crepeau, J.vd.Graaf and A. Tapp: *Committed Oblivious Transfer and Private Multiparty Computation*, Proc. of Crypto 95, Springer Verlag LNCS series.

[16] D. Dolev, C. Dwork, and M. Naor, *Non-malleable cryptography*, Proc. ACM STOC '91, pp. 542–552.

[17] I. Damgård and J. Nielsen: *Universally Composable Efficient Multiparty Computation from Threshold Homomorphic Encryption*, Proc. of Crypto 2003, Springer Verlag LNCS.

[18] M. Fitzi, U. Maurer: *Efficient Byzantine agreement secure against general adversaries*, Proc. Distributed Computing DISC '98.

[19] R. Gennaro, M. Rabin, T. Rabin, *Simplified VSS and Fast-Track Multiparty Computations with Applications to Threshold Cryptography*, in Proc of ACM PODC'98, pp. 101–111.

[20] O. Goldreich, S. Micali and A. Wigderson: *How to Play Any Mental Game or a Completeness Theorem for Protocols with Honest Majority*, Proc. of ACM STOC '87, pp. 218–229.

[21] M. Hirt, U. Maurer: *Complete Characterization of Adversaries Tolerable in General Multiparty Computations*, Proc. ACM PODC'97, pp. 25–34.

[22] M. Karchmer, A. Wigderson: *On Span Programs*, Proc. of Structure in Complexity, 1993.

[23] J.Kilian: *Founding Cryptography on Oblivious Transfer*, Proceedings of the Twentieth Annual ACM Symposium on Theory of Computing, pages 20-31, Chicago, Illinois, 2-4 May 1988.

[24] S. Micali and P. Rogaway:*Secure Computation*, Manuscript, Preliminary version in Proceedings of Crypto 91.

[25] Nielsen: *Protocol Security in the Cryptographic Model*, PhD thesis, Dept. of Comp. Science, Aarhus University, 2003.

[26] T. P. Pedersen: *Non-Interactive and Information-Theoretic Secure Verifiable Secret Sharing*, Proc. CRYPTO '91, Springer Verlag LNCS, vol. 576, pp. 129–140.

[27] P. Pudlák, J. Sgall: *Algebraic Models of Computation and Interpolation for Algebraic Proof Systems* Proc. Feasible Arithmetic and Proof Complexity, Springer Verlag LNCS series.

[28] T. Rabin: *Robust Sharing of Secrets when the Dealer is Honest or Cheating*, J. ACM, 41(6):1089-1109, November 1994.

[29] T. Rabin, M. Ben-Or: *Verifiable Secret Sharing and Multiparty Protocols with Honest majority*, Proc. ACM STOC '89, pp. 73–85.

[30] A. Shamir: *How to Share a Secret*, Communications of the ACM 22 (1979) 612–613.

[31] M. van Dijk: *Secret Key Sharing and Secret Key Generation*, Ph.D. Thesis, Eindhoven University of Technology, 1997.

Foundations of Modern Cryptography

Giovanni Di Crescenzo

1. Introduction

The need for cryptography has been recognized since ancient times. One of its main goals, private communication in the presence of adversary, is traced back to the ancient Roman empire, whose emperor Julius Ceasar used to communicate to his allies by replacing each letter in his message with the third next letter in the alphabet.

Classical cryptography went on until the end of last century focusing on the art of designing and breaking secrecy codes. Modern cryptography has significantly enlarged its scope to the rigorous analysis of any system that is potentially subject to malicious threats and the design of solution that can guarantee the system to withstand such threats. As a consequence, many goals have been added to that of private communication in the presence of adversary, and cryptography has moved from an engineering art built on a number of heuristic techniques to a scientific discipline based on mathematically rigorous design requirements, solution techniques and correctness proofs.

We present here an introduction to some basic topics in the foundation of modern cryptography; specifically: one-way functions, pseudo-random generators, pseudo-random functions and zero-knowledge protocols.

2. One-Way Functions

Modern cryptography is based on the existence of computational problems that are "efficiently" solved by intended users and that can be associated with related computational problems that are conjectured to be "not efficiently" solvable by adversaries. Then the actual execution of the cryptographic protocol by its users is feasible while violating its security property by an adversary is not. The notion of "efficiency" is formulated according to the analogue notion in complexity theory; that is, an algorithm is efficient if it runs in polynomial time in a security parameter (typically specified by the length of its input). Consequently, the notion of a computational problem being not efficiently solvable is formulated by requiring that no algorithm running in polynomial time can solve the problem. We note that these notions are "asymptotic". In particular, a typical security requirement of a system may ask that a certain computational problem cannot be solved by a polynomial time algorithm for "sufficiently large" values of the security parameter.

Although at first sight it is clear that some hardness assumption is required to prove the security of a cryptographic scheme, it is not immediately clear which is the best assumption. Ideally, one would like to prove the security of a cryptographic scheme by assuming that $P \neq NP$, or, at least, that $BPP \neq NP$, since one admits efficient computation to be augmented with probabilistic choices. However, such an assumption would only guarantee that a problem is not efficiently solvable by an adversary in its worst case, while it could be solvable, for instance, in the majority of the cases (which would still be quite far from acceptable in a typical cryptographic application). Therefore an assumption referring to hardness of a problem in an average case sense seems to be needed. We do not know if the assumption that $BPP \neq NP$ implies the existence of languages that are hard on average, but, regardless of that, it seems that even the latter assumption may not suffice. This is because in a cryptographic protocol it would be desirable that honest parties can feasibly run the protocol and are therefore able to generate instances of problems that are hard on average from the point of view of adversaries. Roughly speaking, this implies the requirement of a method to efficiently generate hard on average instances that can be solved efficiently by whoever generates them but inefficiently from someone else. The definition of *one-way functions* precisely satisfies this requirement, by defining, informally, functions that can be computed efficiently, but for which no polynomial time algorithm can invert with non-negligible success an image of the function computed on a randomly chosen input. Since their original proposal, one-way function have played a crucial role in the development of modern cryptography, to the point that all other cryptographic primitives and applications are studied in relationship to one-way function, and central questions are if a specific cryptographic primitive or protocol can be constructed from any one-way function (and viceversa).

As of today, numerous primitives and protocols have been introduced in the literature and there exists a complex structure of relationships between them. In particular, many important primitives such as pseudo-random functions, pseudo-random generators and zero-knowledge proofs can be constructed from any one-way functions (and viceversa). On the other hand several other cryptographic protocols have been proved secure under probably stronger assumptions than the mere existence of one-way functions, and seem to require such stronger assumptions.

2.1. Definitions

We start with some preliminary definitions.

An *algorithm* is a Turing machine, an *efficient algorithm* and an *adversary* are probabilistic polynomial-time algorithms. By the expression $x \leftarrow y$ we denote the possibly random *process* of (1) uniformly and independently choosing element x from set y, or (2) uniformly and independently drawing x according to distribution y, or (3) setting object x equal to object y, or (4) setting object x equal to the output of the (possibly probabilistic) algorithm y (in which case we specify also the input to y). By $\text{Prob}[R_1; \ldots; R_n : E]$ we denote the probability of event E, after the ordered execution of possibly random processes $R_1, \ldots, R_n$.

We define negligible functions as functions that tend to zero smaller than any inverse of a polynomial.

Definition 1. *A function δ is* negligible *if for all positive constants c there exists an integer n_c such that $\delta(n) < n^{-c}$, for all $n \geq n_c$.*

Intuitively, events with a negligible probability should not be noticed by probabilistic polynomial-time algorithms when the input sizes are large enough. We now are ready to formally define one-way functions.

Definition 2. *A function $f : \{0,1\}^* \to \{0,1\}^*$ is* one-way *if*

1. *there exists an efficient algorithm C that, on input x, returns $f(x)$;*
2. *for any efficient algorithm A, the following probability is negligible in n:*

$$\text{Prob}[x \leftarrow \{0,1\}^n; y \leftarrow f(x); x' \leftarrow A(1^n, y) : f(x') = f(x)].$$

We also define collections of one-way functions.

Definition 3. *A collection of functions $F = \{f_n : n \in \mathcal{N}, f_n : \{0,1\}^n \to \{0,1\}^n\}$ is* one-way *if*

1. *there exists an efficient algorithm C that, on input n, x, returns $f_n(x)$, and if*
2. *for any efficient algorithm A, the following probability is negligible in n:*

$$\text{Prob}[x \leftarrow \{0,1\}^n; y \leftarrow f_n(x); x' \leftarrow A(1^n, y) : f_n(x') = f_n(x)].$$

It is possible to prove that one-way functions exist if and only if collections of one-way functions exist. We note that the definition of one-way function essentially implies that almost all inputs to the function produce an output that is hard to invert. A natural relaxation of this intuition is that only a large fraction of the inputs produce inputs that are hard to invert. These functions are called "weak one-way" and will be discussed later in greater detail.

We now recall the definition of "trapdoor" functions as one-way function with the additional property that there exists some information that allows its owner (and only her) to invert the function.

Definition 4. *A* trapdoor function *$f : \{0,1\}^* \to \{0,1\}^*$ is a one-way function for which there exists an efficient algorithm E and a polynomial p such that, for any n, there exists a string t_n such that $|t_n| \leq p(n)$ and for all $x \in \{0,1\}^*$, $E(f(x), t_n) = x'$ and $f(x) = f(x')$.*

Definition 5. *A collection of trapdoor functions $F = \{f_n : n \in \mathcal{N}, f_n : \{0,1\}^n \rightarrow \{0,1\}^n\}$ is a collection of one-way functions for which there exists an efficient algorithm E and a polynomial p such that, for any n, there exists a string t_n such that $|t_n| \leq p(n)$ and for all $x \in \{0,1\}^n$, $E(1^n, f_n(x), t_n) = x'$ and $f_n(x) = f_n(x')$.*

We note that not all collection of one-way functions may be collections of trapdoor one-way functions, and, given the current state of the art, it seems unlikely that one can construct a collection of trapdoor functions from any collection of one-way functions (without making stronger hardness assumptions).

2.2. Candidates from Number Theory

Proving the existence of a one-way function implies a proof that $P \neq NP$, currently the biggest open question in Theoretical Computer Science. Several candidates for one-way functions have been provided in the literature; and many of these are to-day widely believed to satisfy the previous definition (where the belief is essentially based on the fact that many years of researches have not produced an efficient algorithm inverting such functions). Number theory has proved to be a source of several problems that appear to be "hard" and therefore provide good candidates for both collections of one-way functions and collections of trapdoor functions. We will consider some of these problems here. Specifically, we consider the problems of "factoring composite integers" and "computing discrete logarithms modulo primes" in order to construct candidates for collections of one-way functions, and the problem of "computing square or higher-order roots modulo composites", to construct candidates for collections of trapdoor functions.

Factoring. We define two collection of functions based on the multiplications of natural numbers. First, we define the collection of functions $IM_1 = \{f1_n : n \in \mathcal{N}, f1_n : \{0,1\}^n \rightarrow \{0,1\}^n\}$, where $f1_n(p,q) = x$, p, q are interpreted as positive integers of length $n/2$ and x is computed as their product over the set of natural integers $\mathcal{N}$. Then we define $IM_2 = \{f2_n : n \in \mathcal{N}, f2_n : \{0,1\}^n \rightarrow \{0,1\}^n\}$, where $f2_n(r) = x$, where r is used to uniquely determine two primes p, q of length $n/2$ and x is computed as their product over $\mathcal{N}$.

We first note that the product of two positive integers can be computed in polynomial (in fact, quadratic) time. The problem underlying both problems of inverting IM_1 and IM_2, well known as *factoring*, is one of the most fascinating in elementary number theory and most studied today in computational number theory and cryptography. Considerations about the hardness of computing the factorization of large integers are attributed, for instance, to Gauss. After numerous studies, Integer Multiplication, in its definition IM_1 seems a good candidate for a collection of "weak" one-way functions (to be formally defined later); this is because there is certainly a large fraction of inputs for which the function $f1_n$ can be efficiently inverted. Consider, as a simple example, the case in which p and q can be themselves factored as the product of small primes. Then a simple algorithm that tests divisibility can discover each single factor one at a time by checking many small primes. However, if this is not the case, then several research effort have only produced algorithms that run in time superpolynomial in the size of the input, The asymptotically fastest algorithms known today are variations on the so-called 'random squares algorithm' [38], a probabilistic algorithm with running time $L(n)^{\sqrt{2}}$, for $L(n) = e^{\sqrt{\log n \log \log n}}$. Specifically, various versions of the 'number field sieve' are proved, under certain assumptions, to factor integers in expected time

$$e^{((c+o(1))(\log n)^{1/3}(\log \log n)^{2/3})},$$

for some constant c [40, 1]. This state of affairs leads to the belief that IM_2 seems a good candidate for a collection of one-way functions.

Discrete Logarithm. Let p be a prime. Then the multiplicative group $(Z_p^*, \cdot \bmod p)$, where $Z_p^* = \{z : z < p, (z,p) = 1\}$, is cyclic; that is, it can be written as $Z_p^* = \{g^i : i = 1, \ldots, p-1\}$, for some generator g. We define the following collection of functions $EXP = \{f_n : n \in \mathcal{N}, f_n : \{0,1\}^n \to \{0,1\}^n\}$, where $f_n(p,g,x) = (a,b,c)$, where $a = p$, $b = g$ and $c = g^x \bmod p$. We can easily restrict the function so that p is a prime, and g is a generator of the multiplicative group Z_p^*.

We first remark that generating a prime p of a pre-specified length can be done in polynomial time due to a recent breakthrough result and so can the operation of exponentiation modulo a prime (that is, computing $c = g^x \bmod p$), through repeated squaring operations. Computing a generator g of the multiplicative group Z_p^* can be done in expected polynomial time as a random element of Z_p^* can be tested in polynomial time and the density of generators in Z_p^* is high.

The problem of inverting EXP, well known as *computing discrete logarithms*, is another important problem in elementary number theory and well studied today in computational number theory and cryptography. Several efforts have been devoted to trying to solve this problem, the best culminating in the 'index calculus algorithm' that solves the problem in expected running time $L(p)^{\sqrt{2}}$. Consequently, EXP seems a good candidate for a collection of one-way functions.

The RSA Function. We now introduce a candidate for a collection of trapdoor functions. Define the following collection of functions $RSA = \{f_n : n \in \mathcal{N}, f_n : \{0,1\}^n \to \{0,1\}^n\}$, where $f_n(N,e,x) = (a,b,c)$, where $a = N$, $b = e$ and $c = x^e \bmod N$. We are especially interested in the case in which N is the product of two primes p, q, and e is an integer coprime with $\phi(N) = (p-1)(q-1)$. The trapdoor is the factorization p, q of N, and allows to invert the function.

We first remark that generating an integer N as the product of two primes and a value e such that $(e, \phi(N)) = 1$, and computing $c = x^e \bmod N$ can be done in polynomial time. The best algorithm known for inverting RSA consists of factoring N, which is believed to be hard as discussed before.

The Squaring Function. Another candidate for a collection of trapdoor functions is the squaring function; that is, the RSA function, for $e = 2$. For this function, one can prove that the problem of inverting the function is equivalent to factoring N.

2.3. Weak vs. Strong One-Way Functions

We now formally define "weak" one-way functions and will also refer to (previously defined) one-way functions as "strong" one-way functions. Informally, weak one-way function represent a relaxation of one-way function as they only require that no efficient adversary can invert the function for at least a noticeable fraction of the inputs.

Definition 6. *Let p be a polynomial. A function $f : \{0,1\}^* \to \{0,1\}^*$ is a p-weak one-way function if*

1. *there exists an efficient algorithm C that, on input x, returns $f(x)$*
2. *for any efficient algorithm A, it holds that for all sufficiently large n,*

$$\text{Prob}\,[\,x \leftarrow \{0,1\}^n; y \leftarrow f(x); x' \leftarrow A(1^n, y) \,:\, f(x') \neq f(x)\,] \geq 1/p(n).$$

Similarly as before, we can define a *collection of weak one-way functions*. The following theorem was first stated in the oral presentation of [52].

Theorem 1. *A weak one-way function exists if and only if a strong one-way function exists.*

As one-way functions are believed to represent a very minimal notion of cryptographic hardness, this theorem seems to suggest that cryptographic hardness can be amplified from a low (but sufficiently noticeable) level to a high (and sufficiently close to the maximum possible) level.

Proof. We start the proof by recalling the transformation from weak to strong one-way functions from [52]. Intuitively, the strong one-way function is the concatenation of sufficiently many application of the weak one-way function. This is reminiscent of analogue theorems in Information Theory; interestingly, as we will see, the proof of this theorem is significantly harder.

More formally, given a p-weak collection of one-way functions $F = \{f_n : n \in \mathcal{N}\}$, where $f_n : \{0,1\}^n \to \{0,1\}^n$, we define a collection $G = \{g_m : m \in \mathcal{N}\}$, where $g_m : \{0,1\}^m \to \{0,1\}^m$, for $m = 2n^2p(n)$, is defined as

$$g_m(x_1, \ldots, x_{2np(n)}) = (f_n(x_1) \circ \cdots \circ f_n(x_{2np(n)})).$$

We now prove that G is a collection of strong one-way functions. Assume (towards contradiction) that this is not the case. Then there exists an efficient adversary A and a polynomial q such that for infinitely many m, it holds that

$$\text{Prob}\,[\,x \leftarrow \{0,1\}^m; y \leftarrow g_m(x); x' \leftarrow A(1^n, y) \,:\, g_m(x') = g_m(x)\,] \geq 1/q(m).$$

If we present an efficient adversary A' that, using A, can invert f_n with probability at least $1 - 1/p(n)$ then we contradict the assumption that F is a collection of p-weak one-way functions. Consider the following algorithm A'.

Input for Algorithm A': $y \in \{0,1\}^n$, where $y = f_n(x)$, for a randomly chosen x.

Instructions for Algorithm A':

1. repeat $4n^2p(n)q(m)$ times:
      for $i = 1, \ldots, 2np(n)$
            randomly choose $x_j \in \{0,1\}^n$, for $j = 1, \ldots, i-1, i+1, \ldots, 2np(n)$
            compute $y_j = f_n(x_j)$, for $j = 1, \ldots, i-1, i+1, \ldots, m$
            if A successfully inverts $(y_1, \ldots, y_{i-1}, y, y_{i+1}, \ldots, y_{2np(n)})$ then
                  let $(x_1, \ldots, x_{2np(n)}) = A(y_1, \ldots, y_{i-1}, y, y_{i+1}, \ldots, y_{2np(n)})$
                  return: x_j and halt
2. return: 'failure to invert'.

We define the subset BAD $\subseteq \{0,1\}^n$ of x such that the probability, over the randomness used by A', that in a single iteration of its repeat loop A' returns $f_n^{-1}(f_n(x))$ is less than $1/4np(n)q(m)$.

We now show that the probability, over the randomness used by A' and the random choice of x, that A' is not successful is 'essentially' the probability that x is BAD. More precisely, we define event $e(A', x)$ as the event that A' does not invert $y = f_n(x)$, when x is randomly chosen, and A' is run on $f_n(x)$. Then we have that

$$
\begin{aligned}
\mathrm{Prob}[\, e(A', x)\,] \;&=\; \mathrm{Prob}[\, e(A', x) \mid x \in \mathrm{BAD}\,] \cdot \mathrm{Prob}[\, x \in \mathrm{BAD}\,] \\
&\quad + \mathrm{Prob}[\, e(A', x) \mid x \notin \mathrm{BAD}\,] \cdot \mathrm{Prob}[\, x \notin \mathrm{BAD}\,] \\
&\leq\; 1 \cdot \mathrm{Prob}[\, x \in \mathrm{BAD}\,] + (1 - 1/4np(n)q(m))^{4n^2p(n)q(m)} \cdot 1 \\
&\leq\; \mathrm{Prob}[\, x \in \mathrm{BAD}\,] + e^{-n}
\end{aligned}
$$

If we show that $\mathrm{Prob}[\, x \in \mathrm{BAD}\,] \leq 1/2p(n)$ then we have that $\mathrm{Prob}[\, e(A', x)\,] \leq 1/2p(n) + e^{-n} < 1/p(n)$, which brings us to contradicting the assumption that f_n is a weak one-way function. To show that $\mathrm{Prob}[\, x \in \mathrm{BAD}\,] \leq 1/2p(n)$, assume (towards contradiction) that this is not the case. Then let $\vec{x} = (x_1, \ldots, x_{2np(n)})$ and define the event $e(A, \vec{x})$ as the event that A successfully inverts $\vec{y} = g_m(\vec{x})$, when $\vec{x}$ is uniformly chosen. Then we have that the probability of event $e(A, \vec{x})$ is

$$
\begin{aligned}
&=\; \mathrm{Prob}\left[\, e(A, \vec{x}) \mid \bigvee\nolimits_{i=1}^{2np(n)} x_i \in \mathrm{BAD}\,\right] \cdot \mathrm{Prob}\left[\, \bigvee\nolimits_{i=1}^{2np(n)} x_i \in \mathrm{BAD}\,\right] \\
&\quad + \mathrm{Prob}\left[\, e(A, \vec{x}) \mid \bigwedge\nolimits_{i=1}^{2np(n)} x_i \notin \mathrm{BAD}\,\right] \cdot \mathrm{Prob}\left[\, \bigwedge\nolimits_{i=1}^{2np(n)} x_i \notin \mathrm{BAD}\,\right] \\
&\leq\; \Sigma_{i=1}^{2np(n)} \mathrm{Prob}[\, e(A, \vec{x}) \mid x_i \in \mathrm{BAD}\,] \cdot \mathrm{Prob}[\, x_i \in \mathrm{BAD}\,] \\
&\quad + \mathrm{Prob}\left[\, e(A, \vec{x}) \mid \bigwedge\nolimits_{i=1}^{2np(n)} x_i \notin \mathrm{BAD}\,\right] \cdot \mathrm{Prob}\left[\, \bigwedge\nolimits_{i=1}^{2np(n)} x_i \notin \mathrm{BAD}\,\right] \\
&\leq\; (2np(n)) \cdot \left(\frac{1}{4np(n)q(m)}\right) \cdot 1 + 1 \cdot (1 - 1/2p(n))^{2np(n)} \\
&\leq\; \left(\frac{1}{2q(m)}\right) + e^{-n} \\
&<\; \left(\frac{1}{q(m)}\right)
\end{aligned}
$$

which negates our original assumption and therefore gives us a contradiction. $\square$

It has been noted that Yao's construction of a strong one-way function from a weak one-way function is not satisfactory as it significantly increases the size of the input. Perhaps surprisingly, in practical applications, such a large increase in the size of the input can make a supposedly hard function actually easy to invert for all sizes of interest. This is best illustrated with an example. Suppose a one-way function is used in a cryptographic protocol and the amount of resources available to the user evaluating the function is bounded. Specifically, assume that the user can only use 1024-bit input one-way functions and that such functions have been obtained using the above reduction by 32 parallel applications of weak one-way functions on 32-bit inputs. Then the running time necessary to invert the strong one-way function becomes 32 times the running time necessary to invert each weak function, which may be very small given the short input size. This example calls for methods to evaluate the security of reductions between one-way functions, and, in fact, between any two cryptographic primitives (as the same problem can be recast, with appropriate modifications, on other cryptographic primitives as well).

A crucial quantity for evaluating the security of a cryptographic primitive is the amount of memory used by an application of the primitive, and, more specifically, as observed in [32], the amount of *private memory* only. The latter is taken as the *security parameter* of an instance of the primitive. Given an instance f of a primitive P, we denote by A an adversary trying to "break" f, by t a polynomial bounding its running time, by δ a function denoting its success probability, and by R, the function defined as $R(n) = t(n)/\delta(n)$ for all $n \in \mathcal{N}$, the *achievement ratio* of A, n denoting the security parameter.

Given two primitives P1 and P2, using n_1 and n_2 private memory, respectively, we say that a *reduction* from P1 to P2 is a pair of machines (S, A_1) such that:

1. given a description of an instance f of P1, S returns a description of an instance g of P2;
2. given an adversary A_2 running in time $t_2(n_2)$ who breaks g with probability $\delta_2(n_2)$, A_1 is an oracle adversary running in time $t_1(n_1)$, with access to oracle A, who breaks f with probability $\delta_1(n_1)$.

The parameters $t_1, \delta_1, t_2, \delta_2, n_1, n_2$ play an important role into evaluating the strength of the reduction. Specifically, compare the achievement ratios of A_1 and A_2, when both instances have the same private memory n; in general, they might satisfy the following inequality:

$$R_1(n) \leq n^c \cdot R_2(n^\alpha)^\beta,$$

for some constants c, α, β.

We say that a reduction from an instance of primitive P1 to an instance of primitive P2 is

1. *linear-preserving* if $\alpha = \beta = 1$,
2. *polynomially-preserving* if $\alpha = 1$ and $\beta = c > 1$,
3. *slight-preserving* if $\alpha = \beta = c > 1$, for some $c \in \mathcal{N}$.

A linear-preserving reduction is more desirable than a polynomially-preserving one, which in turn is more desirable than a slight-preserving one. The term *security-preserving* is often used in the literature for reductions that are either linear-preserving or polynomially-preserving.

A crucial fact that is often used is that a sufficient condition for a reduction to be security-preserving is that $n_2 = a \cdot n_1$, and $R_1(n) = R_2(n)^\beta$, for some constants $a, \beta > 1$ (in other words, it is enough that the amount of private randomness used by primitive P2 is only a constant times that used by primitive P1, and that the running times and the success probabilities associated with the adversaries are polynomially related).

We note that in Yao's construction of a strong one-way function from any weak one-way function the amount of randomness used by the former is *not* a constant times the amount of randomness used by the latter; in fact, it can be larger even by a large polynomial factor. This motivated researchers to come up with additional constructions that save randomness.

Weak vs. Strong One-Way Permutations:. The construction in [25] is polynomially-preserving and is performed for the case of one-way permutations. Given a $p(n)$-weak collection of one-way permutations $F = \{f_n : n \in \mathcal{N}\}$, where $f_n : \{0,1\}^n \to \{0,1\}^n$, define a collection of one-way permutations $G = \{g_m : m \in \mathcal{N}\}$, where $g_m : \{0,1\}^m \to \{0,1\}^m$, for $m = n + O(p(n))$, is defined as a repeated application of the following two steps: one execution of the permutation f_n on a portion of the input of size n, and one random step on an expander graph having vertex set $\{0,1\}^n$. At the end the final node reached on the expander is returned in output together with the input portion used to choose the random steps on the expander. Later, in [32], additional constructions have been given for security-preserving reductions between weak and strong one-way permutations, some of which being linear-preserving. In particular, the paper [32] formalizes and uses the important observation that the security of a function can be parameterized by the private input only (rather than both private and public).

Weak vs. Strong One-Way Regular Functions. The construction in [15] is polynomially-preserving and is performed for the case of one-way regular functions; that is, functions for which each image has the same number of preimages. The construction in [15] uses pairwise independent hash functions and is obtained by iterating several times an atomic function. Specifically, let $H_{2n,n}$ be the set of pairwise independent hash functions that can described with $4n$ bits. Given a p-weak collection of one-way functions $F = \{f_n : n \in \mathcal{N}\}$, where $f_n : \{0,1\}^n \to \{0,1\}^n$, define the collection of functions $AG = \{ag_n : n \in \mathcal{N}\}$, where $ag_n : \{0,1\}^{2n} \times \{0,1\}^{4n} \to \{0,1\}^{2n}$ is defined as

$$ag_n(a, b; h_n) = (f_n(b), h_n(a \circ b)),$$

for all $a, b \in \{0,1\}^n$ and $h_n \in H_{2n,n}$, and the symbol $\circ$ denotes concatenation. Then the final collection of functions $CG = \{cg_n : n \in \mathcal{N}\}$ is defined as follows.

Input to cg_n: $(a, b; h_0 \ldots, h_{k-1})$, where $a, b \in \{0, 1\}^n$, $h_0, \ldots, h_{k-1} \in H_{2n,n}$, $k = 2n/p(n)$.

Instructions for cg_n:

1. Set $a_0 = a$ and $b_0 = b$.
2. For $i = 0, \ldots, k - 1$,
   set $(a_{i+1}, b_{i+1}) = ag_n(a_i, b_i; h_i)$.
3. Output: (a_k, b_k).

3. Pseudo-Random Generators

As randomness plays a vital role in several areas of computer science, such as cryptography, algorithms and complexity theory, pseudo-random generators are very often crucial tools for the use of randomness in these domains.

Informally, by pseudo-random generators one denotes a deterministic function that, given as input a *short* string of 'random' bits, returns a *longer* string that 'looks random' to an observer with certain 'limited computational resources'.

Real randomness. A first question one may ask is: are there really ways to generate random bits ? This question is currently answered by looking at some natural sources, such as radioactive sources, noise diodes or coins. However, these and similar sources may not be perfect in that they may generate either biased bits (bits for which the probability of 1 is different from the probability of 0) or correlated bits (bits for which the conditional probabilities of 0 and 1 are different). Much research has been devoted to the problem of turning a biased and correlated source into an almost random one. Dealing with bias is not hard; for instance, the well-known Von Neumann's trick suggests to extract bit 0 from pairs 01 returned by the biased source, bit 1 from pairs 10, and discarding pairs 00 and 11. (Note that the resulting source has no bias since the probability of pairs 01 and 10 are identical for any bias.) Dealing with correlation seems harder, and several papers have been proposing interesting techniques that return random sources starting from sources with a certain predefined correlation function. All these techniques turn out to be very helpful in generating random bits from potentially defective natural sources. Therefore, from now on we will assume that there exist effective ways to generate random bits.

Pseudo-random generators outside cryptography. Starting from areas different than cryptography, several methods for pseudo-random generation have been proposed in the past. A classical notion of pseudo-random generators [36], for instance, requires the strings returned by the generator to satisfy certain statistical properties that are also satisfied by really random bits. Examples of such methods are linear feedback shift registers or linear congruential generators. Other methods, motivated by the problem of reducing the randomness required by probabilistic polynomial-time algorithms, only require the strings returned by the generator to hit some large subsets at least once with high probability, or an average number

of times equal to the density of the subset. Some generators with these properties are based on pairwise independent hash functions or permutations, or random
walks on expander graphs. Although useful for their motivating application, these
generators are not strong enough for most cryptographic applications, for which a
new and stronger definition of pseudo-randomness was required.

3.1. Definitions

Two main approaches have been used in defining cryptographically-secure pseudo-
random generators. The first approach [6] required that it would be computationally hard to predict the next bit output by a pseudo-random generator significantly better than by random guessing. (Previously in [48] it had been proposed
a similar test, based on sequences of bits rather than single bits.) Later, another
approach was proposed in [52], requiring that no polynomial-time algorithm could
distinguish the output of a pseudo-random generator from a random string of the
same length. In [52] it was also proved that the two approaches are equivalent; in
other words, a pseudo-random generator that can pass the next-bit test is also a
pseudo-random generator that can pass all polynomial time statistical tests (and
viceversa).

In order to formalize this definition, we will first define the important definitions of polynomial-time indistinguishability (also called computational indistinguishability) between distributions and of pseudo-random distributions. The
definition of polynomial-time indistinguishability captures the intuition of two distributions that cannot be tell apart from any polynomial-time statistical test.

Definition 7. *For any n, let X_n, Y_n be distributions over $\{0,1\}^n$. We say that
the families of distributions $X = \{X_n : n \in \mathcal{N}\}$ and $Y = \{Y_n : n \in \mathcal{N}\}$ are
polynomial-time indistinguishable if for any polynomial-time algorithm A and any
polynomial p, there exists c such that for all $n > c$ it holds that*

$$| \operatorname{Prob}[u \leftarrow X_n : A(u) = 1] - \operatorname{Prob}[u \leftarrow Y_n : A(u) = 1]| < 1/p(n).$$

We note that although in the above definition the algorithm A is given a single
sample from either distribution X_n or distribution Y_n, it has been proved that this
definition is equivalent to one in which A takes as input a polynomial number of
independent samples from either distribution.

Given the above definition, we have that a pseudo-random distribution can
be defined in terms of polynomial-time indistinguishability with the uniform distribution.

Definition 8. *For any n, let X_n be a distribution over $\{0,1\}^n$ and let U_n be the
uniform distribution over $\{0,1\}^n$. We say that the family of distributions $X =
\{X_n : n \in \mathcal{N}\}$ is pseudo-random if it is polynomial-time indistinguishable from
$U = \{U_n : n \in \mathcal{N}\}$.*

We can now formally define pseudo-random generators as functions that expand
the input and induce pseudo-random distributions.

Definition 9. *Let U_n denote the uniform distribution over $\{0,1\}^n$. A deterministic polynomial time computable collection of functions $G = \{G_n : n \in \mathcal{N}\}$, where $G_n : \{0,1\}^n \to \{0,1\}^m$ is a pseudo-random generator if $m > n$ and the family of distributions $DG = \{DG_n : n \in \mathcal{N}\}$, where $DG_n = \{s \leftarrow U_n; r \leftarrow G_n(s) : r\}$, is pseudo-random.*

An important tool that has been crucial for many constructions of pseudo-random generators is that of "hard-core bit" of a function. A hard-core bit is defined for one-way functions as a predicate of an input to the function; the intuition behind this notion is the intention to capture the entire hardness of inverting the one-way function in a single bit.

Definition 10. *A collection of functions $F = \{f_n : n \in \mathcal{N}\}$, where $f_n : \{0,1\}^n \to \{0,1\}^m$, is a collection of boolean predicates if $m = 1$ for all $n \in \mathcal{N}$.*

Definition 11. *Let $F = \{f_n : n \in \mathcal{N}\}$ be a collection of functions. A collection of predicates $B = \{b_n : n \in \mathcal{N}\}$ is a hard-core bit for F if the following holds:*

1. *There exists an efficient algorithm E such that $E(1^n, x) = b_n(x)$ for all $x \in \{0,1\}^n$*
2. *The distribution (induced by B) $DB = \{DB_n : n \in \mathcal{N}\}$, where $DB_n = \{x \leftarrow \{0,1\}^n : b_n(x)\}$ is pseudo-random.*

A deterministic hard-core bit has been presented for collections of one-way functions based on discrete logarithms (the most significant bit) or squaring modulo composite integers (the least significant bit). It has been proved in [27] that for any one-way function the probabilistic predicate returning the inner product of the input with a random string is a hard core bit.

Theorem 2. *For any collection of one-way functions F there exists a probabilistic hard-core bit for F.*

3.2. Constructions

Perhaps surprisingly, hardness (of inverting one-way functions) and pseudo-randomness (of the output of pseudo-random generators) turned out to be very related. A fundamental result in cryptography is the construction of a pseudo-random generator from any one-way function [31]. We cover here the proof of a simpler version of this result: that is, the special case in which the given one-way function is actually a one-way permutation. We divide the proof of this fact in two claims. The first claim shows how to construct pseudo-random generators that expand the input by only one bit from any one-way permutation. The second claim shows how to construct pseudo-random generators expanding the input by an arbitrary polynomial amount from the obtained pseudo-random generator expanding the input by a single bit. We also discuss how to construct a one-way function from any pseudo-random generator.

Claim 1. *If there exists a collection of one-way permutations then there exists a collection of pseudo-random generators $G = \{G_n : n \in \mathcal{N}\}$, where $G_n : \{0,1\}^n \to \{0,1\}^{n+1}$.*

Proof. Given a collection of one-way permutations $F = \{f_n : n \in \mathcal{N}\}$, where $f_n : \{0,1\}^n \to \{0,1\}^n$, we consider the hard-core bit $B = \{b_n : n \in \mathcal{N}\}$ guaranteed by Theorem 2. We then define a collection $G = \{G_n : n \in \mathcal{N}\}$, where $G_n : \{0,1\}^n \to \{0,1\}^{n+1}$ is defined as $G_n(x) = f_n(x) \circ b_n(x)$ for any $x \in \{0,1\}^n$ and would like to prove that G is a collection of pseudo-random generators. Assume by contradiction that this is not the case. Then it holds that there exists an algorithm A and a polynomial p such that the difference

$$|\operatorname{Prob}[\, x \leftarrow U_n; u \leftarrow G_n(x) : A(u) = 1\,] - \operatorname{Prob}[\, u \leftarrow U_{n+1} : A(u) = 1\,]|$$

is at least $1/p(n+1)$. We now define

$$\alpha = \operatorname{Prob}[x \leftarrow U_n : A(f_n(x) \circ b) = 1 \mid b = b_n(x)]$$
$$\beta = \operatorname{Prob}[x \leftarrow U_n : A(f_n(x) \circ b) = 1 \mid b = 1 - b_n(x)].$$

Then we can rewrite the second term $\operatorname{Prob}[\, u \leftarrow U_{n+1} : A(u) = 1\,]$ in the above inequality as $\operatorname{Prob}[\, x \leftarrow U_n; b \leftarrow \{0,1\} : A(f_n(x) \circ b) = 1\,]$ that is equal to $(\alpha+\beta)/2$ after conditioning over $\operatorname{Prob}[\, b = b_n(x)\,]$ and $\operatorname{Prob}[\, b = 1 - b_n(x)\,]$. Also, we see that the first term $\operatorname{Prob}[\, x \leftarrow U_n; u \leftarrow G_n(x) : A(u) = 1\,]$ in the above inequality is equal to α. Therefore we get that $|\alpha - (\alpha - \beta)/2| = |\alpha - \beta|/2$ is $> 1/p(n+1)$. We now construct an algorithm A' that on input $f_n(x)$ tries to compute $b_n(x)$ and we show that it succeeds with probability significantly better than $1/2$.

Input for Algorithm A': $f_n(x)$

Instructions for Algorithm A': $f_n(x)$

1. randomly choose $b \in \{0,1\}$
2. let $d = A(f_n(x) \circ b)$
3. if $d = 1$ then output b else output $1 - b$.

We see that the probability $\operatorname{Prob}[\, x \leftarrow U_n : A(f_n(x)) = 1\,]$ can be computed as $\alpha/2 + (1 - \beta)/2$ after conditioning over $\operatorname{Prob}[\, b = b_n(x)\,]$ and $\operatorname{Prob}[\, b = 1 - b_n(x)\,]$. Finally, observe that $\alpha/2 + (1 - \beta)/2 = 1/2 + (\alpha - \beta)/2 > 1/2 + 1/p(k+1)$. $\square$

Claim 2. *If there exists a collection of one-way permutations then for any polynomial p, there exists a collection of pseudo-random generators $H = \{H_n : n \in \mathcal{N}\}$, where $H_n : \{0,1\}^n \to \{0,1\}^{p(n)}$.*

Proof. The construction of H can be seen as a particular iterated version of the construction of G in Claim 1, and, in turn uses an iterated application of F. Precisely, we define collection $H = \{H_n : n \in \mathcal{N}\}$, where $H_n : \{0,1\}^n \to \{0,1\}^{p(n)}$ is defined as $H_n(x) = b_n(f_n^{p(n)-1}(x)) \circ \cdots \circ b_n(f_n(x)) \circ b_n(x)$, and f^i denotes the i-times iterated application of f, where each application takes as input the output of the previous one. We will prove that H is a collection of pseudo-random generators by using an application of the so-called 'hybrid proof technique' [29].

Assume by contradiction that H_n is not pseudo-random. Then this assumption can be written as saying that there exists a polynomial q and a probabilistic polynomial time algorithm A such that for infinitely many n's, it holds that

$$\Delta = \left| \text{Prob}\left[u \leftarrow H_n : A(u) = 1 \right] - \text{Prob}\left[u \leftarrow U_{p(n)} : A(u) = 1 \right] \right| \geq 1/q(n),$$

where, for any m, by U_m we denote the uniform distribution over $\{0,1\}^m$. Let D_0 denote the distribution induced by H_n on input a randomly chosen n-bit string x. Moreover, for $i = 1, \ldots, p(n)$, let D_i be the distribution that randomly chooses $x \in \{0,1\}^n$, and $r_0, \ldots, r_{i-1} \in \{0,1\}$ and returns $b_n(f_n^{p(n)-1}(x)) \circ \cdots \circ b_n(f_n^i(x)) \circ r_{i-1} \circ \cdots \circ r_0$. Note that $D_{p(n)}$ is equal to the uniform distribution $U_{p(n)}$ over $p(n)$ bits. Then we can rewrite Δ as

$$\Sigma_{i=0}^{p(n)-1} \left| \text{Prob}\left[u \leftarrow D_i : A(u) = 1 \right] - \text{Prob}\left[u \leftarrow D_{i+1} : A(u) = 1 \right] \right|,$$

and since $\Delta \geq 1/q(n)$ we obtain that there exists a $j \in \{0, \ldots, p(n)-1\}$ such that

$$\left| \text{Prob}\left[u \leftarrow D_j : A(u) = 1 \right] - \text{Prob}\left[u \leftarrow D_{j+1} : A(u) = 1 \right] \right| \geq 1/(q(n)p(n)).$$

Then we can construct an algorithm A' that uses A to violate the pseudo-randomness of the collection of generators G from Claim 1.

Input for Algorithm A': $u \in \{0,1\}^{n+1}$, where $u = x \circ b$, for $x \in \{0,1\}^n$ and $b \in \{0,1\}$.

Instructions for Algorithm A':

1. randomly choose $h \in \{1, \ldots, p(n)\}$
2. randomly choose $c_0, \ldots, c_{h-1} \in \{0,1\}$
3. let $y = b_n(f_n^{p(n)-1}(x)) \circ \cdots \circ b_n(f_n^{h+1}(x)) \circ b \circ c_{h-1} \circ \cdots \circ c_0$
4. if $A(y) = 1$ then output 1 else output 0.

Assume $h = j$ (this happens with probability $1/p(n)$). We see that the value y in step 3 is distributed according to D_{j+1} if u is distributed according to U_{n+1} or according to D_j if u is distributed according to G_n. We obtain that

$$\left| \text{Prob}\left[x \leftarrow U_n; u \leftarrow G_n(x) : A'(u) = 1 \right] - \text{Prob}\left[u \leftarrow U_{n+1} : A'(u) = 1 \right] \right|$$

$$\geq \quad (1/p(n)) \cdot \left| \text{Prob}\left[u \leftarrow D_j : A(u) = 1 \right] - \text{Prob}\left[u \leftarrow D_{j+1} : A(u) = 1 \right] \right|$$

$$\geq \quad 1/(q(n) \cdot p^2(n)),$$

from which we derive our desired contradiction. $\square$

An implication of Claim 1 is that any candidate for a a one-way permutation gives rise to a pseudo-random generators via the construction described in the proof of the claim. We note that the construction of a pseudo-random generator starting from a generic one-way permutation uses a probabilistic hard core bit. It is of interest to notice that pseudo-random generators can be constructed also using deterministic hard-core bits. Two of the most important examples are based on squaring modulo composites and discrete logarithms. Specifically, the previously considered squaring function, when defined over $(Z_n^*)^2$, is a one-way permutation, and its hard-core bit is the least significant bit. Moreover, the previously defined exponentiation (modulo primes) function can be used as a one-way permutation and its hard-core bit is its most significant bit.

A pseudo-random generator is itself a one-way function, having different-size domain and range, and can be used to define a one-way function with equal domain and range, by using simple domain padding. An intuition to prove this goes as follows. Assume the function thus constructed is not one-way; then there exists an efficient algorithm that inverts the one-way function with non-negligible probability and for infinitely many input sizes. This algorithm can itself used to distinguish a pseudo-random output from a random string of the same length, as with sufficiently high probability a random string does not belong to the range of the pseudo-random generator and therefore the inverter would not find a preimage for it.

3.3. A Cryptographic Application

An important application of pseudo-random generators is in reducing the amount of random bits required in cryptographic protocols secure against polynomial-time adversaries.

A well-known private-key encryption scheme is the "One-Time Pad", originally invented in [51] in 1918. Assuming Alice and Bob agree on a random key K (a random "pad"); then they can communicate securely (that is, without the eavesdropper Eve obtaining any information about their message) as follows: On input message m, Alice computes the ciphertext $c = m \oplus [K]$ where $[K]$ denotes a substring of K of appropriate length and sends it to Bob. Given c, Bob can recover message m, by decrypting c as $m = c \oplus [K]$. Here, $\oplus$ is the "exclusive OR" operator, and K is at least as large as m. The following two facts make the one-time pad encryption scheme quite remarkable. First, as shown by Shannon, in [49], it holds that encryption scheme such that the ciphertext does not reveal any information about the plaintext (that is, any provably-secure, in the information-theoretic sense, encryption scheme) must satisfy $|K| \geq |m|$. Therefore, one-time pad is optimally secure in an information theoretic sense. Second, the encryption and decryption operations are essentially optimal in terms of time-complexity (being a mere exclusive-or operation). Unfortunately, the length of the key is inappropriate for any practical cryptographic application. Still, one-time pads are widely utilized as atomic components of more elaborate encryption systems by employing pseudo-random generators to generate arbitrarily long sequences of pseudo-random bits (given only a short shared random seed). In this case the resulting pseudo-random sequence is used as a pad. The employment of pseudorandom generators allows the transmission of messages longer than the shared key but, naturally, loses information-theoretic security (its security now relies on the security of the pseudo-random generator). In many practical applications this is an acceptable loss since we assume the adversary runs in polynomial time.

4. Pseudo-Random Functions

Random functions are functions that, on each input, return an output value that is chosen uniformly and independently from any other output. (If called twice on the same imput, however, the function returns the same random output.) Clearly, such functions do not have a short description than their input/output table. This may be too long for practical applications when the input has to be as long as the intended security parameter.

Pseudo-random functions aim to achieve essentially the same effect as random functions, with respect to polynomial time observers, and, yet, at the same time, admit an efficient description. Specifically, pseudo-random functions, are functions that take use a fixed and short random string, the *seed*, and a variable string, the *input*, to produce an *output* string that 'looks' random to a polynomial time observer. Furthermore, the function cannot be distinguished from a random function even if an efficient adversary is able to adaptively repeat the process of choosing an input to the function and obtain the corresponding function's output, for a polynomial number of times. The important requirement for this to be possible is that the seed is randomly chosen and is kept secret from the adversary.

Pseudo-random functions can replace random functions in any cryptographic application where the adversary runs in polynomial time and the function is used in a black-box fashion. When constrasted with pseudo-random generators, we see that pseudo-random functions are even more powerful as they allow efficient direct access to a very long pseudo-random sequence, which cannot even feasibly scanned bit-by-bit. Instead, the output returned by pseudo-random generators is always polynomially longer than the amount of randomness used in the input.

4.1. Definitions

We now proceed with formal definition for pseudo-random functions and permutations. We start by defining oracles and oracle adversaries.

Definition 12. *An* oracle $O = \{O_n : n \in \mathcal{N}\}$ *is a collection of functions* $O_n :$ $\{0,1\}^n \to \{0,1\}^n$. *An efficient algorithm A is an* oracle adversary *if it is given access to oracle O and, on input 1^n, can repeat the following process for a polynomial number of times:*

1. *on input 1^n and $x_1, y_1, \ldots, x_i, y_i \in \{0,1\}^n$, compute x_{i+1}*
2. *set $y_{i+1} = O_n(x_{i+1})$*

An oracle adversary A who is given access to oracle O is also denoted as A^O.

The formal definition of pseudo-random functions is then given as functions that are computationally indistinguishable from random functions from any efficient oracle adversary.

Definition 13. *For any $n \in \mathcal{N}$, let R_n be the set of all functions $r_n : \{0,1\}^n \to \{0,1\}^n$, and let f_n be a function $f_n : \{0,1\}^n \times \{0,1\}^n \to \{0,1\}^n$. Consider the following probabilistic experiment INIT:*

1. *Uniformly choose $r_n \leftarrow R_n$ for each $n \in \mathcal{N}$*
2. *Set $RAND = \{r_n : n \in \mathcal{N}\}$*
3. *Uniformly choose $s \in \{0,1\}^n$ for each $n \in \mathcal{N}$*
4. *Set $f_s = f_n(s, \cdot)$*

5. *Set $REAL = \{f_s : n \in \mathcal{N}\}$*

We say that REAL is a collection of pseudo-random functions if for any efficient oracle adversary A and any polynomial p, there exists c such that for all $n > c$ it holds that

$$\left| \mathrm{Prob} \left[\mathrm{INIT}; O \leftarrow f_s : A^O(1^n) = 1 \right] - \mathrm{Prob} \left[\mathrm{INIT}; O \leftarrow r_n : A^O(1^n) = 1 \right] \right|$$

is $< 1/p(n)$.

The formal definition of pseudo-random permutations is a direct adaptation of the previous definition for functions.

Definition 14. *For any $n \in \mathcal{N}$, let P_n be the set of all permutations $p_n : \{0,1\}^n \to \{0,1\}^n$, and let f_n be a function $f_n : \{0,1\}^n \times \{0,1\}^n \to \{0,1\}^n$ such that for each $s \in \{0,1\}^n$, the function $f_n(s, \cdot)$ is a permutation. Consider the following probabilistic experiment INIT:*

1. *Uniformly choose $p_n \leftarrow P_n$ for each $n \in \mathcal{N}$*
2. *Set $RAND = \{p_n : n \in \mathcal{N}\}$*
3. *Uniformly choose $s \in \{0,1\}^n$ for each $n \in \mathcal{N}$*
4. *Set $f_s = f_n(s, \cdot)$*
5. *Set $REAL = \{f_s : n \in \mathcal{N}\}$*

We say that REAL is a collection of pseudo-random permutations if for any efficient oracle adversary A and any polynomial q, there exists c such that for all $n > c$ it holds that

$$\left| \mathrm{Prob} \left[\mathrm{INIT}; O \leftarrow f_s : A^O(1^n) = 1 \right] - \mathrm{Prob} \left[\mathrm{INIT}; O \leftarrow p_n : A^O(1^n) = 1 \right] \right|$$

is $< 1/q(n)$.

4.2. Constructions

We describe two important constructions of pseudo-random functions and permutations: a construction of a pseudo-random functions from any pseudo-random generator [24] and a construction of a pseudo-random permutation from any pseudo-random function [42]. We also discuss how to construct a one-way function from any pseudo-random function.

The first result we present is the following

Theorem 3. *If there exists a collection of pseudo-random generators then there exists a collection of pseudo-random functions.*

Proof. Let $G = \{G_n : n \in \mathcal{N}\}$ be a collection of pseudo-random generators stretching n bits to $2n$ bits. That is, it holds that $G_n : \{0,1\}^n \to \{0,1\}^{2n}$ for all n. We denote by $G_n^0 : \{0,1\}^n \to \{0,1\}^n$ the function such that $G_n^0(s)$ is equal to the first n bits of $G_n(s)$, for all $s \in \{0,1\}^n$. Similarly, we denote by $G_n^1 : \{0,1\}^n \to \{0,1\}^n$ the function such that $G_n^1(s)$ is equal to the second n bits of $G_n(s)$, for all $s \in \{0,1\}^n$. Then we define a collection of function $F = \{f_s : |s| \in \mathcal{N}\}$, where $f_s : \{0,1\}^n \to \{0,1\}^n$ is defined as

$$f_s(x) = G_n^{x_n}(G_n^{x_{n-1}}(\cdots G_n^{x_2}(G_n^{x_1}(s))\cdots)),$$

for each $x = x_1 \circ \cdots \circ x_n$, and $x_i \in \{0,1\}$, for $i = 1, \ldots, n$.

This construction is also called the 'tree construction' for pseudo-random functions. For each s, consider the following tree T_s: each level of the tree is associated with an application of G_n; on input s, the root computes $G_n(s)$ and branches into two subtrees, returning $G_n^0(s)$ as an input for its left child and $G_n^1(s)$ as an input for its right child; the tree construction then continues recursively for the remaining bits $x_2, \ldots, x_n$ and the leaves of T_s contain all possible 2^n outputs of f_s.

We now show that F is a collection of pseudo-random functions. The proof contains an interesting application of the hybrid proof technique. Assume by contradiction that F is not pseudo-random. Then this assumption can be written as saying that there exists a polynomial q and an efficient oracle adversary A such that for infinitely many n's, it holds that $\Delta = |p_{\mathrm{real}} - p_{\mathrm{rand}}| \geq 1/q(n)$, where

$$p_{\mathrm{real}} \;=\; \mathrm{Prob}\left[\, INIT; O \leftarrow f_s : A^O(1^n) = 1 \,\right]$$
$$p_{\mathrm{rand}} \;=\; \mathrm{Prob}\left[\, INIT; O \leftarrow r_n : A^O(1^n) = 1 \,\right]$$

Also, let p be the polynomial such that A makes at most $p(n)$ queries to O in the above probabilities.

In the sequel to avoid overburden notation we fix $n \in \mathcal{N}$ and a randomly chosen $s \in \{0,1\}^n$. For $i = 0, \ldots, n$, we define hybrid functions g_s^i that differ from f_s only in that they apply i times an independently chosen random function and $n - i$ times generator G_n. Formally, for $i = 0, \ldots, n$, let D_i denote the distribution induced by the following probabilistic experiment INIT':

1. Uniformly choose $r_j^{b_j} \leftarrow R_n$ for $j = 1, \ldots, n$ and $b_j \in \{0,1\}$
2. Uniformly choose $s \in \{0,1\}^n$
3. For each $x \in \{0,1\}^n$ and each $i = 0, \ldots, n$,
   define $g_s^i(x) = G_n^{x_n}(\cdots G_n^{x_{i+1}}(r_i^{x_i}(\cdots (r_1^{x_1}(s)) \cdots)) \cdots)$

Note that p_{real} is equal to $\mathrm{Prob}\left[\, INIT'; O \leftarrow g_s^0 : A^O(1^n) = 1 \,\right]$, and that we can rewrite Δ as at most $|\, p_{\mathrm{rand}} - \mathrm{Prob}\left[\, INIT'; O \leftarrow g_s^n : A^O(1^n) = 1 \,\right]\,| + \Sigma_{i=0}^{n-1} \Delta_i$, where Δ_i is the difference

$$\left|\, \mathrm{Prob}\left[\, INIT'; O \leftarrow g_s^i : A^O(1^n) = 1 \,\right] - \mathrm{Prob}\left[\, INIT'; O \leftarrow g_s^{i+1} : A^O(1^n) = 1 \,\right] \,\right|.$$

We now prove the following

Claim 3. *It holds that*

$$\left|\, p_{\mathrm{rand}} - \mathrm{Prob}\left[\, INIT'; O \leftarrow g_s^n : A^O(1^n) = 1 \,\right] \,\right| \leq 2np(n)^2/2^n.$$

Proof. Note that the function O defined in probability p_{rand} is a random function. Therefore the claim follows from two main observations. Denote by GOOD the event that none of A's queries to g_s^n results in any of the functions $r_i^{x_i}$ being evaluated on two equal inputs. The first observation is that if event GOOD happens then the tuple containing A's queries and replies to such queries by g_s^n is equally distributed to the same tuple when the queries are replied by the random function of experiment p_{rand}. The second observation is that the probability that GOOD does not happen is at most $2np(n)^2/2^n$ as there are at most $p(n)$ queries made by A and each query results in the evaluation of $2n$ random functions $r_i^{x_i}$. □

Given Claim 3, observing that $2np(n)^2/2^n \le 1/2q(n)$ and since by our contradition assumption $\Delta \ge 1/q(n)$, we obtain that there exists a $j \in \{0, \ldots, n-1\}$ such that $\Delta_j \ge 1/2nq(n)$. Then we can construct an adversary B that uses oracle adversary A to violate the pseudo-randomness of the collection of n-bit to $2n$-bit generators G.

Input for Algorithm B: $t : \{0,1\}^n \to \{0,1\}^{2n}$

Instructions for Algorithm B:

1. run INIT′
2. randomly choose $h \in \{1, \ldots, n\}$
3. define $g_s^h(x) = G_n^{x_n}(\cdots G_n^{x_{h+1}}(t(r_{h-1}^{x_{h-1}}(\cdots(r_1^{x_1}(s))\cdots))) \cdots)$
4. set $O = g_s^h$, let $d = A^O(1^n)$ and output: d.

We remark that the functions $r_i^{x_i}$ defined in the above description are implemented as follows: on a new input z, they return an n-bit independently and uniformly chosen string u; on an old input, they return the previously returned output. Note that A can only make polynomially many queries, therefore B only needs to remember a polynomial number of previous outputs.

Assume $h = j$ (this happens with probability $1/n$). We see that in step 3 the function O is equal to g_s^h if t is a random function or to g_s^{h-1} if t is equal to G_n. Then B can contradict the pseudo-randomness of G with respect to multiple samples, and therefore the pseudo-randomness of G. □

The second result we present is the following

Theorem 4. *If there exists a collection of pseudo-random functions then there exists a collection of pseudo-random permutations.*

Proof. Let $F = \{f_s : n \in \mathcal{N}\}$ be a collection of pseudo-random functions, where $f_s : \{0,1\}^n \to \{0,1\}^n$.

The *Feistel transform* FT is defined as follows: On input $(L_0 \circ R_0)$, where $|L_0| = |R_0| = n$, FT returns $(L_1 \circ R_1)$, where $L_1 = R_0$, and $R_1 = L_0 \oplus f_s(R_0)$. Note that this transform is a permutation: given key s and the output $(L_1 \circ R_1)$, one can compute the input $(L_0 \circ R_0)$, where $R_0 = L_1$ and $L_0 = R_1 \oplus f_s(R_0)$. However, it is clearly not pseudo-random: a distinguisher can simply check that $R_0 = L_1$, a condition that always holds for FT but only holds with very small probability for a random permutation over $2n$-bit inputs. Similarly, one can see that the iteration of 2 applications of FT, even using independently chosen atomic pseudo-random functions, is a permutation but is not pseudo-random. It turns out that the 3-round iteration of FT, when using independently chosen atomic pseudo-random functions f_{s1}, f_{s2}, f_{s3}, is both a permutation and is pseudo-random. We call this construction 3FT.

This proof again uses the hybrid proof technique and therefore we only sketch the main ideas of it. Recall that we need to show that an efficient adversary can distinguish only with negligible probability a 3-round iteration of FT, when using independently chosen pseudo-random functions f_{s1}, f_{s2}, f_{s3}, from a random permutation.

For $i = 0, 1, 2, 3$, the intermediate construct D_i in the hybrid argument is defined as the construction 3FT, where the pseudo-random functions in the first i rounds are replaced by a random functions. Then the assumption that 3FT is not pseudo-random can be rephrased by saying that an efficient adversary can distinguish if its oracle is D_0 or a random permutation with probability non-negligible. Then note that D_3 and a random permutation can be distinguished with probability at most $3q(n)^2/2^n$ if $q(n)$ is the upper bound on the number of queries made by the adversary. Then, by an application of the triangle inequality we see that A can distinguish D_i from D_{i+1}, for some $i \in \{0, 1, 2\}$ with non-negligible probability. Now, note that the difference between D_i and D_{i+1} is in the function in the i-th round that is pseudo-random in the former space and random in the latter. Furthermore, the remaining rounds can be efficiently simulated by an algorithm A' that, using A, can distinguish if the oracle she is interacting with is a pseudo-random or random function with non-negligible probability.   □

Since [31] proves that a pseudo-random generator can be constructed from any one-way function, we immediately obtain the following corollaries.

Corollary 1. *If there exists a collection of one-way functions then there exists a collection of pseudo-random functions.*

Corollary 2. *If there exists a collection of one-way functions then there exists a collection of pseudo-random permutations.*   ·

A pseudo-random function $F = \{f_n(s, \cdot) : n \in \mathcal{N}\}$ can be used to define a one-way function $H = \{h_n : n \in \mathcal{N}\}$, where $h_n(x) = f_n(x, 0)$ for any $x \in \{0, 1\}^n$ and any $n \in \mathcal{N}$. H is one-way as otherwise any inverter can be used to compute the key of the pseudo-random function and therefore violate the pseudorandomness of F.

4.3. Examples and Applications

Efficient constructions of pseudo-random functions can be obtained by combining efficient constructions for pseudo-random generators with Theorem 3.

We note that for greater generality we have defined the original 'asymptotic' variant of the notions of pseudo-random functions and permutations. We remark that recently a 'finite' versions of these notions, only considering the case of functions and permutations (rather than collection of them), has received a lot of attention from the literature. (We note that such definitions can be simply derived by the asymptotic by only using functions or permutations f_n, r_n for a fixed n, and parameterizing the distinguishing probability difference.) This has allowed the study of popular finite functions (such as the cryptographic hash function SHA, and the block ciphers DES and AES) in an idealized model where such functions can be assumed to behave as finite pseudo-random functions and used as primitive for more involved constructions. Based on these assumptions, several studies have been made on various aspects of these functions, such as computing upper and lower bounds on the adversary's success probability in distinguishing the constructions from really random oracles.

We briefly review other practical applications of pseudo-random functions, such as dynamic hashing, private-key encryption, message authentication schemes and identification schemes.

Dynamic hashing. As a hashing function $h : \{0,1\}^n \to \{0,1\}^m$, for $m < n$, one can use a pseudo-random function f_s and set $h(x)$ equal to the first m bits of f_s. This makes the hash function more secure in the sense that even if the adversary obtains hashed values $h(x_i)$ of several strings x_i of length n, the adversary still cannot guess $h(y)$ for a new string y.

Private-key encryption. A secure private-key encryption scheme can be constructed from any pseudo-random function. Assume Alice and Bob share a key k. Then, in order to send a message m to Bob, Alice randomly chooses r and sends $(r, f_s(r) \oplus m)$ to Bob. Note that Bob, given s and pair (r, z) received by Alice, can compute $m = z \oplus f_s(r)$. However, an efficient adversary observing the conversation between Alice and Bob, even after seeing polynomially many (r_i, z_i), does not obtain any meaningful information about the messages m_i since she only sees random values r_i and pseudo-random values z_i (that still 'look random' to her).

Message Authentication Schemes. A secure message authentication scheme can be constructed from any pseudo-random function. Assume Alice and Bob share a key k. Then, in order to send a message m to Bob, Alice randomly computes $f_s(m)$ and sends $(m, f_s(m))$ to Bob. Note that Bob, given s and pair (m, z) received by Alice, can verify that $z = f_s(m)$ and therefore believe that the received message m is the same Alice intended to send him. However, an efficient adversary observing the conversation between Alice and Bob, upon seeing (m, z), cannot modify m into a different m' without being detected by Bob, as she cannot produce value $f_s(m')$ (or otherwise she would distinguish $z = f_s(m)$ from a random value).

Client-Server Identification Schemes. A secure client-server identification scheme can be constructed from any pseudo-random function. Assume a client and a server offering some service share a key k. Then, in order to offer a service to her client, the server sends a random message m to the client and gives the service only of she receives in return $f_s(m)$. Note that as for the above message authentication, an adversary, not knowing s, cannot obtain a service from the server as she can produce $f_s(m')$ for some random value m' only with very small probability.

5. Zero-Knowledge Protocols

The seemingly paradoxical notion of Zero-Knowledge Proof Systems, introduced in [30], has received a great amount of attention in both the cryptography and computational complexity literature. Very informally, a zero-knowledge proof is

a method allowing a prover to convince a verifier of a statement without revealing any additional information other than the fact that the theorem is true. In
other words, all the verifier gains by interacting with the prover on input a true
statement is something that the verifier could have generated by herself. While
the two requirements of 'convincing a verifier' and 'yet not revealing anything
else' may seem hard to coexist, zero-knowledge proofs have found rigorous formu-
lations and efficient instantiations in various settings. Furthermore, the general
zero-knowledge methodology of revealing only the necessary minimal information
in communication in the presence of adversaries has become a fundamental tool
having wide applicability throughout cryptography.

5.1. Basic Definitions

We start with some basic notions and definitions, including the definition of interactive protocols of [30].

A language L is a subset of $\{0,1\}^*$. If L is a language, by $\chi_L : \{0,1\}^* \to \{0,1\}$ we denote the *indicator function* for the language L (i.e., $\chi_L(x) = 1$ if and
only if $x \in L$). By GI and GNI we denote the languages of graph isomorphism
and its complement, respectively. NP is the class of languages decidable in non-
deterministic polynomial-time or verifiable in polynomial time. The 'NP proof
system' for a language L consists of two steps: the prover, on input x, sends a
witness w of length polynomial in n to the verifier; the verifier, on input x, w can
run a polynomial time predicate to check that w is a witness of the fact that $x \in L$.
This proof system is *non-interactive*, in the sense that a single message is sent from
the prover to the verifier. Moreover, the verifier runs in deterministic polynomial
time. A binary relation $R(\cdot, \cdot)$ is a boolean predicate over two sets that we will
call respectively the *domain* dom R and the *codomain* codom R of relation R. Any
language in NP can be associated with a polynomial-time relation R_L such that
$R_L(x, w) = 1$ if and only if w is a witness of the fact that $x \in L$. Similarly, one
can define a language L_R associated with a polynomial time relation R.

Interactive Protocols. A *probabilistic Turing machine* is a Turing machine with an
additional read-only tape, called the *random tape* whose content is a sequence of
uniformly and independently distributed bits that can be used to perform probabilistic computation. An *interactive Turing machine* is a probabilistic Turing
machine with two additional read/write tapes: a *input tape* and a *communication
tape*. An *interactive protocol* is a pair of interactive Turing machine sharing the
input and communication tapes. If A and B are two interactive probabilistic Turing machines, by pair (A,B) we denote an interactive protocol. Let x be an input
common to A and B. The *transcript* of an execution of protocol (A,B) on input x,
denoted by $\text{tr}_{(A(y),B(R))}(x)$, where R is the content of B's random tape, and y is
A's private input (if any), is the sequence of messages that are written by A or B
on B's communication tape during such execution.

By A_O we denote algorithm A, when given oracle access to machine O. By
$\text{OUT}_B(\text{tr}_{(A(y),B(R))}(x)) \in \{\text{ACCEPT, REJECT}\}$ we denote B's *output* at the end

of the execution of protocol (A,B) on common input x and where R is B's random tape. We will say that B *accepts (rejects)* x, if $\mathrm{OUT}_B(\mathrm{tr}_{(A(y),B(R))}(x)) = \mathrm{ACCEPT}$ ($\mathrm{OUT}_B(\mathrm{tr}_{(A(y),B(R))}(x)) = \mathrm{REJECT}$). Also, we will say that transcript $\mathrm{tr}_{(A(y),B(R))}(x)$ is *accepting (rejecting)* if B accepts (rejects) x.

We define $A(y)$-$\mathrm{View}_B(x)$, B's view of the interaction with A on input x, as the probability space that assigns to pairs $(R; \mathrm{tr}_{(A(y),B(R))}(x))$ the probability that R is the content of B's random tape and that $\mathrm{tr}_{(A(y),B(R))}(x)$ is the transcript of an execution of protocol (A,B) on input x given that R is B's random tape and y is A's private input (if any).

Let G be a probabilistic Turing machine which is given read-only access to the communication tapes between machines A and B. We define (A,B)-$\mathrm{View}_G(x)$, G's view of the interaction between A and B on input x, as the probability space that assigns to a string $\mathrm{tr}_{(A,B(\cdot))}(x)$ the probability that $\mathrm{tr}_{(A,B(\cdot))}(x)$ is the transcript of some execution of protocol (A,B) on input x.

5.2. Zero-Knowledge Proof Systems of Membership

We start by recalling the formal definition for zero-knowledge proof systems of membership, introduced in [30]. A zero-knowledge proof system of membership is an interactive protocol in which a prover convinces a polynomial time verifier that a string x belongs to a language L. Informally, the requirements for zero-knowledge proof systems of membership are three: completeness, soundness and zero-knowledge. The requirements for interactive proofs of membership are two: completeness and soundness. The completeness requirement states that for any input x in language L, the verifier accepts with overwhelming probability. The soundness requirement states that for any input x not in the language L, the verifier rejects with overwhelming probability. The zero-knowledge requirement can come in three main variants: computational, statistical and perfect zero-knowledge. We will deal with computational and perfect only. The perfect zero-knowledge (resp., computational zero-knowledge) requirement states that for all probabilistic polynomial time verifiers V', the view of V' on input $x \in L$ cannot be distinguished by any algorithm (resp., by any polynomial-time algorithm), from the output of an efficient algorithm, called the 'simulator', on input the same x.

Definition 15. *Let L be a language, and let (P,V) be an interactive protocol, where V runs in polynomial time. We say that a pair (P,V) is an* interactive proof *system of membership for L if*

1. **Completeness.** *For all $x \in L$, $\mathrm{Prob}(\mathrm{OUT}_V(\mathrm{tr}_{(P,V)}(x)) = \mathrm{ACCEPT}) = 1$.*
2. **Soundness.** *For all $x \notin L$, for any Turing machine P',*
   $\mathrm{Prob}(\mathrm{OUT}_V(\mathrm{tr}_{(P',V)}(x)) = \mathrm{ACCEPT}) \leq 1/2$.

We will call the bound $1/2$ in the soundness requirement on the probability that V accepts the *error probability* of the proof system. We remark that by using standard techniques as "sequential composition", such probability can be suitably decreased to, say, 2^{-k}, for any $k \geq 0$ and polynomial in the input size n.

Definition 16. Let L be a language, and let (P,V) be an interactive proof system of membership for L. We say that (P,V) is *computational zero-knowledge* if for each probabilistic polynomial time algorithm V', there exists a polynomial time algorithm S, called the simulator, such that for all $x \in L$ the distributions $S_{V'}(x)$ and $View_{V'}(x)$ are computationally indistinguishable.

Definition 17. *Let L be a language, and let (P, V) be an interactive proof system of membership for L. We say that (P, V) is* perfect zero-knowledge *if for each probabilistic polynomial time algorithm V', there exists a polynomial time algorithm S, called the simulator, such that for all $x \in L$ the following holds:*

1. *$S_{V'}(x) = \perp$ with probability at most $1/2$;*
2. *Conditioned on $S_{V'}(x) \neq \perp$, the two probability distributions $S_{V'}(x)$ and $View_{V'}(x)$ are equal.*

All random self-reducible languages (including graph isomorphism, quadratic residuosity modulo composites and discrete logarithm problems) and their complements have been shown in [30, 28, 50] to have a perfect zero-knowledge proof system of membership. This results have been generalized in [14] to all monotone formulae over random self-reducible languages, and all monotone formulae over complements of random self-reducible languages.

A computational zero-knowledge proof of membership for 3COL. Perhaps the most important result in zero-knowledge protocols is the construction, using commitment schemes, of a zero-knowledge proof system for all languages in NP, due to [28]. An implementation of their protocol using subsequent results gives rise to the following

Theorem 5. *If non-uniform one-way functions exist then there exists a computational zero-knowledge proof system for all languages in NP.*

In order to prove this theorem, we first define commitment schemes.

Informally speaking, a *bit-commitment scheme* (A,B) is a two-phase interactive protocol between two probabilistic polynomial time parties A and B, called the sender and the receiver, respectively, such that the following is true. In the first phase (the commitment phase), A commits to bit b by computing a pair of keys (com, dec) and sending com (the commitment key) to B. Given just σ and the commitment key, the polynomial-time receiver B cannot guess the bit with probability significantly better than $1/2$ (this is the *secrecy* requirement). In the second phase (the decommitment phase) A reveals the bit b and the key dec (the decommitment key) to B. Now B checks whether the decommitment key is valid; if not, B outputs a special string $\perp$, meaning that he rejects the decommitment from A; otherwise, B can efficiently compute the bit b revealed by A and is convinced that b was indeed chosen by A in the first phase (this is the *binding* requirement).

We remark that string commitment schemes can be obtained by independently committing to each bit of the binary string. We also remark that the commitment schemes considered in the literature can be divided in two main types, according to whether the secrecy property holds with respect to computationally bounded adversaries or to unbounded adversaries. A computationally-secret bit-commitment scheme has been constructed under the minimal assumption of the existence of pseudo-random generators (see [43]). A perfectly-secret bit-commitment scheme has been constructed under the assumption of the existence of one-way permutations (see [44]).

As pseudo-random generators have been constructed from any non-uniform one-way functions [31], Theorem 5 is proved if we construct a computational zero-knowledge proof system of membership for an NP-complete language using any commitment scheme. The NP-complete language used in [28] is 3COL, the language of 3-colorable graphs (that is, there exists a function labeling each node of G with one out of three colors such that any two adjacent nodes have been labelled with different colors).

We now informally describe the proof system (P,V) for 3COL. The common input to prover P and verifier V is a graph G and P would like to convince V that $G \in 3COL$. We can divide (P,V) into three messages. First, P computes commitments to the randomly permuted colors of nodes of graph G, and sends its commitments to V. Second, V randomly chooses a "challenge" edge (u, v) and sends it to V. Third, P computes its "answer" message opening the commitments for nodes u, v and showing that the committed colors were different. If this was the case V accepts otherwise V rejects.

A formal description of (P,V) is in Figure 1. We have the following

The Protocol (P,V)

Input to P and V: n-node, m-edge graph G

Input to P: a 3-coloring function $\phi : \{1, \ldots, n\} \to \{1, 2, 3\}$ for G

$P1$: Uniformly choose a permutation ψ over $\{1, 2, 3\}$ and compute function $\rho : \{1, \ldots, n\} \to \{1, 2, 3\}$ as $\rho = \psi \circ \phi$
Compute pairs of commitments/decommitments (com_i, dec_i) of $\rho(i)$, for $i = 1, \ldots, n$

$P \to V$: $com_1, \ldots, com_n$.

$V1$: Uniformly choose edge (u, v), for $u, v \in \{1, \ldots, n\}$

$P \leftarrow V$: u, v.

$P2$: Let dec_u, dec_v be decommitments of com_u, com_v as $\rho(u), \rho(v)$, respectively

$P \to V$: $(\rho(u), dec_u), (\rho(v), dec_v)$.

$V2$: verify that com_u, com_v have been correctly opened as $\rho(u), \rho(v)$
if $\rho(u) \neq \rho(v)$ then return: ACCEPT else return: REJECT.

Fig. 1: A computational zero-knowledge proof system of membership for 3COL

Theorem 6. *The protocol (P, V) is a computational zero-knowledge proof of membership for 3COL.*

Proof. Clearly, V's program can be performed in polynomial time. Now we give a sketch of proof for the requirements of completeness, soundness and computational zero-knowledge.

Completeness. Assume $G \in 3COL$. If P and V behave honestly, then P's verifications in his last step are satisfied with probability 1. This is because P has a 3-coloring ϕ of G and, for any permutation ψ over $\{1, 2, 3\}$ chosen by P, and any adjacent nodes u, v chosen by V, it holds that $\rho(u) = \psi \circ \phi(u) \neq \psi \circ \phi(v) = \rho(v)$.

Soundness. Assume $G \notin 3COL$ and that V behave honestly. Then there is at least one pair of adjacent nodes u', v' in G such that $\rho(u') = \rho(v')$, and $com_{u'}, com_{v'}$ are commitments to $\rho(u'), \rho(v')$, respectively. Consider the event $u = u'$ and $v = v'$. If P reveals $dec_{u'}, dec_{v'}$ then V rejects. On the other hand, by the properties of commitment schemes, a potentially dishonest P can reveal values different from $dec_{u'}, dec_{v'}$ only with negligible probability. Therefore, the probability that V accepts is at most the probability that $u \neq u'$ and $v \neq v'$ plus the probability that P reveals in step P3 different values than the one committed at in step P1. This probability is at most $1 - 1/n^2 + \delta(n)$, for some negligible function δ and can be made exponentially small by performing n^3 independent sequential repetitions of this atomic protocol.

Computational zero-knowledge. An informal sketch on how to construct an expected polynomial time simulator S follows. Recall that S interacts with a verifier V' which may deviate arbitrarily from V's program. S chooses two different colors for some random edge of G and the same color for all other nodes, and sends commitments to all such colors to V', hoping that this particular edge is picked by V', If this does not happen, however, S can "rewind" the program of V' until this event happens, in which case S returns the transcript so obtained.

We note that S needs to try only at most n^2 rewinding attempts on average. Moreover, the output of $S_{V'}$ is computationally indistinguishable from that of a real exection of (P,V'), as the only difference is in the content of the committed values in the first message sent by P. However this difference cannot be observed by a polynomial time distinguisher and therefore the two distributions are computationally indistinguishable. □

The presented zero-knowledge proof system for an NP-complete language has found numerous applications in various areas of cryptography. It has also played an important role in enlarging as much as possible the class of languages having zero-knowledge proof systems of membership, as in the following result, due to [35, 9].

Theorem 7. *If non-uniform one-way functions exist then there exists a computational zero-knowledge proof system of membership for all languages having an interactive proof system of membership.*

We note that the class IP of languages having an interactive proof system of membership has been proved equal to PSPACE in an important result in [47]. It follows then that any language in PSPACE has a zero-knowledge proof system of membership. One may wonder if all languages in PSPACE or NP have a perfect zero-knowledge proof system of membership. It turns out that, as proved in [8, 22, 2], it is very unlikely that all languages in NP have such a proof system (as otherwise the polynomial hierarchy would collapse to its second level). An important consequence of these results is that a way to give evidence that a language is not NP-complete is to construct a perfect zero-knowledge proof system for it.

A perfect zero-knowledge proof of membership for GI. Recall that the language GI is in NP and therefore has a simple proof system of membership: the prover sends an isomorphism between the two input graphs, and the verifier just checks that he indeed received a valid isomorphism. We now present a perfect zero-knowledge proof system for this language from [28], in which the prover does not reveal any information at all about the input graphs, other than the fact that they are isomorphic. Contrarily to the previous computational zero-knowledge proof systems, this result is unconditional in the sense that it does not depend on unproven assumptions, such as the existence of commitment schemes.

We start by informally describing the proof system (P,V) for GI. The common input to prover P and verifier V is a pair of graphs (G_0, G_1) and P would like to convince V that the two graphs are isomorphic, that is, $G_0 \approx G_1$. We can divide (P,V) into three messages. First, P randomly chooses a graph H isomorphic to G_0, and sends its "commitment" message H to V. Second, V randomly chooses a "challenge" bit b and sends it to V. Third, P computes its "answer" message π as an isomorphism between H and G_b and sends it to V, who accepts if and only if π is an isomorphism between H and G_b.

A formal description of (P,V) is in Figure 2. We have the following

The Protocol (P,V)

Input to P and V: (G_0, G_1), where G_0, G_1 are n-node graphs.

Input to P: ϕ, such that $G_1 = \phi(G_0)$

    P1: Uniformly choose a permutation π and compute $H = \pi(G_0)$

$P \to V$: H.

    V1: Uniformly choose bit b

$P \gets V$: b.

    P2: If $b = 0$ then set $\psi = \pi$ otherwise set $\psi = \pi \circ \phi^{-1}$

$P \to V$: ψ.

    V2: if $H = \psi(G_b)$ then return: ACCEPT else return: REJECT.

Fig. 2: A perfect zero-knowledge proof system of membership for GI

Theorem 8. *The protocol (P, V) is a perfect zero-knowledge proof of membership for GI.*

Proof. Clearly, V's program can be performed in polynomial time. Now we prove the three requirements of completeness, soundness and perfect zero-knowledge.

Completeness. Assume $G_0 \approx G_1$. If P and V behave honestly, then P's verifications in his last step are satisfied with probability 1. To see this, consider first the case $b = 0$. In this case V's verification in step V2 is met as $H = \pi(G_0) = \pi(G_b) = \psi(G_b)$. Now, consider the case $b = 1$. Also in this case V's verification in step V2 is met as $H = \pi(G_0) = \pi \circ \phi^{-1}(G_1) = \psi(G_1) = \psi(G_b)$.

Soundness. Assume $G_0 \napprox G_1$ and that V behave honestly. Let H be the graph sent by (a potentially adversary) P in step P1. By the previous assumption H cannot be isomorphic to both G_0 and G_1, but might be isomorphic to one of them, let this graph be G_a. Then P can meet V's verificaton in step V2 only if $a = b$, which happens with probability 1/2. Therefore the probability that V accepts is at most 1/2.

Perfect zero-knowledge. We now show a simulator S. Recall that S interacts with a verifier V' which may deviate arbitrarily from V's program. The basic trick that allows S to produce an accepting conversation between P and V even without knowing a witness for $(G_0, G_1) \in GI$ is that S can "rewind" the verifier until he is as lucky as a dishonest prover.

The simulator S. On input $(G_0, G_1) \in GI$, S will first of all feed V' with a random string of appropriate length. Then S randomly chooses a bit a and a permutation π, computes graph $H = \pi(G_a)$, and sends H to V'. Now, V' sends its random bit b to P. At this point, if $a = b$ then S sets $\psi = \pi$ and returns (H, b, π) and halts; otherwise, he restarts the entire process again, using independently distributed random bits.

We need to show two properties of S: first, S's output is distributed exactly as the output of the protocol; second, S's running time is expected polynomial time.

To see that the first property is satisfied, we start by observing that the messages from V' are clearly equally distributed in both spaces, since they are computed in the same way. The first message from the prover is equally distributed in both spaces since we are assuming that $G_0 \approx G_1$. The second message of the prover is distributed as a random isomorphism between H and G_a in both the transcript of the protocol and the output of the simulator.

To see that the second property is satisfied, we observe that the simulator only executed polynomial time computation and terminates with probability 1/2 at each attempt. Therefore he only needs an expected number of 2 attempts and its total running time is expected polynomial time.   □

5.3. Witness-Indistinguishable Proof Systems of Knowledge

The concept of proof systems of knowledge has been alluded to in [30], developed by [20, 21, 50] and fully formalized in [5]. In this section we recall the definition given in [5], with the additional requirement of witness indistinguishability, introduced in [21]. A witness-indistinguishable proof system of knowledge is an interactive protocol in which, on input a string x, a prover convinces a poly-bounded verifier that he knows a string y such that a polynomial-time relation $R(x, y)$ holds; moreover, for any y_1, y_2, no information is revealed to the verifier about whether the string y used by the prover is equal to y_1 or y_2. Informally, the requirements for witness-indistinguishable proof systems of knowledge are three: non-triviality, extraction and witness-indistinguishability. The non-triviality requirement states that for any input x in the domain of relation R, the verifier accepts with overwhelming probability. The extraction requirement states that there exists an extractor that, for any input x, and interacting with any prover that forces the verifier to accept with 'sufficiently high' probability, is able to compute a string y such that $R(x, y)$ holds, within a 'properly bounded' expected time. The witness-indistinguishability requirement states that for all input $x \in domR$, and for all y_1, y_2 such that $(x, y_1) \in R$ and $(x, y_2) \in R$, the verifier's view when P uses y_1 is identical to the verifier's view when P uses y_2.

Definition 18. *Let P be a probabilistic Turing machine and V a probabilistic polynomial-time Turing machine that share the same input and can communicate with each other. Let R be a two-argument polynomial time relation and err : $\{0,1\}^* \to [0,1]$ be a function. We say that a pair (P,V) is a* WITNESS-INDISTINGUISHABLE PROOF SYSTEM OF KNOWLEDGE *with knowledge error err for relation R if*

1. **Non-Triviality.** *For all $x \in domR$, $\mathrm{Prob}(\mathrm{OUT}_V(\mathrm{tr}_{(P,V)}(x)) = \mathrm{ACCEPT}) = 1$.*
2. **Extraction.** *There exists a probabilistic oracle machine E (called the extractor) such that for all $x \in domR$, and for any Turing machine P', and letting $acc_{P'}(x) = \mathrm{Prob}(\mathrm{OUT}_V(\mathrm{tr}_{(P',V)}(x)) = \mathrm{ACCEPT})$, the following holds: if $acc_{P'}(x) > err(x)$ then,*
   - $\mathrm{Prob}(E_{P'}(x)) = y) \geq 2/3$, *where $(x, y) \in R$.*
   - *The machine E halts within expected time bounded by $\frac{n^c}{(acc_{P'}(x) - err(x))}$, for some constant $c > 0$.*
3. **Witness Indistinguishability.** *For any $x \in domR$, and any y_1, y_2 such that $(x, y_1) \in R$ and $(x, y_2) \in R$, the probability spaces $P(y_1)$-$View_V(x)$ and $P(y_2)$-$View_V(x)$ are equal.*

In [21] it was shown that any zero-knowledge proof of knowledge is also witness-indistinguishable (the converse being not necessarily true). In fact, the concept of witness-indistinguishable proofs is sufficient for many applications. For instance, in some zero-knowledge protocols, 3-round witness-indistinguishable proofs of knowledge are executed as subprotocols, in which the verifier proves the knowledge of some string which certifies that he has computed honestly some previous message.

A Witness-Indistinguishable proof of knowledge for R_{GI}. Define protocol (P,V) as the parallel repetition of n independent executions of the protocol for GI presented in Section 5.2, where n is the size of the input. It has been proved in [26] that protocol (P,V) is *not* zero-knowledge (according to a stronger notion, called "black-box zero-knowledge", unless GI is in BPP, which trivializes the question). Now we show that the protocol (P,V) is a witness-indistinguishable proof of knowledge for R_{GI}. That is, we have the following

Theorem 9. *The protocol (P, V) is a witness-indistinguishable proof of knowledge for R_{GI}.*

Proof. Clearly, V's program can be performed in polynomial time. The non-triviality property directly follows from the completeness of the atomic proof of membership for GI. Now we sketch the proofs of the extraction and perfect witness-indistinguishability of (P,V).

Extraction. This is showed by presenting an extractor E. Recall that E uses as an oracle prover P′ which may deviate arbitrarily from P's program and makes V accept with a certain probability $acc_{P'}(G_0, G_1)$. Intuitively, the trick that E uses to obtain an isomorphism between G_0 and G_1 is that of 'rewinding' the prover in order to ask two different tuples of challenge bits and receive, for at least one copy of the atomic protocol, an answer to challenges 0,1, which reveals the desired isomorphism. We note that if $acc_{P'}(G_0, G_1) > 0$ then one can prove that $\text{Prob}(\text{OUT}_E(\text{tr}_{(P',E)}(G_0, G_1))) = \phi \geq 1 - 2^{-n}$ and that the expected running time of E is a polynomial times the expected number of necessary rewindings of P′. Since E only needs two accepting conversations from P′, the latter number is about $2/acc_{P'}(G_0, G_1)$.

Witness-indistinguishability. Let us observe first that any zero-knowledge proof is also witness-indistinguishable (intuitively, this is because if an adversary can distinguish which witness the prover is using then he can obtain some knowledge he did not know before running the protocol). Therefore a single execution of the atomic protocol for GI is witness-indistinguishable. To prove that a parallel execution of n copies of that protocol is still witness-indistinguishable, we will use again the 'hybrid proof technique' of [29] and contradict the fact that the atomic protocol for GI is witness-indistinguishable. Let ϕ_1, ϕ_2 be two different isomorphisms between G_0 and G_1. Assume, for sake of contradiction, that (P,V) is not witness-indistinguishable. Then there exists an adversary V′ that is able to distinguish with some non-negligible probability a transcript of the protocol when P uses witness ϕ_1 from a transcript of the protocol when P uses witness ϕ_2. Let D_0 (resp., D_n) denote the distribution returning a transcript of an execution of protocol (P,V′), when P is using isomorphism ϕ_1 (resp., ϕ_2). Then the assumption can be written as saying that there exists a polynomial p and a probabilistic polynomial time algorithm V′ such that for infinitely many k's, it holds that

$$\Delta = |\text{Prob}(\text{OUT}_{V'}(D_0) = 1) - \text{Prob}(\text{OUT}_{V'}(D_n) = 1)| \geq 1/p(k).$$

For $i = 1, \ldots, n-1$, we define distribution D_i as the distribution returning a transcript of an execution of protocol (P,V′), where P uses ϕ_2 in the first i parallel executions of the atomic protocol for GI and ϕ_1 in the remaining $n-i$ executions.

Then we can rewrite Δ as

$$\Sigma_{i=0}^{n-1} \, | \, \mathrm{Prob}\,(\mathrm{OUT}_{V'}(D_i) = 1) \, - \, \mathrm{Prob}\,(\,\mathrm{OUT}_{V'}(D_{i+1}) = 1\,)\,|$$

and since $\Delta \geq 1/p(k)$ we obtain that there exists an $i \in \{0, \ldots, n-1\}$ such that

$$|\,\mathrm{Prob}\,(\,\mathrm{OUT}_{V'}(D_i) = 1\,) \, - \, \mathrm{Prob}\,(\,\mathrm{OUT}_{V'}(D_{i+1}) = 1\,)\,| \geq 1/(n \cdot p(k)).$$

This can be used to construct an algorithm that violates the witness-indistinguishability of the atomic protocol for GI, from which a contradiction is derived. □

A Perfect Zero-Knowledge Proof for GNI. An important application of witness indistinguishable proofs of knowledge is in constructing a perfect zero-knowledge proof system for GNI [28]. (Note that GNI is not in NP.)

    We start by informally describing an interactive proof system of membership for GNI. This consists of two messages: on input (G_0, G_1), the verifier randomly chooses a bit b and a graph H isomorphic to G_b, and sends H to the prover. The prover computes b' such that $H \approx G_{b'}$ and sends b' to the verifier that accepts if and only if $b = b'$.

    We note that this proof system is not zero-knowledge as a cheating verifier might send a graph H' for which he does not know if H' is isomorphic to G_0 or G_1 and use the prover's answer to determine that. In order to avoid this problem, the protocol is patched as follows: the verifier, in addition to sending H, also gives a witness-indistinguishable proof of knowledge of an isomorphism between H and one of G_0, G_1. This proof can be obtained as an extension of the previous witness indistinguishable protocol (see [28, 14]).

5.4. Zero-Knowledge Proof Systems of Decision Power

The idea of proving the knowledge of whether a string belongs to a language or not has been given in [20]; a related concept of proving computational power has been introduced in [53]; the formal definition of zero-knowledge proof systems of decision power has first appeared in [16]. Applications of this type of protocols include entity authentication protocols.

    A zero-knowledge proof system of decision power is an interactive protocol in which a prover convinces a poly-bounded verifier that he knows whether a string x belongs to a language L or not, without revealing which is the case, or any other information. Informally, the requirements for zero-knowledge proof systems of decision power are three: verifiability, extraction and zero-knowledge. Verifiability states that the verifier accepts with high probability for any input x, in the language L or not. Extraction states that there exists an extractor that, for any input x, and interacting with any prover that forces the verifier to accept with 'sufficiently high' probability, is able to decide whether $x \in L$ or not, within a 'properly bounded' expected time. This differs from previous work on proofs of knowledge in which the extractor existed only for input in the language and was required to output a string satisfying a polynomial relation with the input. This approach allows to consider even languages above NP. Finally, the zero-knowledge requirement states that for all probabilistic polynomial time verifiers V', the view of V' is efficiently simulatable, and the simulation is correct for all x (in L or not).

Definition 19. *Let P be a probabilistic Turing machine and V a probabilistic polynomial-time Turing machine that share the same input and can communicate with each other. Let L be a language and err : $\{0,1\}^* \to [0,1]$ be a function. We say that a pair (P,V) is a* perfect zero-knowledge proof system of decision power *with knowledge error err for L if*

1. **Verifiability.** *For all x,*

$$\mathrm{Prob}(\mathrm{OUT}_V(\mathrm{tr}_{(\mathrm{P,V})}(x)) = \mathrm{ACCEPT}) = 1.$$

2. **Extraction.** *There exists a probabilistic oracle machine E (called the extractor) such that for all x, and any Turing machine P', and letting $acc_{P'}(x) = \mathrm{Prob}(\mathrm{OUT}_V(\mathrm{tr}_{(\mathrm{P',V})}(x)) = \mathrm{ACCEPT})$, the following holds: if $acc_{P'}(x) > err(x)$ then,*
   - $\mathrm{Prob}(E_{P'}(x)) = \chi_L(x)) \geq 2/3$.
   - *The machine E halts within expected time bounded by $\frac{n^c}{(acc_{P'}(x) - err(x))}$, for some constant $c > 0$.*

3. **Perfect Zero-Knowledge.** *For all probabilistic polynomial-time verifiers V', there exists a polynomial time algorithm S, called the simulator, such that for all x, the following holds:*
   (a) *$S_{V'}(x) = \bot$ with probability at most $1/2$;*
   (b) *Conditioned on $S_{V'}(x) \neq \bot$, the two probability spaces $S_{V'}(x)$ and P-$View_{V'}(x)$ are equal.*

The languages known to have a perfect zero-knowledge proof of decision power are the languages that are known to be random self-reducible, that is, quadratic residuosity [20, 16], graph isomorphism and discrete log [16], and a certain class extending these languages [18].

In principle it might be possible to directly use interactive proof systems of membership in order to construct proof systems of decision power. In particular, consider the following protocol transformation: Given a proof of membership (A,B) for the language OR(L,$\overline{L}$) defined as the set of pairs (x_1, x_2) such that $(x_1 \in L) \vee (x_2 \notin L)$ (in [14] such proofs have been given for GI), derive a protocol (P,V) as (A,B) executed on input (x, x). One would observe that such transformation might be a reasonable approach to construct a proof system of decision power for L. Nevertheless, it turns out that this approach in general fails; that is, the obtained (P,V) fails to be a proof of decision power (an example for this is fully explained in [16]). Therefore we need new techiques to construct these protocols.

A proof of decision power for GI. We start by informally describing the proof system (P,V) from [16]. The common input to prover P and verifier V is a pair of graphs (G_0, G_1). We can divide (P,V) into three basic steps. The first step is done by V; he randomly chooses a bit b and a graph G isomorphic to G_b, and sends it to P. In the second step, V proves to P that graph G has been correctly constructed, using a witness-indistinguishable proof of knowledge, that is, without revealing any information about bit b or the permutation used. In the third step, P checks that V's proof is accepting and then proves to V that he knows an isomorphism between graph G and one of the two input graphs G_0, G_1. V accepts if and only if this proof is convincing.

The implementation of the first step goes as follows. The second and the third step can be implemented in various ways; perhaps, the simpler is to use the same protocol for both steps. Specifically, P and V will run twice a witness-indistinguishable subprotocol (from [30]), where in the first execution (second step of (P,V)) V acts as a prover and P as a verifier, and in the second execution (third step of (P,V)) the roles are reversed. By carefully interleaving such executions, we obtain only 4 rounds of communication between P and V. Let n be an integer and $m = n \log n$; a formal description of (P,V) is in Figure 3. We obtain the following

Theorem 10. *The protocol (P, V) is a perfect zero-knowledge proof of decision power (with decision error 0) for GI.*

Proof. Clearly, V's program can be performed in polynomial time. Now we prove the three requirements of verifiability, extraction and perfect zero-knowledge.

Verifiability. First of all notice that if P and V behave honestly, then P's verifications in his last step are satisfied with probability 1. This implies that with probability 1 the graph G sent by V in his first step is isomorphic to at least one of G_0, G_1. Now, observe that regardless of whether $G_0 \approx G_1$ or not, the prover can compute an isomorphism between G and one of G_0, G_1 and then meet V's verification in the third step of the protocol. Specifically, if $G_0 \approx G_1$ then G is isomorphic to both, and, say, the permutation between G and G_0 can be used to run his program in the third step of the protocol. Instead, if $G_0 \napprox G_1$ then G is isomorphic only to G_b, and then the permutation between G and G_b can be computed by P and used to run his program in the third step of the protocol. Thus, in both cases, V accepts with probability 1.

Extraction. We show an extractor E. Recall that E uses as an oracle prover P' which may deviate arbitrarily from P's program and makes V accept with a certain probability $acc_{P'}(G_0, G_1)$.

The extractor E. On input (G_0, G_1), E starts by running m times a procedure, called Iso-ext, which we now describe.

The procedure Iso-ext takes as input a bit b and returns either a bit v or a special string *fail*. Precisely, each time the procedure is executed, it takes as input a uniformly and independently chosen bit b_i. The procedure starts by repeatedly running the program of the verifier V interacting with P' until an accepting conversation is obtained. In this conversation P' has received a graph G chosen by the procedure as isomorphic to G_b; also, P' has sent some pairs of graphs (D_{i0}, D_{i1}) and answered correctly to V's questions represented by bits e_i. Then the procedure Iso-ext rewinds P' until after his first step. Now, V's second round is run again by sending some uniformly chosen e'_i instead of the bits e_i sent before (here the

The Protocol (P,V)

Input to P and V: (G_0, G_1), where G_0, G_1 are n-node graphs.

$V1$: Uniformly choose bit b and a permutation π and set $G = \pi(G_b)$;
       for $i = 1, \ldots, m$,
             uniformly choose bit a_i and two permutations η_{i0}, η_{i1};
             compute graphs $A_{i0} = \eta_{i0}(G_{a_i})$ and $A_{i1} = \eta_{i1}(G_{1-a_i})$

$P \leftarrow V$: $G, (A_{10}, A_{11}), \ldots, (A_{m0}, A_{m1})$.
   $P1$: For $i = 1, \ldots, m$,
             uniformly choose bits c_i, d_i and permutations ψ_{i0}, ψ_{i1};
             compute graphs $D_{i0} = \psi_{i0}(G_{d_i})$ and $D_{i1} = \psi_{i1}(G_{1-d_i})$

$P \rightarrow V$: $(c_1, \ldots, c_m), (D_{10}, D_{11}), \ldots, (D_{m0}, D_{m1})$.
   $V2$: For $i = 1, \ldots, m$,
             uniformly choose a bit e_i;
             if $c_i = 0$ then set $\sigma_i = (\eta_{i0}, \eta_{i1})$;
             if $c_i = 1$ then set $\sigma_i = \pi \circ \eta_{i,b \oplus a_i}^{-1}$

$P \leftarrow V$: $(e_1, \ldots, e_m), (\sigma_1, \ldots, \sigma_m)$.
   $P2$: For $i = 1, \ldots, m$,
             if $c_i = 0$ then
                 let $\sigma_i = (\eta_{i0}, \eta_{i1})$;
                 check that $A_{i0} = \eta_{i0}(G_{a_i})$, $A_{i1} = \eta_{i1}(G_{1-a_i})$, for some bit a_i;
             if $c_i = 1$ then check that $G = \sigma_i(A_{i0})$ or $G = \sigma_i(G_{i1})$;
             if any of the above verifications is not satisfied then halt;
             if $e_i = 0$ then set $\tau_i = (\psi_{i0}, \psi_{i1})$;
             if $e_i = 1$ then
                 if $G_0 \approx G_1$ then
                     randomly choose a bit g_i;
                     compute a permutation τ_i such that $G = \tau_i(G_{g_i})$;
                 if $G_0 \not\approx G_1$ then
                     compute bit b and permutation π such that $G = \pi(G_b)$;
                     set $\tau_i = \pi \circ \psi_{b \oplus d_i}$.

$P \rightarrow V$: $(\tau_1, \ldots, \tau_m)$.
   $V3$: For $i = 1, \ldots, m$,
             if $e_i = 0$ then
                 let $\tau_i = (\psi_{i0}, \psi_{i1})$;
                 check that $D_{i0} = \psi_{i0}(G_{d_i})$, $D_{i1} = \psi_{i1}(G_{1-d_i})$, for some bit d_i;
             if $e_i = 1$ then check that $G = \tau_i(D_{i0})$ or $G = \tau_i(D_{i1})$.
          If all verifications are successful then output: ACCEPT else output: REJECT. Halt.

Figure 3: A perfect zero-knowledge proof system of decision power for GI

procedure also makes sure that the sequence $(e'_1, \ldots, e'_m)$ is distinct from all previously chosen, including $(e_1, \ldots, e_m)$). This step is repeated until another accepting conversation is obtained. Now, in the case the procedure never finds a second (or even a first) accepting conversation, then it outputs *fail*. If this does not happen, then this implies that P$'$ has given answers to bit e_i and bit e'_i corresponding to the same pair of graphs (D_{i0}, D_{i1}), for $i = 1, \ldots, m$. Since there exists an i such that $e_i \neq e'_i$, from the answers to such two distinct bits, the procedure can easily compute an isomorphism ϕ between G and one of G_0, G_1. In this case the output of procedure Iso-ext will be a bit v such that the isomorphism ϕ obtained by P$'$ is such that $G = \phi(G_v)$.

Now, if procedure Iso-ext has ever output *fail* then E runs an exhaustive search procedure to find a permutation π such that $G_0 = \pi(G_1)$, or a proof that no such permutation exists; if such a permutation is found, then E outputs 1; if not, E outputs 0.

Instead, consider the case procedure Iso-ext never outputs *fail*. As mentioned above, E runs m times the procedure Iso-ext, each time on input a uniformly chosen bit b_i. Then, let v_i be the bit output by the procedure Iso-ext, when given b_i as input, for $i = 1, \ldots, m$. Then E outputs 0 (meaning that the graphs G_0, G_1 are not isomorphic) if $b_i = v_i$, for $i = 1, \ldots, m$, and 1 (meaning that the graphs G_0, G_1 are isomorphic) otherwise.

To prove that the output of E is correct, first of all we observe that if the extractor E outputs because of the search procedure then clearly its output is correct with probability 1. Now we consider the case in which the extractor E outputs after running n times the procedure Iso-ext. First, assume that $G_0 \not\approx G_1$. In this case, in each execution of procedure Iso-ext, E sends a graph G isomorphic to G_{b_i} to P$'$; also, procedure Iso-ext finds an isomorphism between G and exactly one of G_0, G_1, which can only be G_{b_i}. Thus, it holds that $v_i = b_i$, for $i = 1, \ldots, m$, and thus E's output is correct with probability 1. Now, assume that $G_0 \approx G_1$. In this case, in each execution of procedure Iso-ext, E sends a graph G isomorphic to G_b to P$'$, and proves that he knows an isomorphism between G and one of G_0, G_1. Since this proof is witness-indistinguishable, no information is revealed about bit b to any P$'$, and thus the probability that $v_i = b_i$ is exactly $1/2$. This means that the probability that there exists a j such that $b_j \neq v_j$, from which it follows that E's output is correct, is at least $1 - 2^{-m} \geq 1 - 2^{-n}$.

Now, consider the running time of E. The first reason E can output is because of the result of the procedure Iso-ext; in this case the expected running time of E is properly bounded, for the following two reasons: 1) at each iteration such procedure essentially runs the program of verifier V, which is strict polynomial time; 2) the expected number of iterations is at most $2/acc_{P'}(G_0, G_1)$. It follows that E's expected time is at most $poly(n)/acc_{P'}(G_0, G_1)$. Now consider the other case, that is, when E outputs because the result of the search procedure; clearly, this procedure may take exponential time. However, this happens when prover P$'$ makes V accept only in correspondence to one of the sequences $(e_1, \ldots, e_m)$. This implies that in this case the probability $acc_{P'}(G_0, G_1)$ is at most 2^{-m}, and E's expected running time is then $poly(n) \cdot n! \cdot 2^{-m} \leq poly(n)$.

Perfect zero-knowledge. We show a simulator S which satisfies Definition 19. Recall that S interacts with a verifier V' (treated as a black box) which may deviate arbitrarily from V's program.

The simulator S. On input (G_0, G_1), S will first of all feed V' with a random string of appropriate length. Then S obtains the first message from V' and runs P's program to simulate the first message by P. Then he obtains the second message from V', which terminates the proof of knowledge from V'. Now, if this proof is convincing, then S uses this proof to extract the knowledge communicated by V' through this proof. That is, S runs the extractor for the proof of knowledge by V' and obtains a permutation between G and one of G_0, G_1. Then S simulates the last message by P by running P's program, and using the obtained permutation as auxiliary input. Finally S outputs the conversation thus obtained.

We now need to show two properties: first, S's output is distributed exactly as the output of the protocol; second, S's running time is expected polynomial time.

To see that the first property is satisfied, we start by observing that the messages from V' are clearly equally distributed in both spaces, since they are computed in the same way. The first message from the prover is equally distributed in both spaces since S runs algorithm P to compute it. The second message of the prover is also computed by S using algorithm P; here S uses the permutation extracted from the proof of knowledge by V' as his auxiliary-input. Although this auxiliary-input may be different from the one used by P during the protocol, the second message by P has the same distribution, no matter which auxiliary-input is used by V, since P is running a witness-indistinguishable proof of knowledge.

To see that the second property is satisfied, we observe that the simulator computes the first message from the prover, by running P's program which is polynomial time here. Then S runs the extractor for the proof of knowledge by V', which, by properties of proofs of knowledge (see [5]) we know to run in expected polynomial time. Finally, he uses the witness obtained from this extraction to run P's program in polynomial time and simulate the last step of the protocol.   □

5.5. Zero-Knowledge Transfers of Decision

The model for zero-knowledge and result-indistinguishable proofs of decision has been introduced in [23]. A zero-knowledge and result indistinguishable protocol in which a prover convinces a poly-bounded verifier of whether a string x belongs to a language L or not, without revealing which is the case, or any other information to any eavesdropper, and without revealing any other additional information to the verifier. An immediate application of this type of protocols is interactive encryption secure with respect to strong definitions based on languages with such proofs. Here we recall the definition given in [23] for result-indistinguishable proofs of decision. The definition has three requirements. The completeness requirement states that for any input x, with overwhelming probability the verifier accepts and can compute the value $\chi_L(x)$. The correctness requirement states that for any input x and any (possibly dishonest) prover, the probability that the verifier accepts and receives the wrong value $1 - \chi_L(x)$ is negligible. The zero-knowledge requirement states that for all probabilistic polynomial time verifiers V', the view of V' is efficiently simulatable, by a simulator that queries an oracle returning $\chi_L(x)$. Moreover, the simulation is correct for all x (in L or not). The perfect result-indistinguishability requirement states that for all input x, the conversation between prover and verifier is efficiently simulatable.

Definition 20. *Let P be a probabilistic Turing machine and V a probabilistic polynomial-time Turing machine that share the same input and can communicate with each other. Also, let C a probabilistic Turing machine having access to the communication between P and V. Let L be a language. We say that a pair (P,V) is a perfect zero-knowledge and perfectly result-indistinguishable transfer of decision for L if*

1. **Completeness.** *There exists $b \in \{0,1\}$ such that for all x, satisfying $\chi_L(x) = b$, $\mathrm{Prob}(\mathrm{OUT}_V(\mathrm{tr}_{(P,V)}(x)) = (\mathrm{ACCEPT}, \chi_L(x))) = 1$.*
2. **Correctness.** *For all x, and for all P',*
   $\mathrm{Prob}(\mathrm{OUT}_V(\mathrm{tr}_{(P',V)}(x)) = (\mathrm{ACCEPT}, 1 - \chi_L(x))) \leq 1/2$.
3. **Perfect Zero-Knowledge.** *For any Turing machine V', there exists a probabilistic Turing machine $S_{V'}$ (called the V-simulator) running in polynomial-time such that $S_{V'}$, given as input both x and $\chi_L(x)$, returns $\perp$ with probability at most $1/2$, and, conditioned on $S_{V'}(x, \chi_L(x)) \neq \perp$, the probability spaces $\mathrm{P\text{-}View}_{V'}(x)$ and $S_{V'}(x, \chi_L(x))$ are equal.*
4. **Perfect Result-Indistinguishability.** *There exists a probabilistic Turing machine M (called the C-simulator) running in probabilistic polynomial-time such that for all x, the probability spaces $(P,V)\text{-}View_C(x)$ and $M(x)$ are equal.*

The only languages known to have a perfect zero-knowledge transfer of decision power are the specific languages that are known to be random self-reducible, that is, quadratic residuosity [23, 16], graph isomorphism and discrete log [16], and a certain class extending these languages [18].

A transfer of decision for GI. We start by informally describing the proof system (P,V) from [17]. The common input to prover P and verifier V is a pair of graphs (G_0, G_1). We can view (P,V) as made of a sequential composition of $3n$ iterations of an atomic protocol (A,B), which in turn can be divided into three phases. In the first phase B randomly chooses a bit b and a graph G isomorphic to G_b, and sends it to A. In the second phase, B proves to A that graph G has been correctly constructed, without revealing any information about bit b and the permutation chosen. In the third phase, A checks that B's proof is accepting; now, if $G_0 \approx G_1$ then A randomly chooses a bit g; otherwise, if $G_0 \not\approx G_1$ then A computes bit b such that $G \approx G_b$ and sets $g = b$. In both cases A proves in zero-knowledge to B that $G \approx G_g$, and if this proof is not convincing then B rejects. At the end of the $3n$ iterations of protocol (A,B), V accepts if B has never rejected. Furthermore, if in at least n iterations it holds that $b \neq g$, V outputs 1 (meaning that he is convinced that $G_0 \approx G_1$); otherwise V outputs 0 (meaning that he is convinced that $G_0 \napprox G_1$).

The implementation of the first phase of protocol (A,B) is simple. We observe that the second phase can be implemented by using a 'witness-indistingui-shable' subprotocol, as done in [30] in their zero-knowledge proof system of membership for the language of quadratic non-residuosity. In particular, we will use the protocol of [28] used in the middle of a zero-knowledge proof of membership for graph non-isomorphism. Then we observe that the subprotocol in the third phase which allows P to convince V that $G \approx G_g$, for some bit g, can be implemented by using a three-steps protocol, as done in [30] for the language of quadratic residuosity or in [28] for the language of graph isomorphism.

A formal description of (P,V): Let n be an integer and $m = n \log n$. The protocol (P,V) is made of $3n$ sequential repetitions of subprotocol (A,B), which is described in Figure 4. Now, if any verification by B is not satisfied, V outputs: (REJECT), and halts. Otherwise, denote by b_i the bit chosen by V in step V1 of the i-th execution of subprotocol (A,B), and by g_i the bit computed by P in step P2 of the i-th execution of subprotocol (A,B). Then V computes the number s of indices $i \in \{1, \ldots, 3n\}$ such that $b_i = g_i$; if it holds that $s \geq 2n$ then V outputs: (ACCEPT,1); otherwise V outputs: (ACCEPT,0). We obtain the following

Theorem 11. *The protocol (P,V) is a perfectly result-indistinguishable and perfect zero-knowledge transfer of decision for GI.*

The rest of the subsection proves Theorem 11. Clearly, V's program can be performed in polynomial time. Now we prove the requirements in Definition 20.

Completeness. We show that for all pairs (G_0, G_1) of graphs, if P and V follow their protocol, then V accepts and outputs $\chi_{GI}(G_0, G_1)$ with probability greater than $1 - n^{-c}$, for any constant c. We analyze two cases. First assume $G_0 \approx G_1$; now, since V follows his protocol, P will be convinced by V's witness-indistinguishable proof that graph H has been correctly computed. Then H is isomorphic to one of G_0, G_1, and the statement $L_0 \approx L_1$ is true, since L_0 is isomorphic to H and L_1 is isomorphic to a randomly chosen graph between G_0, G_1. Moreover, it holds that $b_i = g_i$ with probability $1/2$, and therefore the number s of indices $i \in \{1, \ldots, 3n\}$ such that $b_i = g_i$ will be at least $2n$ with exponentially small probability (using Chernoff bounds). This guarantees that V outputs (ACCEPT,1) with probability greater than $1 - n^{-c}$, for any constant c. Now, assume $G_0 \napprox G_1$; then P can compute bit b and permutation β such that $H = \beta(G_b)$. This implies that the statement $L_0 \approx L_1$ is true, since L_0 is isomorphic to H and L_1 is chosen isomorphic to G_b. We observe that $b_i = g_i$ for all $i = 1, \ldots, 3n$, and thus V outputs (ACCEPT,0) with probability 1.

Correctness. We show that for any P′ and any input pair (G_0, G_1), the probability that V's output is (ACCEPT,$1 - \chi_{GI}(G_0, G_1)$) is negligible. First, consider

The Protocol (A,B)

Input to A and B: (G_0, G_1), where G_0, G_1 are n-node graphs.

B1: Uniformly choose bit b and a permutation β and compute $H = \beta(G_b)$;
for $j = 1, \ldots, m$,
uniformly choose bit a_j and two permutations α_{j0}, α_{j1};
compute graphs $A_{j0} = \alpha_{j0}(G_{a_j})$ and $A_{j1} = \alpha_{j1}(G_{1-a_j})$.

$A \leftarrow B$: $(H, (A_{10}, A_{11}), \ldots, (A_{m0}, A_{m1}))$.

A1: For $j = 1, \ldots, m$, uniformly choose bit c_j;

$A \rightarrow B$: $(c_1, \ldots, c_m)$.

B2: For $j = 1, \ldots, m$,
if $c_j = 0$ then set $\sigma_j = (\alpha_{j0}, \alpha_{j1})$;
if $c_j = 1$ then set $\sigma_j = \beta \circ \alpha_{j, b \oplus a_j}^{-1}$;

$A \leftarrow B$: $(\sigma_1, \ldots, \sigma_m)$.

A2: For $j = 1, \ldots, m$,
if $c_j = 0$ then
let $\sigma_j = (\eta_{j0}, \eta_{j1})$;
check that $A_{j0} = \eta_{j0}(G_{a_j})$ and $A_{j1} = \eta_{j1}(G_{1-a_j})$, for some bit a_j;
if $c_j = 1$ then check that $H = \sigma_j(A_{j0})$ or $H = \sigma_j(A_{j1})$;
if any of the above verifications is not satisfied then halt;
if $G_0 \approx G_1$ then randomly choose a bit g;
if $G_0 \not\approx G_1$ then
compute bit b and permutation β such that $H = \beta(G_b)$ and set $g = b$;
set $L_0 = H$ and $L_1 = G_g$;
uniformly choose a bit t and a permutation τ and set $T = \tau(L_t)$;

$A \rightarrow B$: $(L_0, L_1), T$.

B3: Uniformly choose a bit l;

$A \leftarrow B$: l.

A3: If $l = t$ then set $\rho = \tau$;
if $l = 1 - t$ then compute ρ such that $T = \rho(L_l)$;

$A \rightarrow B$: ρ.

B4: Check that $T = \rho(L_l)$.

Figure 4: A result-indistinguishable transfer of decision for GI

case $G_0 \approx G_1$ and assume that V accepts. Then notice that V outputs (ACCEPT,0) only when it holds that $g_i = b_i$, for at least $2n$ values of $i \in \{1, \ldots, 3n\}$; however, since $G_0 \approx G_1$, and the subprotocol in the second phase of (P,V) is witness-indistinguishable, bit b_i cannot be computed by any P′ better than by random guessing. Therefore, for any P′, the probability that $g_i = b_i$, for at least $2n$ values of index i, is smaller than n^{-c}, for any constant c (using Chernoff bounds). Now, consider case $G_0 \not\approx G_1$ and assume that V accepts. Then notice that V outputs (AC-CEPT,1) only when it holds that $g_i \neq b_i$, for at least n values of $i \in \{1, \ldots, 3n\}$; however, since $G_0 \not\approx G_1$, the statements $L_{i0} \approx L_{i1}$, for all i such that $g_i \neq b_i$, are all false, and thus the probability that V accepts in this case is smaller than n^{-c}, for all constants c (using Chernoff bounds).

Perfect zero-knowledge. Now we informally describe a simulator $S_{V'}$ such that, for all pairs (G_0, G_1), the probability spaces $S_{V'}(G_0, G_1)$ and $\text{View}_{V'}(G_0, G_1)$ are equal. Since protocol (P,V) is constructed as a sequential repetition of an atomic protocol, it will be enough to describe the program of $S_{V'}$ simulating only such atomic protocol (in this description we will also omit the index of messages denoting the number of iteration).

The algorithm $S_{V'}$. First of all $S_{V'}$ feeds V' with a uniformly chosen random tape R; then he receives from V' graph H and the witness-indistinguishable proof of knowledge certifying that this graph has been correctly constructed (during this proof, $S_{V'}$ acts as a verifier of such proof and can run P's program, since it can be performed in polynomial time). Now, if the proof is not convincing then $S_{V'}$ outputs the conversation obtained so far, and halts. If the proof is convincing and $\chi_{GI}(G_0, G_1) = 1$ then $S_{V'}$ runs the extractor for the proof of knowledge in order to compute bit b and permutation β such that $H = \beta(G_b)$. Now, $S_{V'}$ can compute pair (L_0, L_1) as follows: graph L_0 is computed as done by P in the protocol (i.e., $L_0 = H$), and graph L_1 is computed as uniformly chosen among graphs isomorphic to G_b if $\chi_{GI}(G_0, G_1) = 1$ or isomorphic to G_g for some random bit g, otherwise. Now, the remaining steps of $S_{V'}$ consist of simulating the atomic proof by P that $L_0 \approx L_1$, and can simulated by using the rewinding technique, as follows. First $S_{V'}$ computes a graph T uniformly among those isomorphic to L_t, for some random bit t; then he receives bit l from V'; now, if $t = l$ then $S_{V'}$ sends the permutation between T and L_t, otherwise he rewinds V' until after he has computed graphs L_0, L_1 and tries again until $t = l$. Finally $S_{V'}$ outputs the conversation obtained.

To prove that the perfect zero-knowledge requirement is satisfied, we need to show that algorithm $S_{V'}$ is expected polynomial time, and his output $S_{V'}(G_0, G_1)$ is identically distributed to $\text{View}_{V'}(G_0, G_1)$, for all input pairs (G_0, G_1).

To see that algorithm $S_{V'}$ runs in expected polynomial time, we observe that $S_{V'}$ only runs polynomial-time instructions and the extractor for the proof of knowledge by V', which runs in expected polynomial time. Also, when simulating the third phase of (P,V), the simulation is iterated with probability at most $1/2$.

Now we show that for all pairs (G_0, G_1) and any V', the distributions $S_{V'}(G_0, G_1)$ and $\text{View}_{V'}(G_0, G_1)$ are equal. Clearly the verifier's random tape is uniformly distributed in both spaces and the messages sent by the verifier are computed equally in both spaces. Now, let us consider the messages sent by the prover. It is simple to check that the random bits sent by the prover during the executions of the witness-indistinguishable subprotocol executed in the second phase of (P,V) are also computed in the same way in both spaces. This is true also for graphs L_0, L_1 and for the other messages of P.

Perfect result-indistinguishability. To prove this property, we exhibit an efficient simulator M such that, on input (G_0, G_1), outputs a probability space $M(G_0, G_1)$ which is equal to the view of an observer C of the conversation during the execution of the protocol (P,V) on input G_0, G_1. In this case a description for the atomic protocol (A,B) suffices. Informally, first M simulates the first two phases of (P,V) by executing the same instructions by P and V. That is, he will compute a graph H as $H = \beta(G_b)$ for random b and β, and simulate the witness-indistinguishable proof that H has been correctly constructed, using b, β. Now, the simulator M computes graphs L_0, L_1 as follows: L_0 is set equal to H, and L_1 is uniformly chosen among the graphs isomorphic to G_b. Now, M simulates the proof by P that L_0 is isomorphic to L_1 as follows: he chooses T uniformly among graphs isomorphic to L_t, for some random bit t, sets the message by the verifier equal to bit t, and sets the final message by P equal to the permutation between T and L_t. The probability spaces $M(G_0, G_1)$ and (P,V)-View(G_0, G_1) are equal for any (G_0, G_1).

References

[1] L. Adleman, *Factoring Polynomials using Singular Integers*, TR 90-20, USC, Sep 1990.

[2] W. Aiello and J. Håstad, *Statistical Zero Knowledge Can Be Recognized in Two Rounds*, Journal of Computer and System Sciences, 42, 1991, pp. 327–345.

[3] E. Bach, *How to Generate Random Factored numbers*, SIAM Journal on Computing, vol. 17, n. 2, 1988.

[4] E. Bach and J. Shallit, *Algorithmic Number Theory*, MIT Press, 1996.

[5] M. Bellare and O. Goldreich, *On Defining Proofs of Knowledge*, in Proc. of CRYPTO '92.

[6] M. Blum and S. Micali, *How to Generate Cryptographically Strong Sequence of Pseudo-Random Bits*, SIAM Journal on Computing, vol. 13, no. 4, 1984, pp. 850–864.

[7] D. Boneh and R. Venkatesan, *Breaking RSA may be not harder than Factoring*, Proc. of EUROCRYPT 98.

[8] R. Boppana, J. Hastad, and S. Zachos, *Does co-NP has Short Interactive Proofs ?*, Information Processing Letters, vol. 25, May 1987, pp. 127–132.

[9] M. Ben-Or, O. Goldreich, S. Goldwasser, J. Hastad, J. Kilian, S. Micali, and P. Rogaway, *Everything Provable is Provable in Zero-Knowledge*, in Proc. of CRYPTO 88.

[10] M. Blum, A. De Santis, S. Micali, and G. Persiano, *Non-Interactive Zero-Knowledge*, SIAM Journal of Computing, vol. 20, no. 6, Dec 1991, pp. 1084–1118.

[11] M. Blum, P. Feldman, and S. Micali, *Non-Interactive Zero-Knowledge and Applications*, Proc. of STOC 88.

[12] M. Blum and S. Micali, *How to Generate Cryptographically Strong Sequence of Pseudo-Random Bits*, SIAM J. on Computing, vol. 13, no. 4, 1984, pp. 850–864.

[13] G. Brassard, C. Crépeau, and D. Chaum, *Minimum Disclosure Proofs of Knowledge*, Journal of Computer and System Sciences, vol. 37, no. 2, pp. 156–189.

[14] A. De Santis, G. Di Crescenzo, G. Persiano and M. Yung, *On Monotone Formula Closure of SZK*, in Proc. of FOCS 94.

[15] G. Di Crescenzo and R. Impagliazzo, *Security-Preserving Hardness Amplification for any Regular One-Way Function,* in Proc. of STOC 99.

[16] G. Di Crescenzo, K. Sakurai and M. Yung, *Zero-Knowledge Proofs of Decision Power: New Protocols and Optimal Round-Complexity,* in Proc. of ICICS 98.

[17] G. Di Crescenzo, K. Sakurai and M. Yung, *Result-Indistinguishable Zero-Knowledge Proofs: Increased Power and Constant Round Protocols,* in Proc. of STACS 98.

[18] G. Di Crescenzo, K. Sakurai and M. Yung, *On Zero-Knowledge Proofs: 'From Membership to Decision',* in Proc. of STOC 00.

[19] W. Diffie and M. Hellman, *New Directions in Cryptography,* in IEEE Transaction in Information Theory, 22, 1976.

[20] U. Feige, A. Fiat, and A. Shamir, *Zero-Knowledge Proofs of Identity,* Journal of Cryptology, vol. 1, 1988, pp. 77–94. (previous version STOC 87)

[21] A. Fiat and A. Shamir, *How to Prove yourself: Practical Solutions to Identifications and Signature Problems,* Proc. of CRYPTO 86.

[22] L. Fortnow, *The Complexity of Perfect Zero Knowledge,* STOC 87.

[23] Z. Galil, S. Haber, and M. Yung, *Minimum-Knowledge Interactive Proofs for Decision Problems,* in SIAM Journal on Computing, vol. 18, n.4.

[24] O. Goldreich, S. Goldwasser, and S. Micali, *How to Construct Random Functions,* Journal of the ACM, vol. 33, no. 4, 1986, pp. 792–807.

[25] O. Goldreich, R. Impagliazzo, L. Levin, R. Venkatesan, and D. Zuckerman, *Security-Preserving Amplification of Hardness,* in Proc. of FOCS 90.

[26] O. Goldreich and H. Krawczyk, *On the Composition of Zero-Knowledge Proof Systems,* in Proc. of ICALP 1990.

[27] O. Goldreich and L. Levin, *A Hard-Core Predicate for any One-Way Function,* in Proc. of FOCS 90.

[28] O. Goldreich, S. Micali, and A. Wigderson, *Proofs that Yield Nothing but their Validity or All Languages in NP Have Zero-Knowledge Proof Systems,* Journal of the ACM, vol. 38, n. 1, 1991, pp. 691–729.

[29] S. Goldwasser, and S. Micali, *Probabilistic Encryption,* Journal of Computer and System Sciences, vol. 28, n. 2, 1984, pp. 270–299.

[30] S. Goldwasser, S. Micali, and C. Rackoff, *The Knowledge Complexity of Interactive Proof-Systems,* SIAM Journal on Computing, vol. 18, n. 1, 1989.

[31] J. Hastad, R. Impagliazzo, L. Levin, and M. Luby, *Construction of a Pseudo-Random Generator from any One-Way Function,* SIAM Journal on Computing, vol. 28, n. 4, pp. 1364–1396, 1999.

[32] A. Herzberg and M. Luby, *Public Randomness in Cryptography,* in Proc. of CRYPTO 92.

[33] R. Impagliazzo and M. Luby, *One-Way Functions are Necessary for Complexity-Based Cryptography,* in Proc. of FOCS 89.

[34] R. Impagliazzo and S. Rudich, *Limits on the Provable Consequences of One-Way Permutations,* in Proc. of STOC 91.

[35] R. Impagliazzo and M. Yung, *Direct Minimum Knowledge Computations,* in Proc. of CRYPTO 87.

[36] D. E. Knuth, THE ART OF COMPUTER PROGRAMMING, VOLUME 2: SEMINUMERICAL ALGORITHMS, 3rd edition, Addison-Wesley, 1998.

[37] J. Kobler, U. Schoning, and J. Toran, *The graph isomorphism problem: Its structural complexity*, Progress in Theoretical Computer Science, Birkhauser Ed., 1993.

[38] A. Lenstra and H. Lenstra, *Algorithms in Number Theory*, in Handbook of Theoretical Computer Science, vol. A, chapter 12, Elsevier and MIT Press.

[39] A. Lenstra, H. Lenstra and L. Lovasz, *Factoring Polynomials with Rational Coefficients*, in Matematische Ann., vol. 261, 1982.

[40] A. Lenstra, H. Lenstra, M. Manasse, and J. Pollard, *The Number Field Sieve*, in Proc. of STOC 90.

[41] M. Luby, PSEUDORANDOMNESS AND CRYPTOGRAPHIC APPLICATIONS, Princeton University Press, 1996.

[42] M. Luby and C. Rackoff, *How to Construct a Pseudo-Random Permutation from a Pseudo-Random Function*, in SIAM Journal on Computing, vol. 17, n.2, Aug 1988.

[43] M. Naor, *Bit Commitment using Pseudorandomness*, in Proc. of CRYPTO 91.

[44] M. Naor, R. Ostrovsky, R. Venkatesan, and M. Yung, *Perfectly-Secure Zero-Knowledge Arguments Can be Based on General Complexity Assumptions*, in Proc. of CRYPTO 92.

[45] R. Ostrovsky and A. Wigderson, *One-way Functions are Necessary for Non-Trivial Zero-Knowledge Proofs*, in Proc. of ISTCS 93.

[46] J. Rompel, *One-way Functions are Necessary and Sufficient for Secure Signatures*, in Proc. of STOC 90.

[47] A. Shamir, *IP=PSPACE*, in Proc. of FOCS 90.

[48] A. Shamir, *On the Generation of Cryptographically Strong Pseudo-Random Sequences*, in Proc. of ICALP 81.

[49] Claude E. Shannon, *A Mathematical Theory of Communication*, The Bell System Technical Journal 27, 1948, 379-423, 623-656.

[50] M. Tompa and H. Woll, *Random Self-Reducibility and Zero-Knowledge Interactive Proofs of Possession of Information*, in Proc. of FOCS 87.

[51] G. Vernam, *Secret Signaling Systems*, US Patent, 1919.

[52] A. Yao, *Theory and Applications of Trapdoor Functions,* in Proc. of FOCS 82.

[53] M. Yung, *Zero-Knowledge Proofs of Computational Power*, in Proc. of EUROCRYPT 89.

Provable Security for Public Key Schemes

David Pointcheval

Abstract. Since the appearance of public-key cryptography in the Diffie-Hellman seminal paper, many schemes have been proposed, but many have been broken. Indeed, for a long time, the simple fact that a cryptographic algorithm had withstood cryptanalytic attacks for several years was considered as a kind of validation. But some schemes took a long time before being widely studied, and maybe thereafter being broken.

A much more convincing line of research has tried to provide "provable" security for cryptographic protocols, in a complexity theory sense: if one can break the cryptographic protocol, one can efficiently solve the underlying problem. Unfortunately, this initially was a purely theoretical work: very few practical schemes could be proven in this so-called "standard model" because such a security level rarely meets with efficiency. Ten years ago, Bellare and Rogaway proposed a trade-off to achieve some kind of validation of efficient schemes, by identifying some concrete cryptographic objects with ideal random ones. The most famous identification appeared in the so-called "random-oracle model". More recently, another direction has been taken to prove the security of efficient schemes in the standard model (without any ideal assumption) by using stronger computational assumptions.

In these lectures, we focus on practical asymmetric protocols together with their "reductionist" security proofs, mainly in the random-oracle model. We cover the two main goals that public-key cryptography is devoted to solve: authentication with digital signatures, and confidentiality with public-key encryption schemes.

1. Introduction

Since the beginning of public-key cryptography, with the seminal Diffie-Hellman paper [25], many suitable algorithmic problems for cryptography have been proposed and many cryptographic schemes have been designed, together with more or less heuristic proofs of their security relative to the intractability of the above problems. However, most of those schemes have thereafter been broken.

The simple fact that a cryptographic algorithm withstood cryptanalytic attacks for several years has often been considered as a kind of validation procedure, but some schemes take a long time before being broken. An example is the Chor-Rivest cryptosystem [21, 48], based on the knapsack problem, which took more than 10 years to be totally broken [86], whereas before this attack it was believed to be strongly secure. As a consequence, the lack of attacks at some time should never be considered as a security validation of the proposal.

1.1. Provable Security

A completely different paradigm is provided by the concept of "provable" security. A significant line of research has tried to provide proofs in the framework of complexity theory (*a.k.a.* "reductionist" security proofs [4]): the proofs provide reductions from a well-studied problem (RSA or the discrete logarithm) to an attack against a cryptographic protocol.

At the beginning, people just tried to define the security notions required by actual cryptographic schemes, and then to design protocols which achieve these notions. The techniques were directly derived from the complexity theory, providing polynomial reductions. However, their aim was essentially theoretical. They were indeed trying to minimize the required assumptions on the primitives (one-way functions or permutations, possibly trapdoor, etc) [37, 35, 52, 71] without considering practicality. Therefore, they just needed to design a scheme with polynomial algorithms, and to exhibit polynomial reductions from the basic assumption on the primitive into an attack of the security notion, in an asymptotic way. However, such a result has no practical impact on actual security. Indeed, even with a polynomial reduction, one may be able to break the cryptographic protocol within a few hours, whereas the reduction just leads to an algorithm against the underlying problem which requires many years. Therefore, those reductions only prove the security when very huge (and thus maybe unpractical) parameters are in use, under the assumption that no polynomial time algorithm exists to solve the underlying problem.

1.2. Exact Security and Practical Security

For a few years, more efficient reductions have been expected, under the denominations of either "exact security" [12] or "concrete security" [58], which provide more practical security results. The perfect situation is reached when one manages to prove that, from an attack, one can describe an algorithm against the underlying problem, with almost the same success probability within almost the same amount of time. We have then achieved "practical security".

Unfortunately, in many cases, even just provable security is at the cost of an important loss in terms of efficiency for the cryptographic protocol. Thus some models have been proposed, trying to deal with the security of efficient schemes: some concrete objects are identified with ideal (or black-box) ones.

For example, it is by now usual to identify hash functions with ideal random functions, in the so-called "random-oracle model", informally introduced by Fiat and Shamir [28], and formalized by Bellare and Rogaway [10]. Similarly, block ciphers are identified with families of truly random permutations in the "ideal cipher model" [9]. A few years ago, another kind of idealization was introduced in cryptography, the black-box group [53, 80], where the group operation, in any algebraic group, is defined by a black-box: a new element necessarily comes from the addition (or the subtraction) of two already known elements. It is by now called the "generic model". Recent works [77, 18] even require several ideal models together to provide some new validations.

1.3. Outline of the Notes

In the next section, we explain and motivate more about exact security proofs, and we introduce the notion of the weaker security analyses, the security arguments (in an ideal model, and namely the random-oracle model). Then, we review the formalism of the most important asymmetric primitives: signatures and public-key encryption schemes. For both, we provide some examples, with some security analyses in the "reductionist" sense.

1.4. Related Work

These notes present a survey, based on several published papers, from the author, with often several co-authors: about signature [67, 69, 68, 17, 84], encryption [7, 3, 62, 59, 32, 33] and provably secure constructions [61, 63, 65, 64, 66]. Many other papers are also cited and rephrased, which present efficient provably secure constructions. Among the bibliography list presented at the end, we would like to insist on [10, 11, 12, 22, 82, 83]. We thus refer the reader to the original papers for more details.

2. Security Proofs and Security Arguments

2.1. Computational Assumptions

In both symmetric and asymmetric scenarios, many security notions can not be unconditionally guaranteed (whatever the computational power of the adversary). Therefore, security generally relies on a computational assumption: the existence of one-way functions, or permutations, possibly trapdoor. A one-way function is a function f which anyone can easily compute, but given $y = f(x)$ it is computationally intractable to recover x (or any pre-image of y). A one-way permutation is a bijective one-way function. For encryption, one would like the inversion to be possible for the recipient only: a trapdoor one-way permutation is a one-way permutation for which a secret information (the trapdoor) helps to invert the function on any point.

Given such objects, and thus computational assumptions about the intractability of the inversion without possible trapdoors, we would like that security could be achieved without extra assumptions. The only way to formally prove such a fact is by showing that an attacker against the cryptographic protocol can be used as a sub-part in an algorithm that can break the basic computational assumption.

A partial order therefore exists between computational assumptions (and intractable problems too): if a problem P is more difficult than the problem P' (P' reduces to P, see below) then the assumption of the intractability of the problem P is weaker than the assumption of the intractability of the problem P'. The weaker the required assumption is, the more secure the cryptographic scheme is.

2.2. "Reductionist" Security Proofs

In complexity theory, such an algorithm which uses the attacker as a sub-part in a global algorithm is called a reduction. If this reduction is polynomial, we can say that the attack of the cryptographic protocol is at least as hard as inverting the function: if one has a polynomial algorithm to solve the latter problem, one can polynomially solve the former one. In the complexity theory framework, a *polynomial* algorithm is the formalization of *efficiency*.

Therefore, in order to prove the security of a cryptographic protocol, one first needs to make precise the security notion one wants the protocol to achieve: which adversary's goal one wants to be intractable, under which kind of attack. At the beginning of the 1980's, such security notions have been defined for encryption [35] and signature [37, 38], and provably secure schemes have been suggested. However, those proofs had a theoretical impact only, because both the proposed schemes and the reductions were completely unpractical, yet polynomial. The reductions were indeed efficient (*i.e.* polynomial), and thus a polynomial attack against a cryptosystem would have led to a polynomial algorithm that broke the computational assumption. But the latter algorithm, even polynomial, may require hundreds of years to solve a small instance.

For example, let us consider a cryptographic protocol based on integer factoring. Let us assume that one provides a polynomial reduction from the factorization into an attack. But such a reduction may just lead to a factorization algorithm with a complexity in $2^{25}k^{10}$, where k is the bit-size of the integer to factor. This indeed contradicts the assumption that no-polynomial algorithm exists for factoring. However, on a 1024-bit number ($k = 2^{10}$), it provides an algorithm that requires 2^{125} basic operations, which is much more than the complexity of the best current algorithm, such as NFS [46], which needs less than 2^{100} (see Section 4). Therefore, such a reduction would just be meaningful for numbers above 4096 bits (since with $k = 2^{12}$, $2^{145} < 2^{149}$, where 2^{149} is the estimate effort for factoring a 4096-bit integer with the best algorithm.) Concrete examples are given later.

2.3. Practical Security

Moreover, most of the proposed schemes were unpractical as well. Indeed, the protocols were polynomial in time and memory, but not efficient enough for practical implementation.

For a few years, people have tried to provide both practical schemes, with practical reductions and exact complexity, which prove the security for realistic parameters, under a well-defined assumption: exact reduction in the standard model (which means in the complexity-theoretic framework). For example, under the assumption that a 1024-bit integer cannot be factored with less than 2^{70} basic operations, the cryptographic protocol cannot be broken with less than 2^{60} basic operations. We will see such an example later.

Unfortunately, as already remarked, practical or even just efficient reductions in the standard model can rarely be conjugated with practical schemes. Therefore, one needs to make some hypotheses on the adversary: the attack is generic, independent of the actual implementation of some objects

- hash functions, in the "random-oracle model";
- symmetric block ciphers, in the "ideal-cipher model";
- algebraic groups, in the "generic model".

The "random-oracle model" was the first to be introduced in the cryptographic community [28, 10], and has already been widely accepted. By the way, flaws have been shown in the "generic model" [84] on practical schemes, and the "random-oracle model" is not equivalent to the standard one either. Several gaps have already been exhibited [19, 54, 6]. However, all the counter-examples in the random-oracle model are pathological, counter-intuitive and not natural. Therefore, in the sequel, we focus on security analyses in this model, for real and natural constructions. A security proof in the random-oracle model will at least give a strong argument in favor of the security of the scheme. Furthermore, proofs in the random-oracle model under a weak computational assumption may be of more pratical interest than proofs in the standard model under a strong computational assumption.

2.4. The Random-Oracle Model

As said above, efficiency rarely meets with provable security. More precisely, none of the most efficient schemes in their category have been proven secure in the standard model. However, some of them admit security validations under ideal assumptions: the random-oracle model is the most widely accepted one.

Many cryptographic schemes use a hash function $\mathcal{H}$ (such as MD5 [72] or the American standards SHA-1 [56], SHA-256, SHA-384 and SHA-512 [57]). This use of hash functions was originally motivated by the wish to sign long messages with a single short signature. In order to achieve *non-repudiation*, a minimal requirement on the hash function is the impossibility for the signer to find two different messages providing the same hash value. This property is called *collision-resistance*.

It was later realized that hash functions were an essential ingredient for the security of, first, signature schemes, and then of most cryptographic schemes. In order to obtain security arguments, while keeping the efficiency of the designs that use hash functions, a few authors suggested using the hypothesis that $\mathcal{H}$ behaves like a random function. First, Fiat and Shamir [28] applied it heuristically to provide a signature scheme "as secure as" factorization. Then, Bellare and Rogaway [10, 11, 12] formalized this concept for cryptography, and namely for signature and public-key encryption.

In this model, the so-called "random-oracle model", the hash function can be formalized by an oracle which produces a truly random value for each new query. Of course, if the same query is asked twice, identical answers are obtained. This is precisely the context of relativized complexity theory with "oracles," hence the name.

About this model, no one has ever been able to provide a convincing contradiction to its practical validity, but just theoretical counter-examples on either clearly wrong designs for practical purpose [19], or artificial security notions [54, 6]. Therefore, this model has been strongly accepted by the community, and is considered as a good one, in which security analyses give a good taste of the actual security level. Even if it does not provide a formal proof of security (as in the standard model, without any ideal assumption), it is argued that proofs in this model ensure security of the overall design of the scheme provided that the hash function has no weakness, hence the name "security arguments".

This model can also be seen as a restriction on the adversary's capabilities. Indeed, it simply means that the attack is generic without considering any particular instantiation of the hash functions. Therefore, an actual attack would necessarily use a weakness or a specific feature of the hash function. The replacement of the hash function by another one would rule out this attack.

On the other hand, assuming the tamper-resistance of some devices, such as smart cards, the random-oracle model is equivalent to the standard model, which simply requires the existence of pseudo-random functions [34, 51].

As a consequence, almost all the standards bodies by now require designs provably secure, at least in that model, thanks to the security validation of very efficient protocols.

2.5. The General Framework

Before going into more details of this kind of proofs, we would like to insist on the fact that in the current general framework, we give the adversary complete access to the cryptographic primitive, but as a black-box. It can ask any query of its choice, and the box always answers correctly, in constant time. Such a model does not consider timing attacks [44], where the adversary tries to extract the secrets from the computational time. Some other attacks analyze the electrical energy required by a computation to get the secrets [45], or to make the primitive fail on some computation [13, 16]. They are not captured either by this model.

3. A First Formalism

In this section we describe more formally what a signature scheme and an encryption scheme are. Moreover, we make precise the security notions one wants the schemes to achieve. This is the first imperative step towards provable security.

3.1. Digital Signature Schemes

Digital signature schemes are the electronic version of handwritten signatures for digital documents: a user's signature on a message m is a string which depends on m, on public and secret data specific to the user and —possibly— on randomly chosen data, in such a way that anyone can check the validity of the signature by using public data only. The user's public data are called the *public key*, whereas his secret data are called the *private key*. The intuitive security notion would be the impossibility to forge user's signatures without the knowledge of his private key. In this section, we give a more precise definition of signature schemes and of the possible attacks against them (most of those definitions are based on [38]).

3.1.1. Definitions. A signature scheme $S = (\mathcal{K}, \mathcal{S}, \mathcal{V})$ is defined by the three following algorithms:

- The *key generation algorithm* $\mathcal{K}$. On input 1^k, which is a formal notation for a machine with running time polynomial in k (1^k is indeed k in basis 1), the algorithm $\mathcal{K}$ produces a pair (pk, sk) of matching public and private keys. Algorithm $\mathcal{K}$ is probabilistic. The input k is called the security parameter. The sizes of the keys, or of any problem involved in the cryptographic scheme, will depend on it, in order to achieve an appropriate security level (the expected minimal time complexity of any attack).
- The *signing algorithm* $\mathcal{S}$. Given a message m and a pair of matching public and private keys (pk, sk), $\mathcal{S}$ produces a signature σ. The signing algorithm might be probabilistic.
- The *verification algorithm* $\mathcal{V}$. Given a signature σ, a message m and a public key pk, $\mathcal{V}$ tests whether σ is a valid signature of m with respect to pk. In general, the verification algorithm need not be probabilistic.

3.1.2. Forgeries and Attacks. In this subsection, we formalize some security notions which capture the main practical situations. On the one hand, the **goals** of the adversary may be various:

- Disclosing the private key of the signer. It is the most serious attack. This attack is termed *total break*.
- Constructing an efficient algorithm which is able to sign messages with good probability of success. This is called *universal forgery*.
- Providing a new message-signature pair. This is called *existential forgery*. The corresponding security level is called *existential unforgeability* (EUF).

In many cases the latter forgery, the *existential forgery*, is not dangerous because the output message is likely to be meaningless. Nevertheless, a signature scheme which is existentially forgeable does not guarantee by itself the identity of the signer. For example, it cannot be used to certify randomly looking elements, such as keys. Furthermore, it cannot formally guarantee the non-repudiation property, since anyone may be able to produce a message with a valid signature.

On the other hand, various **means** can be made available to the adversary, helping it into its forgery. We focus on two specific kinds of attacks against signature schemes: the *no-message attacks* and the *known-message attacks* (KMA). In the former scenario, the attacker only knows the public key of the signer. In the latter, the attacker has access to a list of valid message-signature pairs. According to the way this list was created, we usually distinguish many subclasses, but the strongest is definitely the *adaptive chosen-message attack* (CMA), where the attacker can ask the signer to sign any message of its choice, in an adaptive way: it can adapt its queries according to previous answers.

When signature generation is not deterministic, there may be several signatures corresponding to a given message. And then, some notions defined above may become ambiguous [84]. First, in known-message attacks, an existential forgery becomes the ability to forge a fresh message/signature pair that has not been obtained during the attack. There is a subtle point here, related to the context where several signatures may correspond to a given message. We actually adopt the stronger rule that the attacker needs to forge the signature of message, whose signature was not queried. The more liberal rule, which makes the attacker successful when it outputs a second signature of a given message different from a previously obtained signature of the same message, is called *malleability*, while the corresponding security level is called *non-malleability* (NM). Similarly, in adaptive chosen-message attacks, the adversary may ask several times the same message, and each new answer gives it some information. A slightly weaker security model, by now called *single-occurrence adaptive chosen-message attack* (SO-CMA), allows the adversary at most one signature query for each message. In other words the adversary cannot submit the same message twice for signature.

When one designs a signature scheme, one wants to computationally rule out at least existential forgeries, or even achieve non-malleability, under adaptive chosen-message attacks. More formally, one wants that the success probability of any adversary $\mathcal{A}$ with a reasonable time is small, where

$$\mathsf{Succ}_{\mathsf{S}}^{\mathsf{euf}}(\mathcal{A}) = \Pr\left[\ (\mathsf{pk}, \mathsf{sk}) \leftarrow \mathcal{K}(1^k), (m, \sigma) \leftarrow \mathcal{A}^{\mathcal{S}_{\mathsf{sk}}}(\mathsf{pk}) : \mathcal{V}(\mathsf{pk}, m, \sigma) = 1\ \right].$$

We remark that since the adversary is allowed to play an adaptive chosen-message attack, the signing algorithm is made available, without any restriction, hence the oracle notation $\mathcal{A}^{\mathcal{S}_{\mathsf{sk}}}$. Of course, in its answer, there is the natural restriction that, at least, the returned message-signature has not been obtained from the signing oracle $\mathcal{S}_{\mathsf{sk}}$ itself (non-malleability) or even the output message has not been queried (existential unforgeability).

3.2. Public-Key Encryption

The aim of a public-key encryption scheme is to allow anybody who knows the public key of Alice to send her a message that she will be the only one able to recover, granted her private key.

3.2.1. Definitions. A public-key encryption scheme $\mathsf{S} = (\mathcal{K}, \mathcal{E}, \mathcal{D})$ is defined by the three following algorithms:

- The *key generation algorithm* $\mathcal{K}$. On input 1^k where k is the security parameter, the algorithm $\mathcal{K}$ produces a pair $(\mathsf{pk}, \mathsf{sk})$ of matching public and private keys. Algorithm $\mathcal{K}$ is probabilistic.
- The *encryption algorithm* $\mathcal{E}$. Given a message m and a public key pk, $\mathcal{E}$ produces a ciphertext c of m. This algorithm may be probabilistic. In the latter case, we write $\mathcal{E}_{\mathsf{pk}}(m; r)$ where r is the random input to $\mathcal{E}$.
- The *decryption algorithm* $\mathcal{D}$. Given a ciphertext c and the private key sk, $\mathcal{D}_{\mathsf{sk}}(c)$ gives back the plaintext m. This algorithm is necessarily deterministic.

3.2.2. Security Notions. As for signature schemes, the **goals** of the adversary may be various. The first common security notion that one would like for an encryption scheme is *one-wayness* (OW): with just public data, an attacker cannot get back the whole plaintext of a given ciphertext. More formally, this means that for any adversary $\mathcal{A}$, its success in inverting $\mathcal{E}$ without the private key should be negligible over the probability space $\mathbf{M} \times \Omega$, where $\mathbf{M}$ is the message space and Ω is the space of the random coins r used for the encryption scheme, and the internal random coins of the adversary:

$$\mathsf{Succ}_{\mathsf{S}}^{\mathsf{ow}}(\mathcal{A}) = \Pr_{m,r}[(\mathsf{pk}, \mathsf{sk}) \leftarrow \mathcal{K}(1^k) : \mathcal{A}(\mathsf{pk}, \mathcal{E}_{\mathsf{pk}}(m; r)) = m].$$

However, many applications require more from an encryption scheme, namely the *semantic security* (IND) [35], *a.k.a. polynomial security/indistinguishability of encryptions*: if the attacker has some information about the plaintext, for example that it is either "yes" or "no" to a crucial query, any adversary should not learn more with the view of the ciphertext. This security notion requires computational impossibility to distinguish between two messages, chosen by the adversary, which one has been encrypted, with a probability significantly better than one half: its advantage $\mathsf{Adv}_{\mathsf{S}}^{\mathsf{ind}}(\mathcal{A})$, formally defined as

$$2 \times \Pr_{b,r}\left[\begin{array}{l} (\mathsf{pk}, \mathsf{sk}) \leftarrow \mathcal{K}(1^k), (m_0, m_1, s) \leftarrow \mathcal{A}_1(\mathsf{pk}), \\ c = \mathcal{E}_{\mathsf{pk}}(m_b; r) : \mathcal{A}_2(m_0, m_1, s, c) = b \end{array} \right] - 1,$$

where the adversary $\mathcal{A}$ is seen as a 2-stage attacker $(\mathcal{A}_1, \mathcal{A}_2)$, should be negligible.

A later notion is *non-malleability* (NM) [26]. To break it, the adversary, given a ciphertext, tries to produce a new ciphertext such that the plaintexts are meaningfully related. This notion is stronger than the above semantic security, but it is equivalent to the latter in the most interesting scenario [7] (the CCA attacks, see below). Therefore, we will just focus on one-wayness and semantic security.

On the other hand, an attacker can play many kinds of attacks, according to the **available information**: since we are considering asymmetric encryption, the adversary can encrypt any plaintext of its choice, granted the public key, hence the *chosen-plaintext attack* (CPA). It may furthermore have access to additional information, modeled by partial or full access to some oracles:

- A validity-checking oracle which, on input a ciphertext c, answers whether it is a valid ciphertext or not. Such a weak oracle, involved in the so-called reaction attacks [39] or *Validity-Checking Attack* (VCA), had been enough to break some famous encryption schemes [15, 42].

- A plaintext-checking oracle which, on input a pair (m, c), answers whether c encrypts the message m. This attack has been termed the *Plaintext-Checking Attack* (PCA) [59].
- The decryption oracle itself, which on any ciphertext answers the corresponding plaintext. There is of course the natural restriction not to ask the challenge ciphertext to that oracle.

For all these oracles, access may be restricted as soon as the challenge ciphertext is known, the attack is thus said *non-adaptive* since oracle queries cannot depend on the challenge ciphertext, while they depend on previous answers. On the opposite, access can be unlimited and attacks are thus called *adaptive attacks* (w.r.t. the challenge ciphertext). This distinction has been widely used for the chosen-ciphertext attacks, for historical reasons: the *non-adaptive chosen-ciphertext attacks* (CCA1) [52], *a.k.a.* lunchtime attacks, and *adaptive chosen-ciphertext attacks* (CCA2) [71]. The latter scenario which allows adaptively chosen ciphertexts as queries to the decryption oracle is definitely the strongest attack, and will be named the *chosen-ciphertext attack* (CCA).

Furthermore, multi-user scenarios can be considered where related messages are encrypted under different keys to be sent to many people (*e.g.* broadcast of encrypted data). This may provide many useful data for an adversary. For example, RSA is well-known to be weak in such a scenario [40, 79], namely with a small encryption exponent, because of the Chinese Remainders Theorem. But once again, semantic security has been shown to be the appropriate security level, since it automatically extends to the multi-user setting: if an encryption scheme is semantically secure in the classical sense, it is also semantically secure in multi-user scenarios, against both passive [3] and active [5] adversaries.

A general study of these security notions and attacks was conducted in [7], we therefore refer the reader to this paper for more details. See also the summary diagram on Figure 1. However, we can just review the main scenarios we will consider in the following:

- one-wayness under chosen-plaintext attacks (OW-CPA) – where the adversary wants to recover the whole plaintext from just the ciphertext and the public key. This is the weakest scenario.
- semantic security under adaptive chosen-ciphertext attacks (IND-CCA) – where the adversary just wants to distinguish which plaintext, between two messages of its choice, has been encrypted, while it can ask any query it wants to a decryption oracle (except the challenge ciphertext). This is the strongest scenario one can define for encryption (still in our general framework.) Thus, this is our goal when we design a cryptosystem.

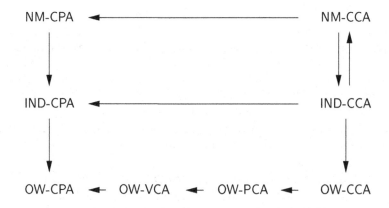

FIGURE 1. Relations between the Security Notions for Asymmetric Encryption

4. The Computational Assumptions

There are two major families in number theory-based public-key cryptography:

1. the schemes based on integer factoring, and on the RSA problem [73];
2. the schemes based on the discrete logarithm problem, and on the Diffie-Hellman problems [25], in any "suitable" group. The first groups in use were cyclic subgroups of $\mathbb{Z}_p^\star$, the multiplicative group of the modular quotient ring $\mathbb{Z}_p = \mathbb{Z}/p\mathbb{Z}$. But many schemes are now converted on cyclic subgroups of elliptic curves, or of the Jacobian of hyper-elliptic curves, with namely the so-called ECDSA [1], the US Digital Signature Standard [55] on elliptic curves.

4.1. Integer Factoring and the RSA Problem

The most famous intractable problem is factorization of integers: while it is easy to multiply two prime integers p and q to get the product $n = p \cdot q$, it is not simple to decompose n into its prime factors p and q.

Currently, the most efficient algorithm is based on sieving on number fields. The Number Field Sieve (NFS) method [46] has a super-polynomial, but sub-exponential, complexity in $\mathcal{O}(\exp((1.923 + o(1))(\ln n)^{1/3}(\ln \ln n)^{2/3}))$. It has been used to establish the main record, in august 1999, by factoring a 155-digit integer (512 bits), product of two 78-digit primes [20]. The factored number, called RSA-155, was taken from the "RSA Challenge List", which is used as a yardstick for the security of the RSA cryptosystem (see later). The latter is used extensively in hardware and software to protect electronic data traffic such as in the SSL (Security Sockets Layer) Handshake Protocol.

This record is very important since 155 digits correspond to 512 bits. And this is the size which is in use in almost all the implementations of the RSA cryptosystem (namely for actual implementations of SSL on the Internet).

RSA-155 =
109417386415705274218097073220403576120\
037329454492059909138421314763499842889\
347847179972578912673324976257528997818\
337970765372440271467435315933543333897
= 102639592829741105772054196573991675900\
716567808038066803341933521790711307779
* 106603488380168454820927220360012878679\
207958575989291522270608237193062808643

Unfortunately, integer multiplication just provides a one-way function, without any possibility to invert the process. No information is known to make factoring easier. However, some algebraic structures are based on the factorization of an integer n, where some computations are difficult without the factorization of n, but easy with it: the finite quotient ring $\mathbb{Z}_n$ which is isomorphic to the product ring $\mathbb{Z}_p \times \mathbb{Z}_q$ if $n = p \cdot q$.

For example, the e-th power of any element x can be easily computed using the *square-and-multiply* method. It consists in using the binary representation of the exponent $e = e_k e_{k-1} \ldots e_0$, computing the successive 2 powers of x (x^{2^0}, x^{2^1}, $\ldots, x^{2^k}$) and eventually to multiply altogether the ones for which $e_i = 1$. However, to compute e-th roots, it seems that one requires to know an integer d such that $ed = 1 \bmod \varphi(n)$, where $\varphi(n)$ is the totient Euler function which denotes the cardinality of the multiplicative subgroup $\mathbb{Z}_n^\star$ of $\mathbb{Z}_n$. In the particular case where $n = pq$, $\varphi(n) = (p-1)(q-1)$. And therefore, $ed - 1$ is a multiple of $\varphi(n)$ which is equivalent to the knowledge of the factorization of n [50]. In 1978, Rivest, Shamir and Adleman [73] defined the following problem:

> **The RSA Problem.** Let $n = pq$ be the product of two large primes of similar size and e an integer relatively prime to $\varphi(n)$. For a given $y \in \mathbb{Z}_n^\star$, compute the modular e-th root x of y (*i.e.* $x \in \mathbb{Z}_n^\star$ such that $x^e = y \bmod n$.)

The Euler function can be easily computed from the factorization of n, since for any $n = \prod p_i^{v_i}$,

$$\varphi(n) = n \times \prod \left(1 - \frac{1}{p_i}\right).$$

Therefore, with the factorization of n (the trapdoor), the RSA problem can be easily solved. But nobody knows whether the factorization is required, and how to do without it either:

> **The RSA Assumption.** For any product of two primes, $n = pq$, large enough, the RSA problem is intractable (presumably as hard as the factorization of n).

4.2. The Discrete Logarithm and the Diffie-Hellman Problems

The setting is quite general: one is given

- a cyclic group $\mathcal{G}$ of prime order q (such as the finite group $(\mathbb{Z}_q, +)$, a subgroup of $(\mathbb{Z}_p^\star, \times)$ for $q|p-1$, of an elliptic curve, etc);
- a generator $\mathbf{g}$ (*i.e.* $\mathcal{G} = \langle \mathbf{g} \rangle$).

We note in bold (such as $\mathbf{g}$) any element of the group $\mathcal{G}$, to distinguish it from a scalar $x \in \mathbb{Z}_q$. But such a $\mathbf{g}$ could be an element in $\mathbb{Z}_p^\star$ or a point of an elliptic curve, according to the setting. Above, we talked about a "suitable" group $\mathcal{G}$. In such a group, some of the following problems have to be hard to solve (using the additive notation):

- the **Discrete Logarithm** problem (**DL**): given $\mathbf{y} \in \mathcal{G}$, compute $x \in \mathbb{Z}_q$ such that $\mathbf{y} = x \cdot \mathbf{g} = \mathbf{g} + \ldots + \mathbf{g}$ (x times), then one writes $x = \log_{\mathbf{g}} \mathbf{y}$.
- the **Computational Diffie-Hellman** problem (**CDH**): given two elements in the group $\mathcal{G}$, $\mathbf{a} = a \cdot \mathbf{g}$ and $\mathbf{b} = b \cdot \mathbf{g}$, compute $\mathbf{c} = ab \cdot \mathbf{g}$. Then one writes $\mathbf{c} = \mathbf{DH}(\mathbf{a}, \mathbf{b})$.
- the **Decisional Diffie-Hellman** Problem (**DDH**): given three elements in the group $\mathcal{G}$, $\mathbf{a} = a \cdot \mathbf{g}$, $\mathbf{b} = b \cdot \mathbf{g}$ and $\mathbf{c} = c \cdot \mathbf{g}$, decide whether $\mathbf{c} = \mathbf{DH}(\mathbf{a}, \mathbf{b})$ (or equivalently, whether $c = ab \bmod q$).

It is clear that they are sorted from the strongest problem to the weakest one. Furthermore, one may remark that they all are "random self-reducible", which means that any instance can be reduced to a uniformly distributed instance: for example, given a specific element $\mathbf{y}$ for which one wants to compute the discrete logarithm x in basis $\mathbf{g}$, one can choose a random $z \in \mathbb{Z}_q$, and compute $\mathbf{z} = z \cdot \mathbf{y}$. The element $\mathbf{z}$ is therefore uniformly distributed in the group, and the discrete logarithm $\alpha = \log_{\mathbf{g}} \mathbf{z}$ leads to $x = \alpha/z \bmod q$. As a consequence, there are only average complexity cases. Thus, the ability to solve a problem for a non-negligible fraction of instances in polynomial time is equivalent to solve any instance in expected polynomial time.

A new variant of the Diffie-Hellman problem has more recently been defined by Tatsuaki Okamoto and the author [60], the so-called *Gap Diffie-Hellman Problem* (**GDH**), where one wants to solve the **CDH** problem with an access to a **DDH** oracle. One may easily remark the following properties about above problems: $\mathbf{DL} \geq \mathbf{CDH} \geq \{\mathbf{DDH}, \mathbf{GDH}\}$, where $A \geq B$ means that the problem A is at least as hard as the problem B. However, in practice, no one knows how to solve any of them without breaking the **DL** problem itself.

Currently, the most efficient algorithms to solve the latter problem depend on the underlying group. For generic groups (for which no specific algebraic property can be used), algorithms have a complexity in the square root of q, the order of the generator $\mathbf{g}$ [78, 70]. For example, on well-chosen elliptic curves only these algorithms can be used. The last record was established in April 2001 on the curve defined by the equation $y^2 + xy = x^3 + x^2 + 1$ over the finite field with 2^{109} elements.

However, for subgroups of $\mathbb{Z}_p^\star$, some better techniques can be applied. The best algorithm is based on sieving on number fields, as for the factorization. The General Number Field Sieve method [41] has a super-polynomial, but sub-exponential, complexity in $\mathcal{O}(\exp((1.923 + o(1))(\ln p)^{1/3}(\ln \ln p)^{2/3}))$. It was used to establish the last record, in April 2001 as well, by computing discrete logarithms in $\mathbb{Z}_p^\star$, for a 120-digit prime p. Therefore, 512-bit primes are still safe enough, as far as the generic attacks cannot be used (the generator must be of large order q, at least a 160-bit prime)

For signature applications, one only requires groups where the **DL** problem is hard, whereas encryption needs trapdoor problems and therefore requires groups where some of the **DH**'s problems are also hard to solve.

5. Digital Signature Schemes

Until 1996, no practical **DL**-based cryptographic scheme has ever been formally studied, but heuristically only. And surprisingly, at the Eurocrypt '96 conference, two opposite studies were conducted on the El Gamal signature scheme [27], the first **DL**-based signature scheme designed in 1985 and depicted on Figure 2.

Initialization $\rightarrow (p, g)$
g a generator of $\mathbb{Z}_p^\star$,   where p is a large prime $\rightarrow (p, g)$
$\mathcal{K}$: **Key Generation** $\rightarrow (y, x)$
private key $x \in \mathbb{Z}_{p-1}^\star$ public key $y = g^x \bmod p$ $\rightarrow (y, x)$
$\mathcal{S}$: **Signature of** $m \rightarrow (r, s)$
K is randomly chosen in $\mathbb{Z}_{p-1}^\star$ $r = g^K \bmod p$ $s = (m - xr)/K \bmod p - 1$ $\rightarrow (r, s)$ is a signature of m
$\mathcal{V}$: **Verification of** (m, r, s)
check whether $g^m \overset{?}{=} y^r r^s \bmod p$ $\rightarrow$ Yes/No

FIGURE 2. The El Gamal Signature Scheme.

Whereas existential forgeries were known for that scheme, it was believed to prevent universal forgeries. The first analysis, from Daniel Bleichenbacher [14], showed such a universal forgery when the generator g is not properly chosen. The second one, from Jacques Stern and the author [67], proved the security against existential forgeries under adaptive chosen-message attacks of a slight variant with a randomly chosen generator g. The latter variant simply replaces the message m by $\mathcal{H}(m, r)$ in the computation, while one uses a hash function $\mathcal{H}$ that is assumed to behave like a random oracle. It is amazing to remark that the Bleichenbacher's attack also applies on our variant. Therefore, depending on the initialization, our variant could be a very strong signature scheme or become a very weak one!

As a consequence, a proof has to be performed in details, with precise assumptions and achievements. Furthermore, the conclusions have to be strictly followed by developers, otherwise the concrete implementation of a secure scheme can be very weak.

5.1. Provable Security

The first *secure* signature scheme was proposed by Goldwasser *et al.* [37] in 1984. It used the notion of claw-free permutations. A pair of permutations (f, g) is said *claw-free* if it is computationally impossible to find a *claw* (x, y), which satisfies $f(x) = g(y)$. Their proposal provided polynomial algorithms with a polynomial reduction between the research of a claw and an existential forgery under an adaptive chosen-message attack. However, the scheme was totally unpractical. What about practical schemes?

5.1.1. The RSA Signature Scheme. Two years after the Diffie-Hellman paper [25], Rivest, Shamir and Adleman [73] proposed the first signature scheme based on the "trapdoor one-way permutation paradigm", using the RSA function: the generation algorithm produces a large composite number $N = pq$, a public key e, and a private key d such that $e \cdot d = 1 \mod \varphi(N)$. The signature of a message m, encoded as an element in $\mathbb{Z}_N^{\star}$, is its e-*th* root, $\sigma = m^{1/e} = m^d \mod N$. The verification algorithm simply checks whether $m = \sigma^e \mod N$.

However, the RSA scheme is not secure by itself since it is subject to existential forgery: it is easy to create a valid message-signature pair, without any help of the signer, first randomly choosing a certificate σ and getting the signed message m from the public verification relation, $m = \sigma^e \mod N$.

5.1.2. The Schnorr Signature Scheme. In 1986 a new paradigm for signature schemes was introduced. It is derived from fair zero-knowledge identification protocols involving a prover and a verifier [36], and uses hash functions in order to create a kind of virtual verifier. The first application was derived from the Fiat–Shamir [28] zero-knowledge identification protocol, based on the hardness of extracting square roots, with a brief outline of its security. Another famous identification scheme [75], together with the signature scheme [76], has been proposed later by Schnorr, based on that paradigm: the generation algorithm produces two large primes p and q, such that $q \geq 2^k$, where k is the security parameter, and $q \mid p - 1$, as well as an element g in $\mathbb{Z}_p^\star$ of order q. It also creates a pair of keys, the private key $x \in \mathbb{Z}_q^\star$ and the public key $y = g^{-x} \bmod p$ The signature of a message m is a triple (r, e, s), where $r = g^K \bmod p$, with a random $K \in \mathbb{Z}_q$, the "challenge" $e = \mathcal{H}(m, r)$ and $s = K + ex \bmod q$. The latter satisfies $r = g^s y^e \bmod p$ with $e = \mathcal{H}(m, r)$, which is checked by the verification algorithm.

The security results for that paradigm have been considered as folklore for a long time but without any formal validation.

5.2. DL-Based Signatures

In our papers [67, 68], with Jacques Stern, we formally proved the above paradigm when $\mathcal{H}$ is assumed to behave like a random oracle. The proof is based on the by now classical *oracle replay technique*: by a polynomial replay of the attack with different random oracles (the $\mathcal{Q}_i$'s are the queries and the ρ_i's are the answers), we allow the attacker to forge signatures that are suitably related. This generic

FIGURE 3. The Oracle Replay Technique

technique is depicted on Figure 3, where the signature of a message m is a triple (σ_1, h, σ_2), with $h = \mathcal{H}(m, \sigma_1)$ which depends on the message and the first part of the signature, both bound not to change for the computation of σ_2, which really relies on the knowledge of the private key. If the probability of fraud is high enough, then with good probability, the adversary is able to answer to many distinct outputs from the $\mathcal{H}$ function, on the input (m, σ_1).

To be more concrete, let us consider the Schnorr signature scheme, which is presented on Figure 4, in any "suitable" cyclic group $\mathcal{G}$ of prime order q, where at least the Discrete Logarithm problem is hard. We expect to obtain two signatures $(\mathbf{r} = \sigma_1, h, s = \sigma_2)$ and $(\mathbf{r}' = \sigma_1', h', s' = \sigma_2')$ of an identical message m such that $\sigma_1 = \sigma_1'$, but $h \neq h'$. Thereafter, we can easily extract the discrete logarithm of the public key:

$$\left. \begin{array}{rcl} \mathbf{r} &=& s \cdot \mathbf{g} + h \cdot \mathbf{y} \\ \mathbf{r} &=& s' \cdot \mathbf{g} + h' \cdot \mathbf{y} \end{array} \right\} \Rightarrow (s - s') \cdot \mathbf{g} = (h' - h) \cdot \mathbf{y},$$

which leads to $\log_{\mathbf{g}} \mathbf{y} = (s - s') \cdot (h' - h)^{-1} \bmod q$.

Initialization (security parameter k) $\to (\mathcal{G}, g, \mathcal{H})$
$\mathbf{g}$ a generator of any cyclic group $(\mathcal{G}, +)$
$\quad$ of order q, with $2^{k-1} \leq q < 2^k$
$\mathcal{H}$ a hash function: $\{0,1\}^* \to \mathbb{Z}_q$
$\to (\mathcal{G}, g, \mathcal{H})$

$\mathcal{K}$: **Key Generation** $\to (\mathbf{y}, x)$
$\quad$ private key $\quad x \in \mathbb{Z}_q^*$
$\quad$ public key $\quad \mathbf{y} = -x \cdot \mathbf{g}$
$\to (\mathbf{y}, x)$
$\mathcal{S}$: **Signature of** $m \to (\mathbf{r}, h, s)$
K is randomly chosen in $\mathbb{Z}_q^*$
$\mathbf{r} = K \cdot \mathbf{g} \qquad h = \mathcal{H}(m, r) \qquad s = K + xh \bmod q$
$\to (\mathbf{r}, h, s)$ is a signature of m
$\mathcal{V}$: **Verification of** (m, r, s)
check whether $h \overset{?}{=} \mathcal{H}(m, \mathbf{r})$
$\qquad$ and $\mathbf{r} \overset{?}{=} s \cdot \mathbf{g} + h \cdot \mathbf{y}$
$\to$ Yes/No

FIGURE 4. The Schnorr Signature Scheme.

5.2.1. General Tools. First, let us recall the "Splitting Lemma" which will be the main probabilistic tool for the "Forking Lemma". It translates the fact that when a subset A is "large" in a product space $X \times Y$, it has many "large" sections.

Lemma 1 (The Splitting Lemma). *Let $A \subset X \times Y$ such that $\Pr[(x, y) \in A] \geq \varepsilon$. For any $\alpha < \varepsilon$, define*

$$B = \left\{ (x, y) \in X \times Y \mid \Pr_{y' \in Y}[(x, y') \in A] \geq \varepsilon - \alpha \right\},$$

then the following statements hold:

(i) $\Pr[B] \geq \alpha$
(ii) $\forall (x, y) \in B, \Pr_{y' \in Y}[(x, y') \in A] \geq \varepsilon - \alpha$.
(iii) $\Pr[B \mid A] \geq \alpha/\varepsilon$.

Proof. In order to prove statement (i), we argue by contradiction, using the notation $\bar{B}$ for the complement of B in $X \times Y$. Assume that $\Pr[B] < \alpha$. Then

$$\varepsilon \leq \Pr[B] \cdot \Pr[A \mid B] + \Pr[\bar{B}] \cdot \Pr[A \mid \bar{B}] < \alpha \cdot 1 + 1 \cdot (\varepsilon - \alpha) = \varepsilon.$$

This implies a contradiction, hence the result.

$\qquad$ Statement (ii) is a straightforward consequence of the definition.

We finally turn to the last assertion, using Bayes' law:

$$\Pr[B \mid A] = 1 - \Pr[\bar{B} \mid A]$$
$$= 1 - \Pr[A \mid \bar{B}] \cdot \Pr[\bar{B}] / \Pr[A] \geq 1 - (\varepsilon - \alpha)/\varepsilon = \alpha/\varepsilon. \qquad \square$$

No-Message Attacks. The following *Forking Lemma* just states that the above oracle replay technique will often success with any good adversary.

Theorem 1 (The Forking Lemma). *Let $(\mathcal{K}, \mathcal{S}, \mathcal{V})$ be a digital signature scheme with security parameter k, with a signature as above, of the form $(m, \sigma_1, h, \sigma_2)$, where $h = \mathcal{H}(m, \sigma_1)$ and σ_2 depends on σ_1 and h only. Let $\mathcal{A}$ be a probabilistic polynomial time Turing machine whose input only consists of public data and which can ask q_h queries to the random oracle, with $q_h > 0$. We assume that, within the time bound T, $\mathcal{A}$ produces, with probability $\varepsilon \geq 7q_h/2^k$, a valid signature $(m, \sigma_1, h, \sigma_2)$. Then, within time $T' \leq 16q_h T/\varepsilon$, and with probability $\varepsilon' \geq 1/9$, a replay of this machine outputs two valid signatures $(m, \sigma_1, h, \sigma_2)$ and $(m, \sigma_1, h', \sigma_2')$ such that $h \neq h'$.*

Proof. We are given an adversary $\mathcal{A}$, which is a probabilistic polynomial time Turing machine with random tape ω. During the attack, this machine asks a polynomial number of questions to the random oracle $\mathcal{H}$. We may assume that these questions are distinct: for instance, $\mathcal{A}$ can store questions and answers in a table. Let $\mathcal{Q}_1, \ldots, \mathcal{Q}_{q_h}$ be the q_h distinct questions and let $\rho = (\rho_1, \ldots, \rho_{q_h})$ be the list of the q_h answers of $\mathcal{H}$. It is clear that a random choice of $\mathcal{H}$ exactly corresponds to a random choice of ρ. Then, for a random choice of $(\omega, \mathcal{H})$, with probability ε, $\mathcal{A}$ outputs a valid signature $(m, \sigma_1, h, \sigma_2)$. Since $\mathcal{H}$ is a random oracle, it is easy to see that the probability for h to be equal to $\mathcal{H}(m, \sigma_1)$ is less than $1/2^k$, unless it has been asked during the attack. So, it is likely that the question (m, σ_1) is actually asked during a successful attack. Accordingly, we define $Ind_\mathcal{H}(\omega)$ to be the index of this question: $(m, \sigma_1) = \mathcal{Q}_{Ind_\mathcal{H}(\omega)}$ (we let $Ind_\mathcal{H}(\omega) = \infty$ if the question is never asked). We then define the sets

$$\mathbf{S} = \left\{ (\omega, \mathcal{H}) \mid \mathcal{A}^\mathcal{H}(\omega) \text{ succeeds } \& \ Ind_\mathcal{H}(\omega) \neq \infty \right\},$$
$$\text{and } \mathbf{S}_i = \left\{ (\omega, \mathcal{H}) \mid \mathcal{A}^\mathcal{H}(\omega) \text{ succeeds } \& \ Ind_\mathcal{H}(\omega) = i \right\} \quad \text{for } i \in \{1, \ldots, q_h\}.$$

We thus call $\mathbf{S}$ the set of the successful pairs $(\omega, \mathcal{H})$. One should note that the set $\{\mathbf{S}_i \mid i \in \{1, \ldots, q_h\}\}$ is a partition of $\mathbf{S}$. With those definitions, we find a lower bound for the probability of success, $\nu = \Pr[\mathbf{S}] \geq \varepsilon - 1/2^k$. Since we did the assumption that $\varepsilon \geq 7q_h/2^k \geq 7/2^k$, then $\nu \geq 6\varepsilon/7$. Let I be the set consisting of the most likely indices i,

$$I = \left\{ i \mid \Pr[\mathbf{S}_i \mid \mathbf{S}] \geq 1/2q_h \right\}.$$

The following lemma claims that, in case of success, the index lies in I with probability at least $1/2$.

Lemma 2.

$$\Pr[Ind_{\mathcal{H}}(\omega) \in I \,|\, \mathbf{S}] \geq \frac{1}{2}.$$

Proof. By definition of the sets $\mathbf{S}_i$, $\Pr[Ind_{\mathcal{H}}(\omega) \in I \,|\, \mathbf{S}] = \sum_{i \in I} \Pr[\mathbf{S}_i \,|\, \mathbf{S}]$. This probability is equal to $1 - \sum_{i \notin I} \Pr[\mathbf{S}_i \,|\, \mathbf{S}]$. Since the complement of I contains fewer than q_h elements, this probability is at least $1 - q_h \times 1/2q_h \geq 1/2$. ☐

We now run the attacker $2/\varepsilon$ times with random ω and random $\mathcal{H}$. Since $\nu = \Pr[\mathbf{S}] \geq 6\varepsilon/7$, with probability greater than $1 - (1 - 6\varepsilon/7)^{2/\varepsilon}$, we get at least one pair $(\omega, \mathcal{H})$ in $\mathbf{S}$. It is easily seen that this probability is lower bounded by $1 - e^{-12/7} \geq 4/5$.

We now apply the Splitting-lemma (Lemma 1, with $\varepsilon = \nu/2q_h$ and $\alpha = \varepsilon/2$) for each integer $i \in I$: we denote by $\mathcal{H}_{|i}$ the restriction of $\mathcal{H}$ to queries of index strictly less than i. Since $\Pr[\mathbf{S}_i] \geq \nu/2q_h$, there exists a subset Ω_i of executions such that,

$$\text{for any } (\omega, \mathcal{H}) \in \Omega_i, \Pr_{\mathcal{H}'}[(\omega, \mathcal{H}') \in \mathbf{S}_i \,|\, \mathcal{H}'_{|i} = \mathcal{H}_{|i}] \;\geq\; \frac{\nu}{4q_h}$$

$$\Pr[\Omega_i \,|\, \mathbf{S}_i] \;\geq\; \frac{1}{2}.$$

Since all the subsets $\mathbf{S}_i$ are disjoint,

$$\Pr_{\omega, \mathcal{H}}\left[(\exists i \in I)\,(\omega, \mathcal{H}) \in \Omega_i \cap \mathbf{S}_i \,|\, \mathbf{S}\right]$$

$$= \Pr\left[\bigcup_{i \in I}(\Omega_i \cap \mathbf{S}_i) \,|\, \mathbf{S}\right] = \sum_{i \in I} \Pr[\Omega_i \cap \mathbf{S}_i \,|\, \mathbf{S}]$$

$$= \sum_{i \in I} \Pr[\Omega_i \,|\, \mathbf{S}_i] \cdot \Pr[\mathbf{S}_i \,|\, \mathbf{S}] \geq \left(\sum_{i \in I} \Pr[\mathbf{S}_i \,|\, \mathbf{S}]\right)/2 \geq \frac{1}{4}.$$

We let β denote the index $Ind_{\mathcal{H}}(\omega)$ corresponding to the successful pair. With probability at least $1/4$, $\beta \in I$ and $(\omega, \mathcal{H}) \in \mathbf{S}_\beta \cap \Omega_\beta$. Consequently, with probability greater than $4/5 \times 1/5 = 1/5$, the $2/\varepsilon$ attacks have provided a successful pair $(\omega, \mathcal{H})$, with $\beta = Ind_{\mathcal{H}}(\omega) \in I$ and $(\omega, \mathcal{H}) \in \mathbf{S}_\beta$. Furthermore, if we replay the attack, with fixed ω but randomly chosen oracle $\mathcal{H}'$ such that $\mathcal{H}'_{|\beta} = \mathcal{H}_{|\beta}$, we know that $\Pr_{\mathcal{H}'}[(\omega, \mathcal{H}') \in \mathbf{S}_\beta \,|\, \mathcal{H}'_{|\beta} = \mathcal{H}_{|\beta}] \geq \nu/4q_h$. Then

$$\Pr_{\mathcal{H}'}[(\omega, \mathcal{H}') \in \mathbf{S}_\beta \text{ and } \rho_\beta \neq \rho'_\beta \,|\, \mathcal{H}'_{|\beta} = \mathcal{H}_{|\beta}]$$

$$\geq \Pr_{\mathcal{H}'}[(\omega, \mathcal{H}') \in \mathbf{S}_\beta \,|\, \mathcal{H}'_{|\beta} = \mathcal{H}_{|\beta}] - \Pr_{\mathcal{H}'}[\rho'_\beta = \rho_\beta] \geq \nu/4q_h - 1/2^k,$$

where $\rho_\beta = \mathcal{H}(\mathcal{Q}_\beta)$ and $\rho'_\beta = \mathcal{H}'(\mathcal{Q}_\beta)$. Using again the assumption that $\varepsilon \geq 7q_h/2^k$, the above probability is lower-bounded by $\varepsilon/14q_h$. We thus replay the attack $14q_h/\varepsilon$ times with a new random oracle $\mathcal{H}'$ such that $\mathcal{H}'_{|\beta} = \mathcal{H}_{|\beta}$, and get another success with probability greater than $1 - (1 - \varepsilon/14q_h)^{14q_h/\varepsilon} \geq 1 - e^{-1} \geq 3/5$.

Finally, after less than $2/\varepsilon + 14q_h/\varepsilon$ repetitions of the attack, with probability greater than $1/5 \times 3/5 \geq 1/9$, we have obtained two signatures $(m, \sigma_1, h, \sigma_2)$ and

$(m', \sigma_1', h', \sigma_2')$, both valid w.r.t. their specific random oracle $\mathcal{H}$ or $\mathcal{H}'$, and with the particular relations

$$\mathcal{Q}_\beta = (m, \sigma_1) = (m', \sigma_1') \text{ and } h = \mathcal{H}(\mathcal{Q}_\beta) \neq \mathcal{H}'(\mathcal{Q}_\beta) = h'. \qquad \square$$

One may have noticed that the mechanics of our reduction depend on some parameters related to the attacker $\mathcal{A}$, namely, its probability of success ε and the number q_h of queries to the random oracle. This induces a lack of uniformity. A uniform version, in expected polynomial time is also possible.

> **Theorem 2 (The Forking Lemma – The Uniform Case).** *Let $(\mathcal{K}, \mathcal{S}, \mathcal{V})$ be a digital signature scheme with security parameter k, with a signature as above, of the form $(m, \sigma_1, h, \sigma_2)$, where $h = \mathcal{H}(m, \sigma_1)$ and σ_2 depends on σ_1 and h only. Let $\mathcal{A}$ be a probabilistic polynomial time Turing machine whose input only consists of public data and which can ask q_h queries to the random oracle, with $q_h > 0$. We assume that, within the time bound T, $\mathcal{A}$ produces, with probability $\varepsilon \geq 7q_h/2^k$, a valid signature $(m, \sigma_1, h, \sigma_2)$. Then there is another machine which has control over $\mathcal{A}$ and produces two valid signatures $(m, \sigma_1, h, \sigma_2)$ and $(m, \sigma_1, h', \sigma_2')$ such that $h \neq h'$, in expected time $T' \leq 84480 T q_h / \varepsilon$.*

Proof. Now, we try to design a machine $\mathbf{M}$ which succeeds in expected polynomial time:

1. $\mathbf{M}$ initializes $j = 0$;
2. $\mathbf{M}$ runs $\mathcal{A}$ until it outputs a successful pair $(\omega, \mathcal{H}) \in \mathbf{S}$ and denotes by N_j the number of calls to $\mathcal{A}$ to obtain this success, and by β the index $Ind_\mathcal{H}(\omega)$;
3. $\mathbf{M}$ replays, at most $140 N_j \alpha^j$ times, $\mathcal{A}$ with fixed ω and random $\mathcal{H}'$ such that $\mathcal{H}'_{|\beta} = \mathcal{H}_{|\beta}$, where $\alpha = 8/7$;
4. $\mathbf{M}$ increments j and returns to 2, until it gets a successful forking.

For any execution of $\mathbf{M}$, we denote by J the last value of j and by N the total number of calls to $\mathcal{A}$. We want to compute the expectation of N. Since $\nu = \Pr[\mathbf{S}]$, and $N_j \geq 1$, then $\Pr[N_j \geq 1/5\nu] \geq 3/4$. We define $\ell = \lceil \log_\alpha q_h \rceil$, so that, $140 N_j \alpha^j \geq 28 q_h / \varepsilon$ for any $j \geq \ell$, whenever $N_j \geq 1/5\nu$. Therefore, for any $j \geq \ell$, when we have a first success in $\mathbf{S}$, with probability greater than $1/4$, the index $\beta = Ind_\mathcal{H}(\omega)$ is in the set I and $(\omega, \mathcal{H}) \in \mathbf{S}_\beta \cap \Omega_\beta$. Furthermore, with probability greater than $3/4$, $N_j \geq 1/5\nu$. Therefore, with the same conditions as before, that is $\varepsilon \geq 7q_h/2^k$, the probability of getting a successful fork after at most $28 q_h / \varepsilon$ iterations at step 3 is greater than $6/7$.

For any $t \geq \ell$, the probability for J to be greater or equal to t is less than $(1 - 1/4 \times 3/4 \times 6/7)^{t-\ell}$, which is less than $\gamma^{t-\ell}$, with $\gamma = 6/7$. Furthermore,

$$E[N \mid J = t] \leq \sum_{j=0}^{j=t} \left(E[N_j] + 140 E[N_j] \alpha^j \right) \leq \frac{141}{\nu} \times \sum_{j=0}^{j=t} \alpha^j \leq \frac{141}{\nu} \times \frac{\alpha^{t+1}}{\alpha - 1}.$$

So, the expectation of N is $E[N] = \sum_t E[N \mid J = t] \cdot \Pr[J = t]$ and then it can be shown to be less than $84480 q_h / \varepsilon$. Hence the theorem. $\qquad \square$

Chosen-Message Attacks. However, this just covers the no-message attacks, without any oracle access. Since we can simulate any zero-knowledge protocol, even without having to restart the simulation because of the honest verifier (*i.e.* the challenge is randomly chosen by the random oracle $\mathcal{H}$) one can easily simulate the signer without the private key:

- one first chooses random $h, s \in \mathbb{Z}_q$;
- one computes $\mathbf{r} = s \cdot \mathbf{g} + h \cdot \mathbf{y}$ and defines $\mathcal{H}(m, \mathbf{r})$ to be equal to h, which is a uniformly distributed value;
- one can output $(\mathbf{r}, h, s)$ as a valid signature of the message m.

This furthermore simulates the oracle $\mathcal{H}$, by defining $\mathcal{H}(m, \mathbf{r})$ to be equal to h. This simulation is almost perfect since $\mathcal{H}$ is supposed to output a random value to any new query, and h is indeed a random value. Nevertheless, if the query $\mathcal{H}(m, \mathbf{r})$ has already been asked, $\mathcal{H}(m, \mathbf{r})$ is already defined, and thus the definition $\mathcal{H}(m, \mathbf{r}) \leftarrow h$ is impossible. But such a situation is very rare, which allows us to claim the following result, which stands for the Schnorr signature scheme but also for any signature derived from a three-round honest verifier zero-knowledge interactive proof of knowledge:

> **Theorem 3.** *Let $\mathcal{A}$ be a probabilistic polynomial time Turing machine whose input only consists of public data. We denote respectively by q_h and q_s the number of queries that $\mathcal{A}$ can ask to the random oracle and the number of queries that $\mathcal{A}$ can ask to the signer. Assume that, within a time bound T, $\mathcal{A}$ produces, with probability $\varepsilon \geq 10(q_s + 1)(q_s + q_h)/2^k$, a valid signature $(m, \sigma_1, h, \sigma_2)$. If the triples (σ_1, h, σ_2) can be simulated without knowing the secret key, with an indistinguishable distribution probability, then, a replay of the attacker $\mathcal{A}$, where interactions with the signer are simulated, outputs two valid signatures $(m, \sigma_1, h, \sigma_2)$ and $(m, \sigma_1, h', \sigma_2')$ such that $h \neq h'$, within time $T' \leq 23q_h T/\varepsilon$ and with probability $\varepsilon' \geq 1/9$.*

A uniform version of this theorem can also be found in [68]. From a more practical point of view, these results state that if an adversary manages to perform an existential forgery under an adaptive chosen-message attack within an expected time T, after q_h queries to the random oracle and q_s queries to the signing oracle, then the discrete logarithm problem can be solved within an expected time less than Cq_hT, for some constant C. This result has been more recently extended to the transformation of any identification scheme secure against passive adversaries into a signature scheme [8].

Brickell, Vaudenay, Yung and the author also extended the *forking lemma* technique [69, 17] to many variants of El Gamal [27] and DSA [55], such as the Korean Standard KCDSA [43]. However, the original El Gamal and DSA schemes were not covered by this study, and are certainly not provably secure, even if no attack has ever been found against DSA.

5.3. RSA-Based Signatures

Unfortunately, with the above signatures based on the discrete logarithm, as any construction using the Fiat-Shamir paradigm, we do not really achieve our goal, because the reduction is costly, since q_h can be huge, as much as 2^{60} in practice. This security proof is meaningful for very large groups only.

5.3.1. FDH-RSA: The Full-Domain Hash Signature.
In 1996, Bellare and Rogaway [12] proposed other candidates, based on the RSA assumption. The first scheme is the by-now classical hash-and-decrypt paradigm (*a.k.a.* the Full-Domain Hash paradigm): as for the basic RSA signature, the generation algorithm produces a large composite number $N = pq$, a public key e, and a private key d such that $e \cdot d = 1 \bmod \varphi(N)$. In order to sign a message m, one first hashes it using a full-domain hash function $\mathcal{H} : \{0,1\}^\star \to \mathbb{Z}_N^\star$, and computes the *e-th* root, $\sigma = \mathcal{H}(m)^d \bmod N$. The verification algorithm simply checks whether the following equality holds, $\mathcal{H}(m) = \sigma^e \bmod N$.

More generally, the Full-Domain Hash signature can be defined as follows, for any trapdoor one-way permutation f:

$\mathcal{K}$: **Key Generation** $\to (f, f^{-1})$
public key $\quad f : X \longrightarrow X$, a trapdoor one-way permutation onto X private key $\quad f^{-1}$ $\to (f, f^{-1})$
$\mathcal{S}$: **Signature of** $m \to \sigma$
$r = \mathcal{H}(m)$ and $\sigma = f^{-1}(r)$ $\to \sigma$ is the signature of m
$\mathcal{V}$: **Verification of** (m, σ)
check whether $f(\sigma) \stackrel{?}{=} \mathcal{H}(m)$ $\to$ Yes/No

FIGURE 5. The FDH Signature.

5.3.2. Security Analysis.
For this scheme, Bellare and Rogaway proved, in the random-oracle model:

> **Theorem 4.** *Let $\mathcal{A}$ be an adversary which can produce, with success probability ε, an existential forgery under a chosen-message attack within a time t, after q_h and q_s queries to the hash function and the signing oracle respectively. Then the permutation f can be inverted with probability ε' within time t' where*
> $$\varepsilon' \geq \frac{\varepsilon}{q_s + q_h + 1} \qquad and \qquad t' \leq t + (q_s + q_h)T_f,$$
> *with T_f the time for an evaluation of f.*

Let us present this proof, using the new formalism introduced by Victor Shoup in [81, 82, 83], and which will be extensively used in these notes. In this technique, we define a sequence $\mathbf{G}_1$, $\mathbf{G}_2$, etc., of modified attack games starting from the actual game $\mathbf{G}_0$. Each of the games operates on the same underlying probability space: the public and private keys of the cryptographic scheme, the coin tosses of the adversary $\mathcal{A}$ and the random oracles. Only the rules defining how the view is computed differ from game to game. To go from one game to another with a slightly different distribution probability, we repeatedly use the following lemma:

Lemma 3. *Let* E_1, E_2 *and* F_1, F_2 *be events defined on a probability space*

$$\Pr[\mathsf{E}_1 \mid \neg\mathsf{F}_1] = \Pr[\mathsf{E}_2 \mid \neg\mathsf{F}_2] \ and \ \Pr[\mathsf{F}_1] = \Pr[\mathsf{F}_2] = \varepsilon \ \Rightarrow \ |\Pr[\mathsf{E}_1] - \Pr[\mathsf{E}_2]| \le \varepsilon.$$

Proof. The proof follows from easy computations:

$$\begin{aligned}
|\Pr[\mathsf{E}_1] - \Pr[\mathsf{E}_2]| &= |\Pr[\mathsf{E}_1 \mid \mathsf{F}_1] \cdot \Pr[\mathsf{F}_1] + \Pr[\mathsf{E}_1 \mid \neg\mathsf{F}_1] \cdot \Pr[\neg\mathsf{F}_1] \\
&\quad - \Pr[\mathsf{E}_2 \mid \mathsf{F}_2] \cdot \Pr[\mathsf{F}_2] - \Pr[\mathsf{E}_2 \mid \neg\mathsf{F}_2] \cdot \Pr[\neg\mathsf{F}_2]| \\
&= |(\Pr[\mathsf{E}_1 \mid \mathsf{F}_1] - \Pr[\mathsf{E}_2 \mid \mathsf{F}_2]) \cdot \varepsilon \\
&\quad + (\Pr[\mathsf{E}_1 \mid \neg\mathsf{F}_1] - \Pr[\mathsf{E}_2 \mid \neg\mathsf{F}_2]) \cdot (1 - \varepsilon)| \\
&= |(\Pr[\mathsf{E}_1 \mid \mathsf{F}_1] - \Pr[\mathsf{E}_2 \mid \mathsf{F}_2]) \cdot \varepsilon| \le \varepsilon. \qquad \square
\end{aligned}$$

Actually, this lemma will not be used in the proofs of the FDH signatures, because all the simulated distributions will remain perfect.

Basic Proof of the FDH Signature. In this proof, we incrementally define a sequence of games starting at the real game $\mathbf{G}_0$ and ending up at $\mathbf{G}_5$. We make a very detailed sequence of games in this proof, since this is the first one. Some steps will be skipped in the other proofs. The goal of this proof is to reduce the inversion of the permutation f on an element y (find x such that $y = f(x)$) to an attack. We are thus given such a random challenge y.

Game $\mathbf{G}_0$:   This is the real attack game, in the random-oracle model, which includes the verification step. This means that the attack game consists in giving the public key to the adversary, and a full access to the signing oracle. When it outputs its forgery, one furthermore checks whether it is actually valid or not. Note that if the adversary asks q_s queries to the signing oracle and q_h queries to the hash oracle, at most $q_s + q_h + 1$ queries are asked to the hash oracle during this game, since each signing query may make such a new query, and the last verification step too. We are interested in the following event: S_0 which occurs if the verification step succeeds (and the signature is new).

$$\mathsf{Succ}^{\mathsf{euf}}_{\mathsf{fdh}}(\mathcal{A}) = \Pr[\mathsf{S}_0]. \tag{1}$$

Game $\mathbf{G}_1$:   In this game, we simulate the oracles, the hash oracle $\mathcal{H}$ and the signing oracle $\mathcal{S}$, and the last verification step, as shown on Figure 6. From this simulation, we easily see that the game is perfectly indistinguishable from the real attack.

$$\Pr[\mathsf{S}_1] = \Pr[\mathsf{S}_0]. \tag{2}$$

$\mathcal{H}$ oracle	For a hash-query $\mathcal{H}(q)$, such that a record $(q, \star, r)$ appears in H-List, the answer is r. Otherwise the answer r is defined according to the following rule: ▶**Rule $\mathcal{H}^{(1)}$** $\quad$	$\quad$ Choose a random element $r \in X$. The record $(q, \perp, r)$ is added to H-List. Note: the second component of the elements of this list will be explained later.
$\mathcal{S}$ oracle	For a sign-query $\mathcal{S}(m)$, one first asks for $r = \mathcal{H}(m)$ to the $\mathcal{H}$-oracle, and then the signature σ is defined according to the following rule: ▶**Rule $\mathcal{S}^{(1)}$** $\quad$	$\quad$ Computes $\sigma = f^{-1}(r)$.
$\mathcal{V}$ oracle	The game ends with the verification of the output (m, σ) from the adversary. One first asks for $r = \mathcal{H}(m)$, and checks whether $r = f(\sigma)$.

FIGURE 6. Simulation of the Attack Game against FDH

Game G_2: Since the verification process is included in the attack game, the output message is necessarily asked to the hash oracle. Let us guess the index c of this (first) query. If the guess failed, we abort the game. Therefore, only a correct guess (event GoodGuess) may lead to a success.

$$\begin{aligned} \Pr[S_2] &= \Pr[S_1 \wedge \mathsf{GoodGuess}] = \Pr[S_1 \mid \mathsf{GoodGuess}] \times \Pr[\mathsf{GoodGuess}] \\ &\geq \Pr[S_1] \times \frac{1}{q_h + q_s + 1}. \end{aligned} \tag{3}$$

Game G_3: We can now simulate the hash oracle, incorporating the challenge y, for which we want to extract the pre-image x by f:

▶**Rule $\mathcal{H}^{(3)}$**
$\quad$ | $\quad$ If this is the c-th query, set $r \leftarrow y$; otherwise, choose a random element $r \in X$. The record $(q, \perp, r)$ is added to H-List.

Because of the random choice for the challenge y, this rule lets the game indistinguishable from the previous one.

$$\Pr[S_3] = \Pr[S_2]. \tag{4}$$

Game G_4: We now modify the simulation of the hash oracle for other queries, which may be used in signing queries:

▶**Rule $\mathcal{H}^{(4)}$**

If this is the c-th query, set $r \leftarrow y$ and $s \leftarrow \perp$; otherwise, choose a random element $s \in X$, and compute $r = f(s)$. The record (q, s, r) is added to H-List.

Because of the permutation property of f, and the random choice for s, this rule lets the game indistinguishable from the previous one.

$$\Pr[\mathsf{S}_4] = \Pr[\mathsf{S}_3]. \tag{5}$$

Game G_5: By now, excepted for the c-th hash query, which will be involved in the forgery (and thus not asked to the signing oracle), the pre-image is known. One can thus simulate the signing oracle without quering f^{-1}:

▶**Rule $\mathcal{S}^{(5)}$**

Lookup for (m, s, r) in H-List, and set $\sigma = s$.

Since the message corresponding to the c-th query cannot be asked to the signing oracle, otherwise it would not be a valid forgery, this rule lets the game indistinguishable from the previous one.

$$\Pr[\mathsf{S}_5] = \Pr[\mathsf{S}_4]. \tag{6}$$

Note that now, the simulation can easily be performed, without any specific computational power or oracle access. Just a few more evaluations of f are done to simulate the hash oracle, and the forgery leads to the pre-image of y:

$$\Pr[\mathsf{S}_5] = \mathsf{Succ}_f^{\mathsf{ow}}(t + (q_h + q_s)T_f). \tag{7}$$

As a consequence, using equations (1), (2), (3), (4), (5), (6) and (7)

$$\begin{aligned}
\mathsf{Succ}_f^{\mathsf{ow}}(t + (q_h + q_s)T_f) &= \Pr[\mathsf{S}_5] = \Pr[\mathsf{S}_3] = \Pr[\mathsf{S}_4] = \Pr[\mathsf{S}_2] \\
&\geq \frac{1}{q_h + q_s + 1} \times \Pr[\mathsf{S}_1] \geq \frac{1}{q_h + q_s + 1} \times \Pr[\mathsf{S}_0].
\end{aligned}$$

And thus,

$$\mathsf{Succ}_{\mathsf{fdh}}^{\mathsf{euf}}(\mathcal{A}) \leq (q_h + q_s + 1) \times \mathsf{Succ}_f^{\mathsf{ow}}(t + (q_h + q_s)T_f). \qquad \square$$

Improved Security Result. This reduction has been thereafter improved [22], thanks to the random self-reducibility of the RSA function. The following result applies as soon as the one-way permutation has some homomorphic property on the group X:

$$f(x \otimes y) = f(x) \otimes f(y).$$

> **Theorem 5.** *Let $\mathcal{A}$ be an adversary which can produce, with success probability ε, an existential forgery under a chosen-message attack within a time t, after q_h and q_s queries to the hash function and the signing oracle respectively. Then the permutation f can be inverted with probability ε' within time t' where*
>
> $$\varepsilon' \geq \frac{\varepsilon}{q_s} \times \exp(-2) \quad \text{and} \quad t' \leq t + (q_s + q_h)T_f,$$
>
> *with T_f the time for an evaluation of f.*

This proof can be performed as the previous one, and thus starts at the real game $\mathbf{G}_0$, then we can use the same simulation as in the game $\mathbf{G}_1$. The sole formal difference in the simulation will be the H-List which elements have one more field, and are thus initially of the form $(q, \perp, \perp, r)$. Things differ much after that, using a real value p between 0 and 1, which will be made precise later. The idea here, is to make any forgery useful for inverting the permutation f, not only a specific (guessed) one. On the other hand, one must still be able to simulate the signing oracle. The probability p will separate the two situations:

Game $\mathbf{G}_2$:   A random coin decides whether we introduce the challenge y in the hash answer, or an element with a known pre-image:

▶**Rule $\mathcal{H}^{(2)}$**

> One chooses a random $s \in X$. With probability p, one sets $r \leftarrow y \otimes f(s)$ and $t \leftarrow 1$; otherwise, $r \leftarrow f(s)$ and $t \leftarrow 0$. The record (q, t, s, r) is added to H-List.

Because of the homomorphic property on the group X of the permutation f, this rule lets the game indistinguishable from the previous one. Note again that elements in H-List contain one more field t than in the previous proof. One may see that $r = y^t \otimes f(s)$.

Game $\mathbf{G}_3$:   For a proportion $1 - p$ of the signature queries, one can simulate the signing oracle without having to invert the permutation f:

▶**Rule $\mathcal{S}^{(3)}$**

> Lookup for (m, t, s, r) in H-List, if $t = 1$ then halt the game, otherwise set $\sigma = s$.

This rule lets the game indistinguishable, unless one signing query fails ($t = 1$), which happens with probability p, for each signature:

$$\Pr[S_3] = (1 - p)^{q_s} \times \Pr[S_2]. \tag{8}$$

Note that now, the simulation can easily be performed, without any specific computational power or oracle access. Just a few more exponentiations are done to simulate the hash oracle, and the forgery (m, σ) leads to the pre-image of y, if ($t = 1$). The latter case holds with probability p. Indeed, (m, t, s, r) can be found in the H-List, and then $r = y^t \otimes f(s) = y \otimes f(s) = f(\sigma)$, which easily leads to the pre-image of y by f:

$$\mathsf{Succ}_f^{\mathsf{ow}}(t + (q_h + q_s)T_f) = p \times \Pr[S_3]. \tag{9}$$

Using equations (1), (2), (8) and (9)

$$
\begin{aligned}
\mathsf{Succ}_f^{\mathsf{ow}}(t + (q_h + q_s)T_f) &= p \times \Pr[S_3] = p \times (1 - p)^{q_s} \times \Pr[S_2] \\
&= p \times (1 - p)^{q_s} \times \Pr[S_1] = p \times (1 - p)^{q_s} \times \Pr[S_0].
\end{aligned}
$$

And thus,

$$\mathsf{Succ}_{\mathsf{fdh}}^{\mathsf{euf}}(\mathcal{A}) \leq \frac{1}{p(1 - p)^{q_s}} \times \mathsf{Succ}_f^{\mathsf{ow}}(t + (q_h + q_s)T_f).$$

Therefore, the success probability of our inversion algorithm is $p(1 - p)^{q_s} \varepsilon$, if ε is the success probability of the adversary. If $q_s > 0$, the latter expression is optimal for $p = 1/(q_s + 1)$. And for this parameter, and a huge value q_s, the success probability is approximately ε/eq_s. It is anyway larger than $\varepsilon/e^2 q_s$ (where $e = \exp(1) \approx 2.17\ldots$).

As far as time complexity is concerned, each random oracle simulation (which can be launched by a signing simulation) requires a modular exponentiation to the power e, hence the result.   □

This is a great improvement since the success probability does not depend anymore on q_h. Furthermore, q_s can be limited by the user, whereas q_h cannot. In practice, one only assumes $q_h \leq 2^{60}$, but q_s can be limited below 2^{30}.

5.3.3. The Probabilistic Signature Scheme. However, one would like to get more, suppressing any coefficient. In their paper [12], Bellare and Rogaway proposed such a better candidate, the Probabilistic Signature Scheme (PSS, see Figure 7): the key generation is still the same, but the signature process involves three hash

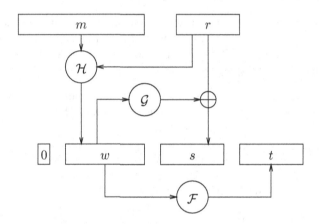

FIGURE 7. Probabilistic Signature Scheme

functions
$$\mathcal{F}: \{0,1\}^{k_2} \to \{0,1\}^{k_0}, \quad \mathcal{G}: \{0,1\}^{k_2} \to \{0,1\}^{k_1},$$
$$\mathcal{H}: \{0,1\}^* \to \{0,1\}^{k_2},$$

where $k = k_0 + k_1 + k_2 + 1$ satisfies $\{0,1\}^{k-1} \subset X \subset \{0,1\}^k$. We remind that f is a trapdoor one-way permutation onto X, with an homomorphic relationship. For each message m to be signed, one chooses a random string $r \in \{0,1\}^{k_1}$. One first computes $w = \mathcal{H}(m,r)$, $s = \mathcal{G}(w) \oplus r$ and $t = \mathcal{F}(w)$. Then one concatenates $y = 0\|w\|s\|t$, where $a\|b$ denotes the concatenation of the bit strings a and b. Finally, one computes the pre-image by f, $\sigma = f^{-1}(y)$. The verification algorithm first computes $y = f(\sigma)$, and parses it as $y = b\|w\|s\|t$. Then, one can get $r = s \oplus \mathcal{G}(w)$, and checks whether $b = 0$, $w = \mathcal{H}(m,r)$ and $t = \mathcal{F}(w)$.

About this PSS construction, Bellare and Rogaway proved the security in the random-oracle model.

Theorem 6. *Let $\mathcal{A}$ be a CMA-adversary against f–PSS which produces an existential forgery within a time t, after q_f, q_g, q_h and q_s queries to the hash functions $\mathcal{F}$, $\mathcal{G}$ and $\mathcal{H}$ and the signing oracle respectively. Then its success probability is upper-bounded by*

$$\mathsf{Succ}_f^{\mathsf{ow}}(t + (q_s + q_h)k_2 \cdot T_f) + \frac{1}{2^{k_2}} + (q_s + q_h) \cdot \left(\frac{q_s}{2^{k_1}} + \frac{q_f + q_g + q_h + q_s + 1}{2^{k_2}} \right),$$

with T_f the time for an evaluation of f.

Proof. First, we assume the existence of an adversary $\mathcal{A}$ that produces an existential forgery with probability ε within time t, after q_f, q_g and q_h queries to the random oracles $\mathcal{F}$, $\mathcal{G}$ and $\mathcal{H}$ and q_s queries to the signing oracle.

Game G_0: This is the real-world attack game. In any game G_n, we denote by S_n the event $\mathcal{V}(\mathsf{pk}, m, \sigma) = 1$, for a new signature σ.

Game G_1: In this game, we make the classical simulation of the random oracles, with random answers for any new query, as shown on Figure 8. This game is clearly identical to the previous one. The $\mathcal{H}$ simulation may seem a bit intricate, but the bit c is never used. It will appear later.

Game G_2: In this game, we introduce the random challenge $y^\star$, for which one is looking for $x^\star$ such that $y^\star = f(x^\star)$. Thus, we replace the random oracle $\mathcal{H}$ by the following simulation, which may abort:

▶**Rule $\mathcal{H}$-New$^{(2)}$**
Choose a random $u \in X$, then if $c = 0$, compute $z = y^\star \otimes f(u)$, otherwise compute $z = f(u)$, until the most significant bit of z is 0, but at most k_2 times (otherwise one aborts the game). Choose a random element $w \in \{0,1\}^{k_2}$. The record $(m, r, c, \perp, w)$ is added in H-List.

Let us remark that the number of calls to $\mathcal{H}$ is upper-bounded by $q_h + q_s$ (direct queries and queries asked by the signing oracle.) This game may only differ from the previous one during some $\mathcal{H}$-simulations, if the simulation aborts because z is still in the bad range, even after the k_2 attempts (event $\mathsf{BadRange}_2$). Using the Lemma 3, noting that

$$\Pr[S_2 \mid \neg\mathsf{BadRange}_2] = \Pr[S_1 \mid \neg\mathsf{BadRange}_2] \text{ and } \Pr[\mathsf{BadRange}_2] \leq \frac{q_h + q_s}{2^{k_2}},$$

one gets

$$|\Pr[S_2] - \Pr[S_1]| \leq \frac{q_h + q_s}{2^{k_2}}. \tag{10}$$

Game G_3: In the above game, one may have noted that z is uniformly distributed in X, because of the permutation property of f, with the conditioning that the most significant bit is 0. One can thus parse it into $0\|w\|s\|t$, where w is uniformly distributed in $\{0,1\}^{k_2}$:

$\mathcal{F}$, $\mathcal{G}$ and $\mathcal{H}$ oracles	Query $\mathcal{F}(w)$: if a record (w, t) appears in F-List, the answer is t. Otherwise the answer t is chosen randomly: $t \in \{0, 1\}^{k_0}$ and the record (w, t) is added in F-List.
	Query $\mathcal{G}(w)$: if a record (w, g) appears in G-List, the answer is g. Otherwise the answer g is chosen randomly: $g \in \{0, 1\}^{k_1}$ and the record (w, g) is added in G-List.
	Query $\mathcal{H}(m, r)$: one first sets $c = 0$ if the query is asked by the signing oracle, and $c = 1$ otherwise (by the adversary directly). If a record $(m, r, \star, \perp, w)$ appears in H-List: **▶Rule $\mathcal{H}$-Old$^{(1)}$** | The answer is w. Otherwise the answer w is defined according to the following rule: **▶Rule $\mathcal{H}$-New$^{(1)}$** | Choose a random element $w \in \{0, 1\}^{k_2}$. The record $(m, r, c, \perp, w)$ is added in H-List. Note: the fourth component of the elements of this list will be explained later.
$\mathcal{S}$ oracle	For a sign-query $\mathcal{S}(m)$, one first chooses a random $r \in \{0, 1\}^{k_1}$ and asks for $w = \mathcal{H}(m, r)$, $s = \mathcal{G}(w) \oplus r$ and $t = \mathcal{F}(w)$. Then one concatenates $y = 0\|w\|s\|t$ and computes the signature σ according to the following rule: **▶Rule $\mathcal{S}^{(1)}$** | Computes $\sigma = f^{-1}(y)$.

FIGURE 8. Simulation of the Attack Game against PSS

▶Rule $\mathcal{H}$-New$^{(3)}$
> Choose a random $u \in X$, then if $c = 0$, compute $z = y^{\star} \otimes f(u)$, otherwise compute $z = f(u)$, until the most significant bit of z is 0, but at most k_2 times (otherwise one aborts the game). Thereafter, z is parsed into $0\|w\|s\|t$, The record (m, r, c, u, w) is added in H-List.

This simulation is thus perfectly indistinguishable, since the additional field u in the H-List is never used. But note that $z = y^{\star c} \otimes f(u)$.

Game G_4: Now, we furthermore anticipate some $\mathcal{F}$ or $\mathcal{G}$ answers, with random numbers, which is the case of the above s and t:

▶**Rule $\mathcal{H}$-New**[(4)]

> Choose a random $u \in X$, then if $c = 0$, compute $z = y^\star \otimes f(u)$, otherwise compute $z = f(u)$, until the most significant bit of z is 0, but at most k_2 times (otherwise one aborts the game). Thereafter, z is parsed into $0\|w\|s\|t$, and one adds the record (w, t) to the F-List and $(w, s \oplus r)$ to the G-List. The record (m, r, c, u, w) is added in H-List.

This game may only differ from the previous one if during some $\mathcal{H}$-simulations, $\mathcal{F}(w)$ or $\mathcal{G}(w)$ have already been defined (either by a direct query, or by a $\mathcal{H}$-simulation.)

$$|\Pr[\mathsf{S}_4] - \Pr[\mathsf{S}_3]| \leq \frac{(q_h + q_s)(q_f + q_g + q_h + q_s)}{2^{k_2}}. \tag{11}$$

Game G_5: Now, we simply abort if the signing oracle makes a $\mathcal{H}(m, r)$-query for some (m, r) that has already been asked to $\mathcal{H}$.

▶**Rule $\mathcal{H}$-Old**[(5)]

> If $c = 0$, then one aborts the game, otherwise the answer is w.

Because of the possible abortion

$$|\Pr[\mathsf{S}_5] - \Pr[\mathsf{S}_4]| \leq q_s(q_h + q_s)/2^{k_1}. \tag{12}$$

Game G_6: In the last game, we replace the signing oracle by an easy simulation, returning the value u involved in the answer $\mathcal{H}(m, r)$, which defines $z = f(u)$:

▶**Rule $\mathcal{S}$**[(6)]

> Look up for (m, r, c, u, w) in H-List, and set $\sigma = u$.

The simulation is perfect since $c = 0$.

The event S_6 means that, at the end of that game, the adversary outputs a valid message/signature (m, σ). The latter satisfies: $y = f(\sigma) = b\|w\|s\|t$. Then one gets $r = s \oplus \mathcal{G}(w)$, and checks whether $b = 0$, $w = \mathcal{H}(m, r)$ and $t = f(w)$. Such a signature is valid

- without having queried $\mathcal{H}(m, r)$, which is possible with probability bounded by 2^{-k_2};
- with $y = y^\star \otimes f(u)$, where $(m, r, 1, u, w) \in$ H-List, and thus one gets $x^\star$.

$$\Pr[\mathsf{S}_6] \leq \mathsf{Succ}_f^{\mathsf{ow}}(t', k) + 2^{-k_2}, \tag{13}$$

where t' is the running time of the adversary, including the time for the simulations: $t' \leq t + (q_s + q_h) \cdot k_2 \cdot T_f$. $\qquad\square$

The important point in this security result is the very tight link between success probabilities, but also the almost linear time of the reduction. Thanks to this exact and efficient security result, RSA–PSS has become the new PKCS #1 v2.1 standard for signature [74]. Another variant has been proposed with message-recovery: PSS-R which allows one to include a large part of the message inside the signature. This makes a signed-message shorter than the size of the signature plus the size of the message, since the latter is inside the former one.

6. Public-Key Encryption

6.1. History

6.1.1. The RSA Encryption Scheme. In the same paper [73] as the RSA signature scheme appeared, Rivest, Shamir and Adleman also proposed a public-key encryption scheme, thanks to the "trapdoor one-way permutation" property of the RSA function: the generation algorithm produces a large composite number $N = pq$, a public key e, and a private key d such that $e \cdot d = 1 \mod \varphi(N)$. The encryption of a message m, encoded as an element in $\mathbb{Z}_N^\star$, is simply $c = m^e \mod N$. This ciphertext can be easily decrypted thanks to the knowledge of d, $m = c^d \mod N$. Clearly, this encryption is OW-CPA, relative to the RSA problem. The determinism makes a plaintext-checking oracle useless. Indeed, the encryption of a message m, under a public key pk is always the same, and thus it is easy to check whether a ciphertext c really encrypts m, by re-encrypting it. Therefore the RSA-encryption scheme is OW-PCA relative to the RSA problem as well.

Because of this determinism, it cannot be semantically secure: given the encryption c of either m_0 or m_1, the adversary simply computes $c' = m_0^e \mod N$ and checks whether $c' = c$. Furthermore, with a small exponent e (e.g. $e = 3$), any security vanishes under a multi-user attack: given $c_1 = m^3 \mod N_1$, $c_2 = m^3 \mod N_2$ and $c_3 = m^3 \mod N_3$, one can easily compute $m^3 \mod N_1 N_2 N_3$ thanks to the Chinese Remainders Theorem, which is exactly m^3 in $\mathbb{Z}$ and therefore leads to an easy recovery of m.

6.1.2. The El Gamal Encryption Scheme. In 1985, El Gamal [27] also designed a public-key encryption scheme based on the Diffie-Hellman key exchange protocol [25]: given a cyclic group $\mathcal{G}$ of order prime q and a generator $\mathbf{g}$, the generation algorithm produces a random element $x \in \mathbb{Z}_q^\star$ as private key, and a public key $\mathbf{y} = x \cdot \mathbf{g}$. The encryption of a message m, encoded as an element $\mathbf{m}$ in $\mathcal{G}$, is a pair $(\mathbf{c} = a \cdot \mathbf{g}, \mathbf{d} = a \cdot \mathbf{y} + \mathbf{m})$, for a random $a \in \mathbb{Z}_q$. This ciphertext can be easily decrypted thanks to the knowledge of x, since

$$a \cdot \mathbf{y} = ax \cdot \mathbf{g} = x \cdot \mathbf{c},$$

and thus $\mathbf{m} = \mathbf{d} - x \cdot \mathbf{c}$. This encryption scheme is well-known to be OW-CPA relative to the Computational Diffie-Hellman problem. It is also semantically secure (against chosen-plaintext attacks) relative to the Decisional Diffie-Hellman problem [85]. For OW-PCA, it relies on the Gap Diffie-Hellman problem [60].

As we have seen above, the expected security level is IND-CCA, whereas the RSA encryption just reaches OW-CPA under the RSA assumption, and the El Gamal encryption achieves IND-CPA under the **DDH** assumption. Can we achieve IND-CCA for practical encryption schemes?

6.2. A First Generic Construction

In [10], Bellare and Rogaway proposed the first generic construction which applies to any trapdoor one-way permutation f onto X. We need two hash functions $\mathcal{G}$ and $\mathcal{H}$:

$$\mathcal{G} : X \longrightarrow \{0,1\}^n \quad \text{and} \quad \mathcal{H} : \{0,1\}^* \longrightarrow \{0,1\}^{k_1},$$

where n is the bit-length of the plaintexts, and k_1 a security parameter. Then the encryption scheme $\mathsf{BR} = (\mathcal{K}, \mathcal{E}, \mathcal{D})$ can be described as follows:

- $\mathcal{K}(1^k)$: specifies an instance of the function f, and of its inverse f^{-1}. The public key pk is therefore f and the private key sk is f^{-1}.
- $\mathcal{E}_{\mathsf{pk}}(m; r)$: given a message $m \in \{0,1\}^n$, and a random value $r \overset{R}{\leftarrow} X$, the encryption algorithm $\mathcal{E}_{\mathsf{pk}}$ computes

$$a = f(r), \qquad b = m \oplus \mathcal{G}(r) \quad \text{and} \quad c = \mathcal{H}(m, r),$$

and outputs the ciphertext $y = a\|b\|c$.
- $\mathcal{D}_{\mathsf{sk}}(a\|b\|c)$: thanks to the private key, the decryption algorithm $\mathcal{D}_{\mathsf{sk}}$ extracts

$$r = f^{-1}(a), \quad \text{and next} \quad m = b \oplus \mathcal{G}(r).$$

If $c = \mathcal{H}(m, r)$, the algorithm returns m, otherwise it returns "Reject."

About this construction, one can prove:

Theorem 7. *Let $\mathcal{A}$ be a* CCA-*adversary against the semantic security of the above encryption scheme* BR. *Assume that $\mathcal{A}$ has advantage ε and running time τ and makes q_d, q_g and q_h queries to the decryption oracle, and the hash functions $\mathcal{G}$ and $\mathcal{H}$, respectively. Then*

$$\mathsf{Succ}_f^{\mathsf{ow}}(\tau') \geq \frac{\varepsilon}{2} - \frac{2q_d}{2^{k_1}} - \frac{q_h}{2^n},$$
$$\text{with} \quad \tau' \leq \tau + (q_g + q_h) \cdot T_f,$$

where T_f denotes the time complexity for evaluating f.

Proof. In the following we use starred letters (r^*, a^*, b^*, c^* and y^*) to refer to the challenge ciphertext, whereas unstarred letters (r, a, b, c and y) refer to the ciphertext asked to the decryption oracle.

Game $\mathbf{G}_0$:   A pair of keys $(\mathsf{pk}, \mathsf{sk})$ is generated using $\mathcal{K}(1^k)$. Adversary A_1 is fed with pk, the description of f, and outputs a pair of messages (m_0, m_1). Next a challenge ciphertext is produced by flipping a coin b and producing a ciphertext $y^* = a^*\|b^*\|c^*$ of m_b. This ciphertext comes from a random $r^* \overset{R}{\leftarrow} X$ and $a^* = f(r^*)$, $b^* = m_b \oplus \mathcal{G}(r^*)$ and $c^* = \mathcal{H}(m_b, r^*)$. On input y^*, A_2 outputs bit b'. In both stages, the adversary is given additional access to the decryption oracle $\mathcal{D}_{\mathsf{sk}}$. The only requirement is that the challenge ciphertext y^* cannot be queried from the decryption oracle.

We denote by S_0 the event $b' = b$ and use a similar notation S_i in any $\mathbf{G}_i$ below. By definition, we have

$$\Pr[S_0] = \frac{1}{2} + \frac{\varepsilon}{2}. \tag{14}$$

Game $\mathbf{G}_1$:   In this game, one makes the classical simulation of the random oracles, with random answers for any new query, as shown on Figure 9. This game is clearly identical to the previous one.

<table>
<tr>
<td rowspan="2" style="writing-mode: vertical-lr">$\mathcal{G}, \mathcal{H}$ Oracles</td>
<td>Query $\mathcal{G}(r)$: if a record (r, g) appears in G-List, the answer is g. Otherwise the answer g is chosen randomly: $g \in \{0, 1\}^n$ and the record (r, g) is added in G-List.</td>
</tr>
<tr>
<td>Query $\mathcal{H}(m, r)$: if a record (m, r, h) appears in H-List, the answer is h. Otherwise the answer h is chosen randomly: $h \in \{0, 1\}^{k_1}$ and the record (m, r, h) is added in H-List.</td>
</tr>
</table>

<table>
<tr>
<td style="writing-mode: vertical-lr">$\mathcal{D}$ Oracle</td>
<td>Query $\mathcal{D}_{\mathsf{sk}}(a\|b\|c)$: one applies the following rules:
▶**Rule Decrypt−R$^{(1)}$**
     | Compute $r = f^{-1}(a)$;
Then, compute $m = b \oplus \mathcal{G}(r)$, and finally,
▶**Rule Decrypt−H$^{(1)}$**
     | If $c = \mathcal{H}(m, r)$, one returns m, otherwise one returns "Reject."</td>
</tr>
</table>

<table>
<tr>
<td style="writing-mode: vertical-lr">Challenger</td>
<td>For two messages (m_0, m_1), flip a coin b and set $m^\star = m_b$.
▶**Rule Chal−Hash$^{(1)}$**
     Choose randomly $r^\star$, then set
     $a^\star = f(r^\star)$,
     $g^\star = \mathcal{G}(r^\star)$, $b^\star = m^\star \oplus g^\star$,
     $c^\star = \mathcal{H}(m^\star, r^\star)$.
Then, output $y^\star = a^\star \| b^\star \| c^\star$.</td>
</tr>
</table>

FIGURE 9. Formal Simulation of the IND-CCA Game against the BR Construction

Game $\mathbf{G}_2$:   In this game, one randomly chooses $h^+ \xleftarrow{R} \{0, 1\}^{k_1}$, and uses it instead of $\mathcal{H}(m^\star, r^\star)$.

▶**Rule Chal−Hash$^{(2)}$**
     | The value $h^+ \xleftarrow{R} \{0, 1\}^{k_1}$ has been chosen ahead of time, choose randomly $r^\star$, then set $a^\star = f(r^\star)$, $g^\star = \mathcal{G}(r^\star)$, $b^\star = m^\star \oplus g^\star$, and $c^\star = h^+$.

The two games $\mathbf{G}_2$ and $\mathbf{G}_1$ are perfectly indistinguishable unless $(m^\star, r^\star)$ is asked for $\mathcal{H}$, either by the adversary or the decryption oracle. But the latter case is not possible, otherwise the decryption query would be the challenge ciphertext. More generally, we denote by AskR_2 the event that $r^\star$ has been asked to $\mathcal{G}$ or to $\mathcal{H}$, by the adversary. We have:

$$|\Pr[\mathsf{S}_2] - \Pr[\mathsf{S}_1]| \leq \Pr[\mathsf{AskR}_2]. \tag{15}$$

Game $\mathbf{G}_3$:   We start modifying the simulation of the decryption oracle, by rejecting any ciphertext $(a\|b\|c)$ for which the corresponding (m, r) has not been queried to $\mathcal{H}$:

▶**Rule Decrypt−H**$^{(3)}$

> Look up in H-List for (m, r, c). If such a triple does not exist,
> then output "Reject", otherwise output m.

Such a simulation differs from the previous one if the value c has been correctly guessed, by chance:

$$|\Pr[\mathsf{S}_3] - \Pr[\mathsf{S}_2]| \leq \frac{q_d}{2^{k_1}} \qquad |\Pr[\mathsf{AskR}_3] - \Pr[\mathsf{AskR}_2]| \leq \frac{q_d}{2^{k_1}}. \tag{16}$$

Game $\mathbf{G}_4$:   In this game, one randomly chooses $r^+ \xleftarrow{R} X$ and $g^+ \xleftarrow{R} \{0,1\}^n$, and uses r^+ instead of $r^\star$, as well as g^+ instead of $\mathcal{G}(r^\star)$.

▶**Rule Chal−Hash**$^{(4)}$

> The three values $r^+ \xleftarrow{R} X$, $g^+ \xleftarrow{R} \{0,1\}^n$ and $h^+ \xleftarrow{R} \{0,1\}^{k_1}$ have been chosen ahead of time, then set $a^\star = f(r^+)$, $b^\star = m^\star \oplus g^+$, $c^\star = h^+$.

The two games $\mathbf{G}_4$ and $\mathbf{G}_3$ are perfectly indistinguishable unless $r^\star$ is asked for $\mathcal{G}$, either by the adversary or the decryption oracle. The former case has already been cancelled in the previous game, in AskR_3. The latter case does not make any difference since either $\mathcal{H}(m, r^\star)$ has been queried by the adversary, which falls in AskR_3, or the ciphertext is rejected in both games. We have:

$$\Pr[\mathsf{S}_4] = \Pr[\mathsf{S}_3] \qquad \Pr[\mathsf{AskR}_4] = \Pr[\mathsf{AskR}_3]. \tag{17}$$

In this game, $m^\star$ is masked by g^+, a random value which never appears anywhere else. Thus, the input to $\mathcal{A}_2$ follows a distribution that does not depend on b. Accordingly:

$$\Pr[\mathsf{S}_4] = \frac{1}{2}. \tag{18}$$

Game $\mathbf{G}_5$:   Finally, one randomly chooses $a^+ \xleftarrow{R} X$, which implicitly defines a random r^+ in X. Actually, a^+ is the given random challenge for which one is looking for the pre-image r^+.

▶**Rule Chal−Hash**$^{(5)}$

> The three values $a^+ \xleftarrow{R} X$, $g^+ \xleftarrow{R} \{0,1\}^n$ and $h^+ \xleftarrow{R} \{0,1\}^{k_1}$ have been chosen/given ahead of time, then set $a^\star = a^+$, $b^\star = m^\star \oplus g^+$, $c^\star = h^+$.

The two games $\mathbf{G}_5$ and $\mathbf{G}_4$ are perfectly indistinguishable, thanks to the permutation property of f.

Game $\mathbf{G}_6$:   In the simulation of the decryption oracle, we may reject even earlier, if the corresponding r has not been queried to $\mathcal{G}$:

▶**Rule Decrypt$-\mathbf{R}^{(6)}$**

Look up in G-List for (r, g) such that $a = f(r)$. If no r is found, then output "Reject".

Such a simulation differs from the previous one if the value (m, r) has been queried to $\mathcal{H}$, while $\mathcal{G}(r)$ is unpredictable, and thus $m = \mathcal{G}(r) \oplus b$ is unpredictable too:

$$| \Pr[\mathsf{AskR}_6] - \Pr[\mathsf{AskR}_5] | \leq \frac{q_h}{2^n}. \tag{19}$$

One may now note that the event AskR_6 leads to the pre-image of a^+ by f in the queries asked to $\mathcal{G}$ and $\mathcal{H}$, by the adversary. By checking all of them, one gets it:

$$\Pr[\mathsf{AskR}_6] \leq \mathsf{Succ}_f^{\mathsf{ow}}(\tau + (q_g + q_h)T_f). \tag{20}$$

$\square$

6.3. OAEP: the Optimal Asymmetric Encryption Padding.

6.3.1. Description. The problem with the above generic construction is the high over-head. When one encrypts with a trapdoor one-way permutation onto X, one could hope the ciphertext to be an element in X, without anything else. In 1994, Bellare and Rogaway proposed such a more compact generic conversion [11], in the random-oracle model, the "Optimal Asymmetric Encryption Padding" (OAEP, see Figure 10), obtained from a trapdoor one-way permutation f onto $\{0, 1\}^k$, whose

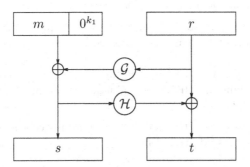

FIGURE 10. Optimal Asymmetric Encryption Padding

inverse is denoted by f^{-1}. We need two hash functions $\mathcal{G}$ and $\mathcal{H}$:

$$\mathcal{G} : \{0, 1\}^{k_0} \longrightarrow \{0, 1\}^{k - k_0} \quad \text{and} \quad \mathcal{H} : \{0, 1\}^{k - k_0} \longrightarrow \{0, 1\}^{k_0},$$

for some k_0. We also need n and k_1 which satisfy $k = n + k_0 + k_1$. Then the encryption scheme OAEP $= (\mathcal{K}, \mathcal{E}, \mathcal{D})$ can be described as follows:

- $\mathcal{K}(1^k)$: specifies an instance of the function f, and of its inverse f^{-1}. The public key pk is therefore f and the private key sk is f^{-1}.
- $\mathcal{E}_{\sf pk}(m;r)$: given a message $m \in \{0,1\}^n$, and a random value $r \stackrel{R}{\leftarrow} \{0,1\}^{k_0}$, the encryption algorithm $\mathcal{E}_{\sf pk}$ computes

$$s = (m\|0^{k_1}) \oplus \mathcal{G}(r) \quad \text{and} \quad t = r \oplus \mathcal{H}(s),$$

and outputs the ciphertext $c = f(s,t)$.
- $\mathcal{D}_{\sf sk}(c)$: thanks to the private key, the decryption algorithm $\mathcal{D}_{\sf sk}$ extracts

$$(s,t) = f^{-1}(c), \quad \text{and next} \quad r = t \oplus \mathcal{H}(s) \quad \text{and} \quad M = s \oplus \mathcal{G}(r).$$

If $[M]_{k_1} = 0^{k_1}$, the algorithm returns $[M]^n$, otherwise it returns "Reject."

In the above description, $[M]_{k_1}$ denotes the k_1 least significant bits of M, while $[M]^n$ denotes the n most significant bits of M.

6.3.2. About the Security. Paper [11] includes a proof that, provided f is a one-way trapdoor permutation, the resulting OAEP encryption scheme is both semantically secure and weakly plaintext-aware. This implies the semantic security against indifferent chosen-ciphertext attacks, also called security against lunchtime attacks (IND-CCA1). Indeed, the *Weak Plaintext-Awareness* means that the adversary cannot produce a new valid ciphertext, until it has seen any valid one, without knowing (awareness) the plaintext. This is more formally defined by the existence of a plaintext-extractor which, on input a ciphertext and the list of the query-answers of the random oracles, outputs the corresponding plaintext. This plaintext-extractor is thus enough for simulating the decryption oracle, but in the first step of the attack only. We briefly comment on the intuition behind (weak) plaintext-awareness. When the plaintext-extractor receives a ciphertext c, then:

- either s has been queried to $\mathcal{H}$ and r has been queried to $\mathcal{G}$, in which case the extractor finds the cleartext by inspecting the two query lists G-List and H-List,
- or else the decryption of (s,t) remains highly random and there is little chance to meet the redundancy 0^{k_1}: the plaintext extractor can safely declare the ciphertext invalid.

The argument collapses when the plaintext-extractor receives additional valid ciphertexts, since this puts additional implicit constraints on $\mathcal{G}$ and $\mathcal{H}$. These constraints cannot be seen by inspecting the query lists. Hence the requirement of a stronger notion of *plaintext-awareness*. In [7], Bellare, Desai, Rogaway and the author defined such a stronger notion which extends the previous *awareness* of the plaintext even after having seen valid ciphertexts. But such a plaintext-awareness notion had never been studied for OAEP, while it was still widely admitted.

Shoup's Counter-Example. In his papers [82, 83], Shoup showed that it was quite unlikely to extend the results of [11] to obtain adaptive chosen-ciphertext security, under the sole one-wayness of the permutation. His counter-example made use of the ad hoc notion of an *XOR-malleable* trapdoor one-way permutation: for such permutation f_0, one can compute $f_0(x \oplus a)$ from $f_0(x)$ and a, with non-negligible probability.

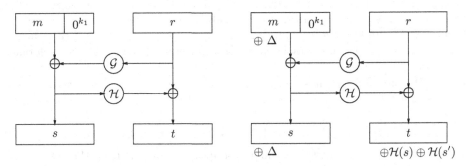

FIGURE 11. Shoup's attack.

Let f_0 be such an XOR-malleable permutation. Define f by $f(s\|t) = s\|f_0(t)$. Clearly, f is also a trapdoor one-way permutation. However, it leads to a malleable encryption scheme as we now show. Start with a challenge ciphertext $y = f(s\|t) = s\|u$, where $s\|t$ is the output of the OAEP transformation on the redundant message $m\|0^{k_1}$ and the random string r (see Figure 11),

$$s = \mathcal{G}(r) \oplus (m\|0^{k_1}), \qquad t = \mathcal{H}(s) \oplus r \quad \text{and} \quad u = f_0(t).$$

Since f is the identity on its leftmost part, we know s, and can define $\Delta = \delta\|0^{k_1}$, for any random string δ, and $s' = s \oplus \Delta$. We then set $t' = \mathcal{H}(s') \oplus r = t \oplus (\mathcal{H}(s) \oplus \mathcal{H}(s'))$. The XOR-malleability of f_0 allows one to obtain $u' = f_0(t')$ from $u = f_0(t)$ and $\mathcal{H}(s) \oplus \mathcal{H}(s')$, with significant probability. Finally, $y' = s'\|u'$ is a valid ciphertext of $m' = m \oplus \delta$, built from $r' = r$, since:

$$t' = f_0^{-1}(u') = t \oplus (\mathcal{H}(s) \oplus \mathcal{H}(s')) = \mathcal{H}(s') \oplus r, \qquad r' = \mathcal{H}(s') \oplus t' = r$$

and

$$s' \oplus \mathcal{G}(r') = \Delta \oplus s \oplus \mathcal{G}(r) = \Delta \oplus (m\|0^{k_1}) = (m \oplus \delta)\|0^{k_1}.$$

Note that the above definitely contradicts adaptive chosen-ciphertext security: asking the decryption of y' after having received the ciphertext y, an adversary obtains m' and easily recovers the actual cleartext m from m' and δ. Also note that Shoup's counter-example exactly stems from where the intuition developed at the end of the previous section failed: a valid ciphertext y' was created without querying the oracle at the corresponding random seed r', using in place the implicit constraint on $\mathcal{G}$ coming from the received valid ciphertext y.

Using methods from relativized complexity theory, Shoup [82, 83] built a non-standard model of computation, where there exists an XOR-malleable trapdoor one-way permutation. As a consequence, it is very unlikely that one can prove the IND-CCA security of the OAEP construction, under the sole one-wayness of the underlying permutation. Indeed, all methods of proof currently known still apply in relativized models of computation.

6.3.3. The Actual Security of OAEP. Shoup [82, 83] furthermore provided a specific proof for RSA with public exponent 3. However, there is little hope of extending this proof for higher exponents. Hopefully, Fujisaki, Okamoto, Stern and the author provided a general security analysis, but under a stronger assumption about the underlying permutation [32, 33]. Indeed, we prove that the scheme is IND-CCA in the random-oracle model [10], relative to the *partial-domain* one-wayness of permutation f.

Partial-Domain One-Wayness. Let us first introduce this new computational assumption. Let f be a permutation $f : \{0,1\}^k \longrightarrow \{0,1\}^k$, which can also be written as

$$f : \{0,1\}^{n+k_1} \times \{0,1\}^{k_0} \longrightarrow \{0,1\}^{n+k_1} \times \{0,1\}^{k_0},$$

with $k = n + k_0 + k_1$. In the original description of OAEP from [11], it is only required that f is a trapdoor one-way permutation. However, in the following, we consider two additional related problems, namely partial-domain one-wayness and set partial-domain one-wayness:

- Permutation f is (τ, ε)-one-way if any adversary $\mathcal{A}$ whose running time is bounded by τ has success probability $\mathsf{Succ}_f^{\mathsf{ow}}(\mathcal{A})$ upper-bounded by ε, where

$$\mathsf{Succ}_f^{\mathsf{ow}}(\mathcal{A}) = \Pr_{s,t}[\mathcal{A}(f(s,t)) = (s,t)].$$

- Permutation f is (τ, ε)-partial-domain one-way if any adversary $\mathcal{A}$ whose running time is bounded by τ has success probability $\mathsf{Succ}_f^{\mathsf{pd\text{-}ow}}(\mathcal{A})$ upper-bounded by ε, where

$$\mathsf{Succ}_f^{\mathsf{pd\text{-}ow}}(\mathcal{A}) = \Pr_{s,t}[\mathcal{A}(f(s,t)) = s].$$

- Permutation f is $(\ell, \tau, \varepsilon)$-set partial-domain one-way if any adversary $\mathcal{A}$, outputting a set of ℓ elements within time bound τ, has success probability $\mathsf{Succ}_f^{\mathsf{s\text{-}pd\text{-}ow}}(\mathcal{A})$ upper-bounded by ε, where

$$\mathsf{Succ}_f^{\mathsf{s\text{-}pd\text{-}ow}}(\mathcal{A}) = \Pr_{s,t}[s \in \mathcal{A}(f(s,t))].$$

We denote by $\mathsf{Succ}_f^{\mathsf{ow}}(\tau)$ (resp. $\mathsf{Succ}_f^{\mathsf{pd\text{-}ow}}(\tau)$ and $\mathsf{Succ}_f^{\mathsf{s\text{-}pd\text{-}ow}}(\ell, \tau)$) the maximal success probability $\mathsf{Succ}_f^{\mathsf{ow}}(\mathcal{A})$ (resp. $\mathsf{Succ}_f^{\mathsf{pd\text{-}ow}}(\mathcal{A})$ and $\mathsf{Succ}_f^{\mathsf{s\text{-}pd\text{-}ow}}(\mathcal{A})$). The maximum ranges over all adversaries whose running time is bounded by τ. In the third case, there is an obvious additional restriction on this range from the fact that $\mathcal{A}$ outputs sets with ℓ elements. It is clear that for any τ and $\ell \geq 1$,

$$\mathsf{Succ}_f^{\mathsf{s\text{-}pd\text{-}ow}}(\ell, \tau) \geq \mathsf{Succ}_f^{\mathsf{pd\text{-}ow}}(\tau) \geq \mathsf{Succ}_f^{\mathsf{ow}}(\tau).$$

Note that, by randomly selecting an element in the set returned by an adversary to the set partial-domain one-wayness, one breaks partial-domain one-wayness with probability $\mathsf{Succ}_f^{\mathsf{s\text{-}pd\text{-}ow}}(\mathcal{A})/\ell$. This provides the following inequality

$$\mathsf{Succ}_f^{\mathsf{pd\text{-}ow}}(\tau) \geq \mathsf{Succ}_f^{\mathsf{s\text{-}pd\text{-}ow}}(\ell, \tau)/\ell.$$

However, for specific choices of f, more efficient reductions may exist. Also, in some cases, all three problems are polynomially equivalent. This is the case for the RSA permutation [73], hence the global security result for RSA-OAEP.

6.3.4. The Proof of Security. In the following we use starred letters ($r^\star$, $s^\star$, $t^\star$ and $y^\star$) to refer to the challenge ciphertext, whereas unstarred letters (r, s, t and y) refer to the ciphertext asked to the decryption oracle.

The Intuition. Referring to our description of the intuition behind the original OAEP proof of security, given above, we can carry a more subtle analysis by distinguishing the case where s has not been queried from oracle $\mathcal{H}$ from the case where r has not been queried from $\mathcal{G}$. If s is not queried, then $\mathcal{H}(s)$ is random and uniformly distributed and r is necessarily defined as $t \oplus \mathcal{H}(s)$. This holds even if s matches with the string $s^\star$ coming from the valid ciphertext $y^\star$. There is a minute probability that $t \oplus \mathcal{H}(s)$ is queried from $\mathcal{G}$ or equals $r^\star$. Thus, $\mathcal{G}(r)$ is random: there is little chance that the redundancy 0^{k_1} is met and the extractor can safely reject.

We claim that r cannot match with $r^\star$, unless $s^\star$ is queried from $\mathcal{H}$. This is because $r^\star = t^\star \oplus \mathcal{H}(s^\star)$ equals $r = t \oplus \mathcal{H}(s)$ with minute probability. Thus, if r is not queried, then $\mathcal{G}(r)$ is random and we similarly infer that the extractor can safely reject. The argument fails only if $s^\star$ is queried.

Thus rejecting when it cannot combine elements of the lists G-List and H-List so as to build a pre-image of y, the plaintext-extractor is only wrong with minute probability, unless $s^\star$ has been queried by the adversary. This seems to show that OAEP leads to an IND-CCA encryption scheme if it is difficult to invert f "partially", which means: given $y^\star = f(s^\star \| t^\star)$, find $s^\star$.

The Strategy. Based on the intuition just described, we can formally prove that applying OAEP encoding to a trapdoor permutation which is difficult to partially invert, leads to an IND-CCA encryption scheme, hence the *partial-domain one-wayness*, which expresses the fact that the above partial inversion problem is difficult.

Chosen-ciphertext security is actually addressed, by turning the intuition explained above into a formal argument, involving a restricted variant of plaintext-awareness (where the list C of ciphertexts is limited to only one ciphertext, the challenge ciphertext $y^\star$).

Theorem 8. *Let $\mathcal{A}$ be a* CCA*-adversary against the semantic security of the encryption scheme* OAEP. *Assume that $\mathcal{A}$ has advantage ε and running time τ and makes q_d, q_g and q_h queries to the decryption oracle, and the hash functions $\mathcal{G}$ and $\mathcal{H}$, respectively. Then*

$$\mathsf{Succ}_f^{\text{s-pd-ow}}(q_h, \tau') \ \geq \ \frac{\varepsilon}{2} - \left(\frac{2(q_d + 2)(q_d + 2q_g)}{2^{k_0}} + \frac{3q_d}{2^{k_1}} \right),$$

$$\text{with} \quad \tau' \ \leq \ \tau + q_g \cdot q_h \cdot (T_f + \mathcal{O}(1)),$$

where T_f denotes the time complexity for evaluating f.

6.3.5. The Plaintext-Extractor.

Description. In order to prove the security against adaptive chosen-ciphertext attacks, it is necessary to simulate calls to a decryption oracle. As in the original paper [11], we design a plaintext-extractor (which is actually the same). But the analysis is more intricate because the success probability of the extractor cannot be estimated unconditionally but only relatively to some computational assumption. When the plaintext-extractor receives a ciphertext c, then:

- either s has been queried to $\mathcal{H}$ and r has been queried to $\mathcal{G}$, in which case the extractor finds the cleartext by inspecting the two query lists G-List and H-List. One indeed looks up for $(\gamma, \mathcal{G}_\gamma) \in$ G-List and $(\delta, \mathcal{H}_\delta) \in$ H-List. For such a pair, one defines $\sigma = \delta$, $\theta = \gamma \oplus \mathcal{H}_\delta$, $\mu = \mathcal{G}_\gamma \oplus \delta$, and checks whether $c = f(\sigma, \theta)$. If $[\mu]_{k_1} = 0^{k_1}$, then the tailing part is the plaintext.
- or else the decryption of (s, t) remains highly random and there is little chance to meet the redundancy 0^{k_1}: the plaintext extractor can safely declare the ciphertext invalid.

Comments. One can easily check that the output of the plaintext-extractor is uniquely defined, regardless of the ordering of the lists. To see this, observe that since f is a permutation, the value of $\sigma = s$ is uniquely defined and so is δ. Keep in mind that the G-List and H-List correspond to input-output pairs for the functions $\mathcal{G}$ and $\mathcal{H}$, and at most one output is related to a given input. This makes $\mathcal{H}_\delta$ uniquely defined as well. Similarly, $\theta = t$ is uniquely defined, and thus γ and $\mathcal{G}_\gamma$: at most one μ may be selected, which is output depending on whether $[\mu]_{k_1} = 0^{k_1}$ or not.

Furthermore, if both r and s have been queried by the adversary, the plaintext-extractor perfectly simulates the decryption oracle.

6.3.6. Proof.

In the following, $y^\star$ is the challenge ciphertext, obtained from the encryption oracle. Since we have in mind using the plaintext-extractor instead of the decryption oracle, trying to contradict semantic security, we assume that $y^\star$ is a ciphertext of m_b and denote by $r^\star$ its random seed. We have

$$r^\star = \mathcal{H}(s^\star) \oplus t^\star \quad \text{and} \quad \mathcal{G}(r^\star) = s^\star \oplus (m_b \| 0^{k_1}).$$

In what follows, all unstarred variables refer to the decryption queries.

We now present a proof with games which sequentially discard all cases for which the above plaintext-extractor may fail.

Game G_0: A pair of keys (pk, sk) is generated using $\mathcal{K}(1^k)$. Adversary A_1 is fed with pk, the description of f, and outputs a pair of messages (m_0, m_1). Next a challenge ciphertext is produced by flipping a coin b and producing a ciphertext $y^\star$ of m_b. This ciphertext comes from a random $r^\star \xleftarrow{R} \{0, 1\}^{k_0}$ and $s^\star$ and $t^\star$ such that $y^\star = f(s^\star, t^\star)$, where $s^\star = (m_b \| 0^{k_1}) \oplus \mathcal{G}(r^\star)$ and $t^\star = r^\star \oplus \mathcal{H}(s^\star)$. On input $y^\star$, A_2 outputs bit b'. In both stages, the adversary is given additional access to the decryption oracle $\mathcal{D}_{sk}$. The only requirement is that the challenge ciphertext cannot be queried from the decryption oracle.

We denote by S_0 the event $b' = b$ and use a similar notation S_i in any $\mathbf{G}_i$ below. By definition, we have

$$\Pr[S_0] = \frac{1}{2} + \frac{\varepsilon}{2}. \tag{21}$$

Game $\mathbf{G}_1$:   In this game, one makes the classical simulation of the random oracles, with random answers for any new query, as shown on Figure 12. This game is clearly identical to the previous one.

Game $\mathbf{G}_2$:   In this game, one randomly chooses $r^+ \overset{R}{\leftarrow} \{0,1\}^{k_0}$ and $g^+ \overset{R}{\leftarrow} \{0,1\}^{k-k_0}$, and uses r^+ instead of $r^\star$, as well as g^+ instead of $\mathcal{G}(r^\star)$.

▶**Rule Chal−Hash**[(2)]

> The two values $r^+ \overset{R}{\leftarrow} \{0,1\}^{k_0}$, $g^+ \overset{R}{\leftarrow} \{0,1\}^{k-k_0}$ have been chosen ahead of time, then set $r^\star = r^+$, $g^\star = g^+$, $s^\star = M^\star \oplus g^+$, $h^\star = \mathcal{H}(s^\star)$, $t^\star = r^+ \oplus h^\star$.

The two games $\mathbf{G}_2$ and $\mathbf{G}_1$ are perfectly indistinguishable unless $r^\star$ is asked for $\mathcal{G}$, either by the adversary or by the decryption oracle. We define this event AskG_2. We have:

$$| \Pr[S_2] - \Pr[S_1] | \leq \Pr[\mathsf{AskG}_2]. \tag{22}$$

In this game, g^+ is used in (s,t) but does not appear in the computation since $\mathcal{G}(r^+)$ is not defined to be equal to g^+. Thus, the input to $\mathcal{A}_2$ follows a distribution that does not depend on b. Accordingly:

$$\Pr[S_2] = \frac{1}{2}. \tag{23}$$

Game $\mathbf{G}_3$:   We start dealing with the decryption oracle, which has remained perfect up to this game, but using the ability to invert f. We first make the decryption oracle reject all ciphertexts c such that the corresponding r value has not been previously queried from $\mathcal{G}$ by the adversary.

▶**Rule Decrypt−SnoR**[(3)]

> $g = \mathcal{G}(r)$, $M = 1^k$.

This new rule leads to a **Reject** since the 0^{k_1} is not verified. This makes a difference only if c is a valid ciphertext, while $\mathcal{G}(r)$ has not been asked. Since $\mathcal{G}(r)$ is uniformly distributed, equality $[s \oplus \mathcal{G}(r)]_{k_1} = 0^{k_1}$ happens with probability $1/2^{k_1}$. Summing up for all decryption queries, we get

$$|\Pr[\mathsf{AskG}_3] - \Pr[\mathsf{AskG}_2]| \leq \frac{q_d}{2^{k_1}}. \tag{24}$$

Note that we cannot remove the query $\mathcal{G}(r)$ from this rule, even if it would not change anything in the simulation of the output of this decryption. However, it would remove a pair (r,g) from G-List, which could be $r^\star$ itself, and this would have a non-negligible impact on the event AskG_3.

$\mathcal{G}, \mathcal{H}$ Oracles	Query $\mathcal{G}(r)$: if a record (r, g) appears in G-List, the answer is g. Otherwise the answer g is chosen randomly: $g \in \{0,1\}^{k-k_0}$ and the record (r, g) is added in G-List.
	Query $\mathcal{H}(s)$: if a record (s, h) appears in H-List, the answer is h. Otherwise the answer h is chosen randomly: $h \in \{0,1\}^{k_0}$ and the record (s, h) is added in H-List.

$\mathcal{D}$ Oracle	Query $\mathcal{D}_{\mathsf{sk}}(c)$:the value M is defined according to the following rules:
	▶**Rule Decrypt−Init**[(1)]
	$\quad\mid\quad$ Compute $(s, t) = f^{-1}(c)$;
	Look up for $(s, h) \in$ H-List:
	$\quad$ • if the record is found, compute $r = t \oplus h$.
	$\qquad$ Look up for $(r, g) \in$ G-List:
	$\qquad$ − if the record is found
	$\qquad\quad$ ▶**Rule Decrypt−SR**[(1)]
	$\qquad\qquad\mid\quad h = \mathcal{H}(s), \quad r = t \oplus h,$
	$\qquad\qquad\quad\ g = \mathcal{G}(r), \quad M = s \oplus g.$
	$\qquad$ − otherwise
	$\qquad\quad$ ▶**Rule Decrypt−SnoR**[(1)]
	$\qquad\qquad\mid\quad$ same as rule **Decrypt−SR**[(1)].
	$\quad$ • otherwise
	$\qquad$ ▶**Rule Decrypt−noS**[(1)]
	$\qquad\quad\mid\quad$ same as rule **Decrypt−SR**[(1)].
	If $[M]_{k_1} = 0^{k_1}$, one returns $m = [M]^n$, otherwise one returns "Reject."

Challenger	For two messages (m_0, m_1), flip a coin b and set $m^\star = m_b$, $M^\star = m^\star \| 0^{k_1}$.
	▶**Rule Chal−Hash**[(1)]
	$\quad\mid\quad$ Choose randomly $r^\star$, then set
	$\qquad g^\star = \mathcal{G}(r^\star), \quad s^\star = M^\star \oplus g^\star,$
	$\qquad h^\star = \mathcal{H}(s^\star), \quad t^\star = r^\star \oplus h^\star.$
	▶**Rule Chal−Output**[(1)]
	$\quad\mid\quad$ Compute and output $y^\star = f(s^\star, t^\star)$.

FIGURE 12. Formal Simulation of the IND-CCA Game against OAEP

Game G_4:   We now make the decryption oracle reject all ciphertexts c such that the corresponding s value has not been previously queried from $\mathcal{H}$ by the adversary.

▶**Rule Decrypt−noS$^{(4)}$**
$$\left|\begin{array}{ll} h = \mathcal{H}(s), & r = t \oplus h, \\ g = \mathcal{G}(r), & M = 1^k. \end{array}\right.$$

This makes a difference only if y is a valid ciphertext, while $\mathcal{H}(s)$ has not been asked. First, since $r = \mathcal{H}(s) \oplus t$ is uniformly distributed, it has been queried from $\mathcal{G}$ with probability less than $(q_g + q_d)/2^{k_0}$. Then, if $\mathcal{G}(r)$ has not been queried, the redundancy is satisfied with probability less than $1/2^{k_1}$. Summing up for all decryption queries, we get

$$|\Pr[\mathsf{AskG_4}] - \Pr[\mathsf{AskG_3}]| \le \frac{q_d(q_g + q_d)}{2^{k_0}} + \frac{q_d}{2^{k_1}}. \tag{25}$$

Game G_5:   Here, we can make the first formal modification in the previous game since, whatever the h-value is, the message M is 1^k, and g and h are never revealed:

▶**Rule Decrypt−noS$^{(5)}$**
$$\left|\begin{array}{ll} h = \mathcal{H}(s), & M = 1^k. \end{array}\right.$$

This will just postpone the definition of $\mathcal{G}(r)$ and also remove one pair (r, g) from G-List. The latter removal may have some impact:

- on the simulation of a later decryption c', if $r' = r$ was found in the previous game, but that is no longer in the list. A rule **Decrypt−SR** is thus replaced by the rule **Decrypt−SnoR**, which means that $g' = g$ was just defined in the modified rule, and never revealed (by any means: no information is leaked.) Therefore, the probability for M' to satisfy the redundancy was 2^{-k_1};
- the removed r could be r^*, but this is $t \oplus \mathcal{H}(s)$, for $s \notin$ H-List. Such a case is bounded by 2^{-k_0}.

Summing up for all decryption queries, we get

$$|\Pr[\mathsf{AskG_5}] - \Pr[\mathsf{AskG_4}]| \le q_d \times \left(\frac{1}{2^{k_0}} + \frac{1}{2^{k_1}}\right). \tag{26}$$

Game G_6:   We follow in making formal modifications:

▶**Rule Decrypt−noS$^{(6)}$**
$$\left|\begin{array}{l} M = 1^k. \end{array}\right.$$

This will postpone the definition of $\mathcal{H}(s)$, and also remove the pair (s, h) from H-List. The latter removal may have some impact on the simulation of a later decryption c': if $s' = s$ was found in the previous game, but that is no longer in the list:

- a rule **Decrypt−SnoR** is replaced by the rule **Decrypt−noS** (which just cancels r' from G-List), which means that $h' = h$ was just defined in the modified rule, and never revealed. The probability for r' to be equal to r^* is 2^{-k_0}.
- a rule **Decrypt−SR** is replaced by the rule **Decrypt−noS**, which means that $h' = h$ was just defined in the modified rule, and never revealed. The probability for $r' = t' \oplus h'$ to be in G-List was less than $q_g/2^{k_0}$, which is an upper-bound of this case to appear.

In both cases, the decryption is anyway still the same. Summing up for all decryption queries, we get

$$|\Pr[\mathsf{AskG_6}] - \Pr[\mathsf{AskG_5}]| \leq \frac{q_d(q_g + 1)}{2^{k_0}}. \tag{27}$$

Furthermore, in the decryption simulation, when both r and s have been asked, no new query occurs:

▶**Rule Decrypt$-$SR$^{(6)}$**
| $M = s \oplus g$.

As a consequence, the new decryption simulation makes no new $\mathcal{H}$-query.

Game G_7: We now define $s^\star$ independently of anything else, as well as $\mathcal{H}(s^\star)$, by randomly choosing $s^+ \overset{R}{\leftarrow} \{0,1\}^{k-k_0}$ and $h^+ \overset{R}{\leftarrow} \{0,1\}^{k_0}$, and using s^+ instead of $s^\star$, as well as h^+ instead of $\mathcal{H}(s^\star)$. The only change is that $s^\star = s^+$ instead of $M^\star \oplus g^+$, which in some sense defines $g^+ = M^\star \oplus s^+$ but we do not need it. The game obeys the following rule:

▶**Rule Chal$-$Hash$^{(7)}$**
| The three values $r^+ \overset{R}{\leftarrow} \{0,1\}^{k_0}$, $s^+ \overset{R}{\leftarrow} \{0,1\}^{k-k_0}$ and $h^+ \overset{R}{\leftarrow}$
| $\{0,1\}^{k_0}$ have been chosen ahead of time, then set $s^\star =$
| s^+, $\quad t^\star = r^+ \oplus h^+$.

The two games G_7 and G_6 are perfectly indistinguishable unless $s^\star$ is asked for $\mathcal{H}$ by the adversary, or used by the decryption oracle. The former event is denoted $\mathsf{AskH_7}$, while the latter makes a difference only if the rule **Decrypt$-$SR$^{(6)}$** was used, with an accepted ciphertext, or the rule **Decrypt$-$SnoR$^{(6)}$** was used, with $r = r^\star$ (because this rule becomes **Decrypt$-$noS$^{(6)}$**, where no $\mathcal{G}(r)$ query is done, since it could have been $r^\star$, and thus made the event AskG happen.)

We thus insist here on that the event $\mathsf{AskH_7}$ denotes the fact that $s^\star$ is asked for $\mathcal{H}$ by the adversary, whereas the event AskG denotes the fact that $r^\star$ is asks for $\mathcal{G}$ by the adversary or the decryption oracle/simulation.

Let us briefly deal with the bad cases:

- the rule **Decrypt$-$SR$^{(6)}$** was used, with an accepted ciphertext. This means that there exists a valid ciphertext $c = f(s^\star \| t)$ that is queried to the decryption oracle, with the corresponding r queried to $\mathcal{G}$, where $r = t \oplus \mathcal{H}(s^\star) = t \oplus t^\star \oplus r^+$, and r^+ is a random value.
- the rule **Decrypt$-$SnoR$^{(6)}$** was used, with $r = r^+$, where r^+ is a random value.

$$|\Pr[\mathsf{AskG_7}] - \Pr[\mathsf{AskG_6}]| \leq \Pr[\mathsf{AskH_7}] + \frac{q_d(q_g + q_d)}{2^{k_0}} + \frac{q_d}{2^{k_0}}. \tag{28}$$

In this new game, $r^\star = t^\star \oplus h^+$ is uniformly distributed, and independent of the adversary's view, since h^+ is never revealed:

$$\Pr[\mathsf{AskG_7}] \leq \frac{q_g + q_d}{2^{k_0}}, \tag{29}$$

where q_g and q_d denote the number of queries asked by the adversary to $\mathcal{G}$, or to the decryption oracle, respectively. As a consequence,

$$\Pr[\mathsf{AskG_2}] \le \frac{3q_d}{2^{k_1}} + \frac{(2q_d + 1)(q_g + q_d)}{2^{k_0}} + \frac{q_d(q_g + 3)}{2^{k_0}} + \Pr[\mathsf{AskH_7}] \qquad (30)$$

Game G_8:   Finally, we define $s^\star$ and $t^\star$ independently of anything else, by randomly choosing $s^+ \overset{R}{\leftarrow} \{0,1\}^{k-k_0}$ and $t^+ \overset{R}{\leftarrow} \{0,1\}^{k_0}$:

▶**Rule Chal–Hash$^{(8)}$**

$\qquad\Big|\quad$ The two values $s^+ \overset{R}{\leftarrow} \{0,1\}^{k-k_0}$ and $t^+ \overset{R}{\leftarrow} \{0,1\}^{k_0}$ have
$\qquad\Big|\quad$ been chosen ahead of time, then set $s^\star = s^+, \quad t^\star = t^+$.

The two games G_8 and G_7 are perfectly indistinguishable.

Game G_9:   We now completely manufacture the challenge ciphertext: we randomly choose $y^+ \overset{R}{\leftarrow} \{0,1\}^k$, and simply set $y^\star = y^+$, ignoring the encryption algorithm altogether. This implicitly defines s^+ and t^+, because of the permutation property of f. Actually, y^+ is the given random challenge for which one is looking for the partial pre-image s^+.

▶**Rule Chal–Hash$^{(9)}$**

$\qquad\Big|\quad$ Do nothing.

▶**Rule Chal–Output$^{(9)}$**

$\qquad\Big|\quad$ The challenge $y^+ \overset{R}{\leftarrow} \{0,1\}^k$ has been given ahead of time,
$\qquad\Big|\quad$ then set and output $y^\star = y^+$.

The distribution of $y^\star$ remains the same: due to the fact that f is a permutation, the previous method defining $y^\star = f(s^\star\|t^\star)$, with $s^\star = s^+$ and $t^\star = t^+$ was already generating a uniform distribution over the k-bit elements.

Game G_{10}:   Before concluding, one may remark that the new simulation of the decryption oracle is exactly the way the plaintext-extractor previously explained would operate, with some extra but unuseful $\mathcal{G}$-queries. Since we do not care anymore about the event $\mathsf{AskG_{10}}$, they can be simplified:

▶**Rule Decrypt–SR$^{(10)}$**

$\qquad\Big|\quad M = s \oplus g$.

▶**Rule Decrypt–SnoR$^{(10)}$**

$\qquad\Big|\quad M = 1^k$.

▶**Rule Decrypt–noS$^{(10)}$**

$\qquad\Big|\quad M = 1^k$.

Finally, simply outputting the list of queries to $\mathcal{H}$ during this game, one gets

$$\Pr[\mathsf{AskH_{10}}] \le \mathsf{Succ}_f^{\mathsf{s\text{-}pd\text{-}ow}}(q_h, \tau'). \qquad (31)$$

To conclude the proof of Theorem 8, one just has to comment on the running time τ'. Although the plaintext-extractor is called q_d times, there is no q_d multiplicative factor in the bound for τ'. This comes from a simple bookkeeping argument. Instead of only storing the lists G-List and H-List, one stores an additional structure consisting of tuples $(\gamma, \mathcal{G}_\gamma, \delta, \mathcal{H}_\delta, y)$. A tuple is included only for $(\gamma, \mathcal{G}_\gamma) \in$ G-List and $(\delta, \mathcal{H}_\delta) \in$ H-List. For such a pair, one defines $\sigma = \delta$, $\theta = \gamma \oplus \mathcal{H}_\delta$, $\mu = \mathcal{G}_\gamma \oplus \delta$, and computes $y = f(\sigma, \theta)$. If $[\mu]_{k_1} = 0^{k_1}$, one stores the tuple $(\gamma, \mathcal{G}_\gamma, \delta, \mathcal{H}_\delta, y)$. The cumulative cost of maintaining the additional structure is $q_g \cdot q_h \cdot (T_f + \mathcal{O}(1))$ but, handling it to the plaintext-extractor allows one to output the expected decryption of y, by table lookup, in constant time. Of course, a time-space tradeoff is possible, giving up the additional table, but raising the computing time to $q_d \cdot q_g \cdot q_h \cdot (T_f + \mathcal{O}(1))$. □

6.3.7. Particular Case: RSA–OAEP. Theorem 8 unfortunately requires a very strong assumption on the trapdoor permutation: the partial-domain one-wayness. Hopefully, in [33], we furthermore proved that for RSA, this is not a stronger assumption than the classical RSA assumption:

Lemma 4. *Let $\mathcal{A}$ be an algorithm that outputs a q-set containing $k - k_0$ of the most significant bits of the e-th root of its input (partial-domain RSA, for any modulus N, which $2^{k-1} < N < 2^k$ and $k > 2k_0$), within time bound t, with probability ε. There exists an algorithm that solves the RSA problem (N, e) with success probability ε', within time bound t' where*

$$\varepsilon' \geq \varepsilon \times (\varepsilon - 2^{2k_0 - k + 6}), \qquad t' \leq 2t + q^2 \times \mathcal{O}(k^3).$$

Combining this lemma with the previous general security result about OAEP, one gets

---

Theorem 9. *Let $\mathcal{A}$ be a CCA–adversary against the "semantic security" of RSA–OAEP (where the modulus is k-bit long, $k > 2k_0$), with running time bounded by t and advantage ε, making q_d, q_g and q_h queries to the decryption oracle, and the hash functions G and H, respectively. Then the RSA problem can be solved with probability ε' greater than*

$$\frac{\varepsilon^2}{4} - \varepsilon \cdot \left(\frac{2(q_d + 2)(q_d + 2q_g)}{2^{k_0}} + \frac{3q_d}{2^{k_1}} + \frac{32}{2^{k - 2k_0}} \right)$$

within time bound $t' \leq 2t + q_h \cdot (q_h + 2q_g) \times \mathcal{O}(k^3)$.

There is actually a slight inconsistency in piecing together the two above results, coming from the fact that RSA is not a permutation over k-bit strings. Research papers usually ignore the problem. Of course, standards have to cope with it. Observe that one may decide only to encode a message of $n - 8$ bits, where n is $k - k_0 - k_1$ as before, as is done in the PKCS #1 standard. The additional redundancy leading bit can be treated the same way as the 0^{k_1} redundancy, especially with respect to decryption. However, this is not enough since $G(r)$ might still carry the string $(s\|t)$ outside the domain of the RSA encryption function. An easy way out is to start with another random seed if this happens. On average, 256 trials will be enough.

This security result does not achieve the practical security, because of the expensive reduction. In [33], we improved the reduction cost, with a more intricate proof. More precisely:

Theorem 10. *Let $\mathcal{A}$ be a* CCA–*adversary against the "semantic security" of* RSA–OAEP *(where the modulus is k-bit long, $k > 2k_0$), with running time bounded by t and advantage ε, making q_d, q_g and q_h queries to the decryption oracle, and the hash functions G and H, respectively. Then the RSA problem can be solved with probability ε' greater than*

$$\varepsilon^2 - 2\varepsilon \cdot \left(\frac{2q_d q_g + q_d + q_g}{2^{k_0}} + \frac{2q_d}{2^{k_1}} + \frac{32}{2^{k-2k_0}} \right)$$

within time bound $t' \le 2t + q_h \cdot (q_h + 2q_g) \times \mathcal{O}(k^3)$.

Unfortunately, the reduction is still very expensive, and is thus meaningful for huge moduli only, more than 4096-bit long. Indeed, the RSA inverter we can build, thanks to this reduction, has a complexity at least greater than $q_h \cdot (q_h + 2q_g) \times \mathcal{O}(k^3)$. As already remarked, the adversary can ask up to 2^{60} queries to the hash functions, and thus this overhead in the inversion is at least 2^{151}. However, current factoring algorithms can factor up to 4096 bit-long integers within this number of basic operations (see [47] for complexity estimates of the most efficient factoring algorithms).

Anyway, the formal proof shows that the global design of OAEP is sound, and that it is still probably safe to use it in practice (*e.g.* in PKCS #1 v2.0, while being very careful during the implementation [49]).

6.4. REACT: a Rapid Enhanced-security Asymmetric Cryptosystem Transform

Unfortunately, there is no hope to use OAEP with any **DL**-based primitive, because of the "permutation" requirement. The OAEP construction indeed requires the primitive to be a permutation (trapdoor partial-domain one-way), which is the case of the RSA function. However, the only trapdoor problem known in the **DL**-setting is the Diffie-Hellman problem, and it does not provide any bijection. Thus, first Fujisaki and Okamoto [30] proposed a generic conversion from any IND-CPA scheme into an IND-CCA one, in the random-oracle model. While applying this conversion to the above El Gamal encryption (see Section 6.1), one obtains an IND-CCA encryption scheme relative to the **DDH** problem. Later, independently, Fujisaki and Okamoto [31] and the author [62] proposed better generic conversions since they apply to any OW-CPA scheme to make it into an IND-CCA one, still in the random-oracle model.

This high security level is just at the cost of two more hashings for the new encryption algorithm, as well as two more hashings but one re-encryption for the new decryption process.

6.4.1. Description. The re-encryption cost is the main drawback of these conversions for practical purposes. Therefore, Okamoto and the author tried and succeeded in providing a conversion that is both secure and efficient [59]: REACT, for "Rapid Enhanced-security Asymmetric Cryptosystem Transform". It is actually quite similar to the BR construction, excepted that it applies to any trapdoor one-way function, not permutations only.

$\mathcal{K}'$: **Key Generation** $\rightarrow$ (pk, sk)
(pk, sk) $\leftarrow \mathcal{K}(1^k)$ $\rightarrow$ (pk, sk)
$\mathcal{E}'$: **Encryption of** $m \in \mathbf{M}' = \{0,1\}^\ell \rightarrow (a,b,c)$
$R \in \mathbf{M}$ and $r \in \mathbf{R}$ are randomly chosen $a = \mathcal{E}_{\mathsf{pk}}(R;r) \qquad b = m \oplus \mathcal{G}(R) \qquad c = \mathcal{H}(R,m,a,b)$ $\rightarrow (a,b,c)$ is the ciphertext
$\mathcal{D}'$: **Decryption of** (a,b,c)
Given $a \in \mathbf{C}$, $b \in \{0,1\}^\ell$ and $c \in \{0,1\}^\kappa$ $R = \mathcal{D}_{\mathsf{sk}}(a) \qquad m = b \oplus \mathcal{G}(R)$ if $c = \mathcal{H}(R,m,a,b)$ and $R \in \mathbf{M} \rightarrow m$ is the plaintext   (otherwise, "Reject: invalid ciphertext")

FIGURE 13. Rapid Enhanced-security Asymmetric Cryptosystem Transform REACT $= (\mathcal{K}', \mathcal{E}', \mathcal{D}')$

The latter conversion is indeed very efficient in many senses:

- the computational overhead is just the cost of two hashings for *both* encryption and decryption
- if one can break IND-CCA of the resulting scheme with an expected time T, one can break OW-PCA of the basic scheme within almost the same amount of time, with a low overhead (not as with OAEP). It thus provides a *practical* security result.

Let us describe this generic conversion REACT [59] on any encryption scheme $\mathsf{S} = (\mathcal{K}, \mathcal{E}, \mathcal{D})$

$$\mathcal{E} : \mathbf{PK} \times \mathbf{M} \times \mathbf{R} \rightarrow \mathbf{C}, \qquad \mathcal{D} : \mathbf{SK} \times \mathbf{C} \rightarrow \mathbf{M},$$

where $\mathbf{PK}$ and $\mathbf{SK}$ are the sets of the public and private keys, $\mathbf{M}$ is the messages space, $\mathbf{C}$ is the ciphertexts space and $\mathbf{R}$ is the random coins space. One should remark that $\mathbf{R}$ may be small and even empty, with a deterministic encryption scheme, such as RSA. But in many other cases, such as the El Gamal encryption, it is as large as $\mathbf{M}$. We also need two hash functions $\mathcal{G}$ and $\mathcal{H}$,

$$\mathcal{G} : \mathbf{M} \rightarrow \{0,1\}^\ell, \qquad \mathcal{H} : \mathbf{M} \times \{0,1\}^\ell \times \mathbf{C} \times \{0,1\}^\ell \rightarrow \{0,1\}^\kappa,$$

where κ is the security parameter, while ℓ denotes the size of the messages to encrypt. The REACT conversion is depicted on Figure 13.

6.4.2. Security Result. About this construction, one can prove:

Theorem 11. *Let $\mathcal{A}$ be a* CCA*-adversary against the semantic security of the encryption scheme* REACT $= (\mathcal{K}', \mathcal{E}', \mathcal{D}')$. *Assume that $\mathcal{A}$ has advantage ε and running time τ and makes q_d, q_g and q_h queries to the decryption oracle, and the hash functions $\mathcal{G}$ and $\mathcal{H}$, respectively. Then*

$$\mathsf{Succ}_{\mathsf{S}}^{\mathsf{ow-pca}}(\tau') \ \geq \ \frac{\varepsilon}{2} - \frac{2q_d}{2^\kappa} - \frac{q_h}{2^\ell},$$

$$\text{with} \quad \tau' \ \leq \ \tau + (q_g + q_h) \cdot T_{\mathsf{pca}},$$

where T_{pca} denotes the times required by the PCA *oracle to answer any query.*

Proof. In the following we use starred letters ($r^\star$, $a^\star$, $b^\star$, $c^\star$ and $y^\star$) to refer to the challenge ciphertext, whereas unstarred letters (r, a, b, c and y) refer to the ciphertext asked to the decryption oracle.

Game $\mathbf{G}_0$: A pair of keys (pk, sk) is generated using $\mathcal{K}(1^k)$. Adversary A_1 is fed with pk, and outputs a pair of messages (m_0, m_1). Next a challenge ciphertext is produced by flipping a coin b and producing a ciphertext $y^\star = a^\star \| b^\star \| c^\star$ of m_b. This ciphertext comes from random $R^\star \overset{R}{\leftarrow} \mathbf{M}$ and $r^\star \overset{R}{\leftarrow} \mathbf{R}$ and $a^\star = \mathcal{E}_{\mathsf{pk}}(R^\star, r^\star)$, $b^\star = m_b \oplus \mathcal{G}(R^\star)$ and $c^\star = \mathcal{H}(R^\star, m_b, a^\star, b^\star)$. On input $y^\star$, A_2 outputs bit b'. In both stages, the adversary is given additional access to the decryption oracle $\mathcal{D}'_{\mathsf{sk}}$. The only requirement is that the challenge ciphertext cannot be queried from the decryption oracle.

We denote by S_0 the event $b' = b$ and use a similar notation S_i in any $\mathbf{G}_i$ below. By definition, we have

$$\Pr[\mathsf{S}_0] = \frac{1}{2} + \frac{\varepsilon}{2}. \tag{32}$$

Game $\mathbf{G}_1$: In this game, one makes the classical simulation of the random oracles, with random answers for any new query, as shown on Figure 14. This game is clearly identical to the previous one.

Game $\mathbf{G}_2$: In this game, one randomly chooses $h^+ \overset{R}{\leftarrow} \{0,1\}^\kappa$, and uses it instead of $\mathcal{H}(R^\star, m^\star, a^\star, b^\star)$.

▶**Rule Chal−Hash$^{(2)}$**

> The value $h^+ \overset{R}{\leftarrow} \{0,1\}^\kappa$ has been chosen ahead of time, choose randomly $R^\star$ and $r^\star$, then set
> $a^\star = \mathcal{E}_{\mathsf{pk}}(R^\star, r^\star)$, $\quad g^\star = \mathcal{G}(R^\star)$, $\quad b^\star = m^\star \oplus g^\star$, $\quad c^\star = h^+$.

The two games $\mathbf{G}_2$ and $\mathbf{G}_1$ are perfectly indistinguishable unless $(R^\star, m^\star, a^\star, b^\star)$ is asked for $\mathcal{H}$, either by the adversary or the decryption oracle. But the latter case is not possible, otherwise the decryption query would be the challenge ciphertext itself. More generally, we denote by AskR_2 the event that $R^\star$ has been asked to $\mathcal{G}$ or to $\mathcal{H}$, by the adversary. We have:

$$|\Pr[\mathsf{S}_2] - \Pr[\mathsf{S}_1]| \leq \Pr[\mathsf{AskR}_2]. \tag{33}$$

$\mathcal{G}, \mathcal{H}$ Oracles	Query $\mathcal{G}(r)$: if a record (r, g) appears in G-List, the answer is g. Otherwise the answer g is chosen randomly: $g \in \{0, 1\}^\ell$ and the record (r, g) is added in G-List.
	Query $\mathcal{H}(R, m, a, b)$: if a record (R, m, a, b, h) appears in H-List, the answer is h. Otherwise the answer h is chosen randomly: $h \in \{0, 1\}^\kappa$ and the record (R, m, a, b, h) is added in H-List.

$\mathcal{D}'$ Oracle	Query $\mathcal{D}'_{\mathsf{sk}}(a\|b\|c)$: one applies the following rules:
	▶**Rule Decrypt$-\mathbf{R}^{(1)}$**
	$\qquad$ Compute $R = \mathcal{D}_{\mathsf{sk}}(a)$;
	Then, compute $m = b \oplus \mathcal{G}(R)$, and finally,
	▶**Rule Decrypt$-\mathbf{H}^{(1)}$**
	$\qquad$ If $c = \mathcal{H}(R, m, a, b)$, one returns m, otherwise one returns "Reject."

Challenger	For two messages (m_0, m_1), flip a coin b and set $m^\star = m_b$.
	▶**Rule Chal$-\mathbf{Hash}^{(1)}$**
	$\qquad$ Choose randomly $R^\star$ and $r^\star$, then set
	$\qquad a^\star = \mathcal{E}_{\mathsf{pk}}(R^\star, r^\star),$
	$\qquad g^\star = \mathcal{G}(R^\star), \quad b^\star = m^\star \oplus g^\star,$
	$\qquad c^\star = \mathcal{H}(R^\star, m^\star, a^\star, b^\star).$
	Then, output $y^\star = a^\star\|b^\star\|c^\star$.

FIGURE 14. Formal Simulation of the IND-CCA Game against REACT

Game G_3: We start modifying the simulation of the decryption oracle, by rejecting any ciphertext $(a\|b\|c)$ for which the corresponding (R, m, a, b) has not been queried to $\mathcal{H}$:

▶**Rule Decrypt$-\mathbf{H}^{(3)}$**
$\qquad$ Look up in H-List for (R, m, a, b, c). If such a triple does not
$\qquad$ exist, then output "Reject", otherwise output m.

Such a simulation differs from the previous one if the value c has been correctly guessed, by chance:

$$|\Pr[S_3] - \Pr[S_2]| \le \frac{q_d}{2^\kappa} \qquad |\Pr[\mathsf{AskR}_3] - \Pr[\mathsf{AskR}_2]| \le \frac{q_d}{2^\kappa}. \qquad (34)$$

Game G_4: In this game, one randomly chooses $R^+ \xleftarrow{R} \mathbf{M}$ and $r^+ \xleftarrow{R} \mathbf{R}$, and $g^+ \xleftarrow{R} \{0, 1\}^\ell$, and uses R^+ instead of $R^\star$, r^+ instead of $r^\star$, as well as g^+ instead of $\mathcal{G}(R^\star)$.

▶**Rule Chal−Hash**[(4)]

> The four values $R^+ \overset{R}{\leftarrow} \mathbf{M}$, $r^+ \overset{R}{\leftarrow} \mathbf{R}$, $g^+ \overset{R}{\leftarrow} \{0,1\}^\ell$ and
> $h^+ \overset{R}{\leftarrow} \{0,1\}^\kappa$ have been chosen ahead of time, then set
> $a^\star = \mathcal{E}_{\mathsf{pk}}(R^+, r^+), \quad b^\star = m^\star \oplus g^+, \quad c^\star = h^+.$

The two games $\mathbf{G}_4$ and $\mathbf{G}_3$ are perfectly indistinguishable unless $R^\star$ is asked for $\mathcal{G}$, either by the adversary or the decryption oracle. The former case has already been cancelled in the previous game in AskR_3. The latter case makes no difference since either $\mathcal{H}(R^\star, m, a, b)$ has been queried by the adversary, which falls in AskR_3, or the ciphertext is rejected in both games. We have:

$$\Pr[\mathsf{S}_4] = \Pr[\mathsf{S}_3] \qquad \Pr[\mathsf{AskR}_4] = \Pr[\mathsf{AskR}_3]. \qquad (35)$$

In this game, $m^\star$ is masked by g^+, a random value which never appears anywhere else. Thus, the input to $\mathcal{A}_2$ follows a distribution that does not depend on b. Accordingly:

$$\Pr[\mathsf{S}_4] = \frac{1}{2}. \qquad (36)$$

Game $\mathbf{G}_5$: Finally, one chooses $a^+ \overset{R}{\leftarrow} \mathbf{C}$, according the following distribution: $R^+ \overset{R}{\leftarrow} \mathbf{M}, r^+ \overset{R}{\leftarrow} \mathbf{R}, a^+ \leftarrow \mathcal{E}_{\mathsf{pk}}(R^+, r^+)$. This implicitly defines one pair (R^+, r^+), but the latter is unknown to the simulator.

▶**Rule Chal−Hash**[(5)]

> The three values $a^+ \overset{R}{\leftarrow} \mathbf{C}$, $g^+ \overset{R}{\leftarrow} \{0,1\}^\ell$ and $h^+ \overset{R}{\leftarrow}$
> $\{0,1\}^\kappa$ have been chosen/given ahead of time, then set
> $a^\star = a^+, \quad b^\star = m^\star \oplus g^+, \quad c^\star = h^+.$

The two games $\mathbf{G}_5$ and $\mathbf{G}_4$ are perfectly indistinguishable.

Game $\mathbf{G}_6$: In the simulation of the decryption oracle, we may reject even earlier, if the corresponding R has not been queried to $\mathcal{G}$:

▶**Rule Decrypt−R**[(6)]

> Look up in G-List for (R, g) such that $R = \mathcal{D}_{\mathsf{sk}}(a)$ (using
> the PCA-oracle). If no R is found, then output "Reject".

Note that this game differs from the analogous one for the first generic construction BR, because the encryption function is not deterministic, as was the permutation f. Such a simulation differs from the one in the previous game if the value (R, m, a, b) has been queried to $\mathcal{H}$, while $\mathcal{G}(R)$ is unpredictable, and thus $m = \mathcal{G}(R) \oplus b$ in unpredictable too:

$$|\Pr[\mathsf{AskR}_6] - \Pr[\mathsf{AskR}_5]| \le \frac{q_h}{2^\ell}. \qquad (37)$$

One may now note that the event AskR_6 leads to the plaintext R^+ of a^+ by $\mathbf{S}$ in the queries asked to $\mathcal{G}$ and $\mathcal{H}$. By checking all of them, one gets it:

$$\Pr[\mathsf{AskR}_6] \le \mathsf{Succ}_{\mathsf{S}}^{\mathsf{ow-pca}}(\tau + (q_g + q_h)T_{\mathsf{pca}}). \qquad (38)$$

$\square$

This construction is very generic, and achieves practical security.

6.4.3. Hybrid REACT. In this REACT conversion, one can even improve efficiency, replacing the one-time pad [87] by any symmetric encryption scheme: indeed, we have computed some $b = m \oplus K$, where $K = \mathcal{G}(R)$ can be seen as a session key used in a one-time pad encryption scheme. But one could use any symmetric

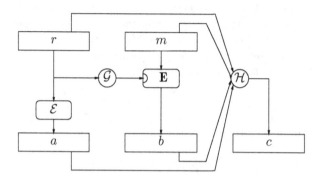

FIGURE 15. Hybrid Rapid Enhanced-security Asymmetric Cryptosystem Transform

encryption scheme $(\mathbf{E}, \mathbf{D})$ that is just semantically secure (under no plaintext nor ciphertext attacks). Indeed, the one-time pad achieves perfect semantic security, against this kind of very weak attacks. But one can tolerate some imperfection. Anyway, most of the candidates to the AES process (the call for symmetric encryption schemes, from the NIST, to become the new international standard), and the AES itself (the winner), resisted to more powerful attacks, and thus can be considered strongly secure in our scenario. Therefore, plaintexts of any size could be encrypted using this conversion (see Figure 15), with a very high speed rate.

7. Conclusion

Recently, Cramer and Shoup proposed the first schemes, for both encryption [23] and signature [24], with formal security proofs in the standard model (without any ideal assumption). The encryption scheme achieves IND-CCA under the sole **DDH** assumption, which says that the **DDH** problem is intractable. The signature scheme prevents existential forgeries, even against adaptive chosen-message attacks, under the Strong RSA assumption [2, 29], which claims the intractability of the Flexible RSA problem:

> Given an RSA modulus N and any $y \in \mathbb{Z}_N^*$, produce x and a prime integer e such that $y = x^e \bmod N$.

Both schemes are very nice because they are the first efficient schemes with formal security proofs in the standard model, but under stronger computational assumptions. We have not presented them, nor the reductions either, which can be found in the original papers. Actually, they are intricate and pretty expensive. Indeed, the complexity of the reductions make them meaningful for large parameters only.

Furthermore, as already noted, no ideal assumptions (such as the random-oracle model) are required, but stronger computational assumptions are needed: the final decision for the best for practical use is not easy.

Moreover, even if the schemes are much more efficient than previous proposals in the standard model, they are still more than twice as expensive as the schemes presented along this paper, in the random-oracle model. This is enough to rule them out from most of the practical applications. Indeed, everybody wants security, but only if it is quite transparent. Therefore, provable security must not decrease efficiency. It is the reason why strong security arguments (which are in an ideal model, but this can be seen as realistic restrictions on the adversary's capabilities) for efficient schemes have a more practical impact than security proofs in the standard model for less efficient schemes.

Of course, quite efficient schemes with formal security proofs are still the target, and thus an exciting challenge.

Acknowledgment

These notes are based on several of my papers, for both signature [67, 68] and encryption [7, 59, 32, 33], written in collaboration with many co-authors, and I would like to take the opportunity of thanking all of them: Mihir Bellare, Anand Desai, Eiichiro Fujisaki, Tatsuaki Okamoto, Philip Rogaway and Jacques Stern. I also warmly thank Benoît Chevallier-Mames, Javier Herranz Sotoca and Duong Hieu Phan for their useful comments on preliminary versions of these notes.

References

[1] American National Standards Institute. Public Key Cryptography for the Financial Services Industry: The Elliptic Curve Digital Signature Algorithm. ANSI X9.62-1998. January 1999.

[2] N. Barić and B. Pfitzmann. Collision-Free Accumulators and Fail-Stop Signature Schemes without Trees. In *Eurocrypt '97*, LNCS 1233, pages 480–484. Springer-Verlag, Berlin, 1997.

[3] O. Baudron, D. Pointcheval, and J. Stern. Extended Notions of Security for Multicast Public Key Cryptosystems. In *Proc. of the 27th ICALP*, LNCS 1853, pages 499–511. Springer-Verlag, Berlin, 2000.

[4] M. Bellare. Practice-Oriented Provable Security. In *ISW '97*, LNCS 1396. Springer-Verlag, Berlin, 1997.

[5] M. Bellare, A. Boldyreva, and S. Micali. Public-key Encryption in a Multi-User Setting: Security Proofs and Improvements. In *Eurocrypt '00*, LNCS 1807, pages 259–274. Springer-Verlag, Berlin, 2000.

[6] M. Bellare, A. Boldyreva, and A. Palacio. A Separation between the Random-Oracle Model and the Standard Model for a Hybrid Encryption Problem, 2003. Cryptology ePrint Archive 2003/077.

[7] M. Bellare, A. Desai, D. Pointcheval, and P. Rogaway. Relations among Notions of Security for Public-Key Encryption Schemes. In *Crypto '98*, LNCS 1462, pages 26–45. Springer-Verlag, Berlin, 1998.

[8] M. Bellare and A. Palacio. GQ and Schnorr Identification Schemes: Proofs of Security against Impersonation under Active and Concurrent Attacks. In *Crypto '02*, LNCS 2442, pages 162–177. Springer-Verlag, Berlin, 2002.

[9] M. Bellare, D. Pointcheval, and P. Rogaway. Authenticated Key Exchange Secure Against Dictionary Attacks. In *Eurocrypt '00*, LNCS 1807, pages 139–155. Springer-Verlag, Berlin, 2000.

[10] M. Bellare and P. Rogaway. Random Oracles Are Practical: a Paradigm for Designing Efficient Protocols. In *Proc. of the 1st CCS*, pages 62–73. ACM Press, New York, 1993.

[11] M. Bellare and P. Rogaway. Optimal Asymmetric Encryption – How to Encrypt with RSA. In *Eurocrypt '94*, LNCS 950, pages 92–111. Springer-Verlag, Berlin, 1995.

[12] M. Bellare and P. Rogaway. The Exact Security of Digital Signatures – How to Sign with RSA and Rabin. In *Eurocrypt '96*, LNCS 1070, pages 399–416. Springer-Verlag, Berlin, 1996.

[13] E. Biham and A. Shamir. Differential Fault Analysis of Secret Key Cryptosystems. In *Crypto '97*, LNCS 1294, pages 513–525. Springer-Verlag, Berlin, 1997.

[14] D. Bleichenbacher. Generating El Gamal Signatures without Knowing the Secret Key. In *Eurocrypt '96*, LNCS 1070, pages 10–18. Springer-Verlag, Berlin, 1996.

[15] D. Bleichenbacher. A Chosen Ciphertext Attack against Protocols based on the RSA Encryption Standard PKCS #1. In *Crypto '98*, LNCS 1462, pages 1–12. Springer-Verlag, Berlin, 1998.

[16] D. Boneh, R. DeMillo, and R. Lipton. On the Importance of Checking Cryptographic Protocols for Faults. In *Eurocrypt '97*, LNCS 1233, pages 37–51. Springer-Verlag, Berlin, 1997.

[17] E. Brickell, D. Pointcheval, S. Vaudenay, and M. Yung. Design Validations for Discrete Logarithm Based Signature Schemes. In *PKC '00*, LNCS 1751, pages 276–292. Springer-Verlag, Berlin, 2000.

[18] D. R. L. Brown and D. B. Johnson. Formal Security Proofs for a Signature Scheme with Partial Message Recovery. In *CT – RSA '01*, LNCS 2020, pages 126–142. Springer-Verlag, Berlin, 2001.

[19] R. Canetti, O. Goldreich, and S. Halevi. The Random Oracles Methodology, Revisited. In *Proc. of the 30th STOC*, pages 209–218. ACM Press, New York, 1998.

[20] S. Cavallar, B. Dodson, A. K. Lenstra, W. Lioen, P. L. Montgomery, B. Murphy, H. te Riele, K. Aardal, J. Gilchrist, G. Guillerm, P. Leyland, J. Marchand, F. Morain, A. Muffett, Ch. Putnam, Cr. Putnam, and P. Zimmermann. Factorization of a 512-bit RSA Modulus. In *Eurocrypt '00*, LNCS 1807, pages 1–18. Springer-Verlag, Berlin, 2000.

[21] B. Chor and R. L. Rivest. A Knapsack Type Public Key Cryptosystem based on Arithmetic in Finite Fields. In *Crypto '84*, LNCS 196, pages 54–65. Springer-Verlag, Berlin, 1985.

[22] J.-S. Coron. On the Exact Security of Full-Domain-Hash. In *Crypto '00*, LNCS 1880, pages 229–235. Springer-Verlag, Berlin, 2000.

[23] R. Cramer and V. Shoup. A Practical Public Key Cryptosystem Provably Secure against Adaptive Chosen Ciphertext Attack. In *Crypto '98*, LNCS 1462, pages 13–25. Springer-Verlag, Berlin, 1998.

[24] R. Cramer and V. Shoup. Signature Scheme based on the Strong RSA Assumption. In *Proc. of the 6th CCS*, pages 46–51. ACM Press, New York, 1999.

[25] W. Diffie and M. E. Hellman. New Directions in Cryptography. *IEEE Transactions on Information Theory*, IT–22(6):644–654, November 1976.

[26] D. Dolev, C. Dwork, and M. Naor. Non-Malleable Cryptography. *SIAM Journal on Computing*, 30(2):391–437, 2000.

[27] T. El Gamal. A Public Key Cryptosystem and a Signature Scheme Based on Discrete Logarithms. *IEEE Transactions on Information Theory*, IT–31(4):469–472, July 1985.

[28] A. Fiat and A. Shamir. How to Prove Yourself: Practical Solutions of Identification and Signature Problems. In *Crypto '86*, LNCS 263, pages 186–194. Springer-Verlag, Berlin, 1987.

[29] E. Fujisaki and T. Okamoto. Statistical Zero Knowledge Protocols to Prove Modular Polynomial Relations. In *Crypto '97*, LNCS 1294, pages 16–30. Springer-Verlag, Berlin, 1997.

[30] E. Fujisaki and T. Okamoto. How to Enhance the Security of Public-Key Encryption at Minimum Cost. In *PKC '99*, LNCS 1560, pages 53–68. Springer-Verlag, Berlin, 1999.

[31] E. Fujisaki and T. Okamoto. Secure Integration of Asymmetric and Symmetric Encryption Schemes. In *Crypto '99*, LNCS 1666, pages 537–554. Springer-Verlag, Berlin, 1999.

[32] E. Fujisaki, T. Okamoto, D. Pointcheval, and J. Stern. RSA–OAEP is Secure under the RSA Assumption. In *Crypto '01*, LNCS 2139, pages 260–274. Springer-Verlag, Berlin, 2001. Also appeared as *RSA–OAEP is Still Alive* in the Cryptology ePrint Archive 2000/061. November 2000.
Available from http://eprint.iacr.org/.

[33] E. Fujisaki, T. Okamoto, D. Pointcheval, and J. Stern. RSA–OAEP is Secure under the RSA Assumption. *Journal of Cryptology*, 17(2):81–104, 2004.

[34] O. Goldreich, S. Goldwasser, and S. Micali. How to Construct Random Functions. *Journal of the ACM*, 33(4):792–807, 1986.

[35] S. Goldwasser and S. Micali. Probabilistic Encryption. *Journal of Computer and System Sciences*, 28:270–299, 1984.

[36] S. Goldwasser, S. Micali, and C. Rackoff. The Knowledge Complexity of Interactive Proof Systems. In *Proc. of the 17th STOC*, pages 291–304. ACM Press, New York, 1985.

[37] S. Goldwasser, S. Micali, and R. Rivest. A "Paradoxical" Solution to the Signature Problem. In *Proc. of the 25th FOCS*, pages 441–448. IEEE, New York, 1984.

[38] S. Goldwasser, S. Micali, and R. Rivest. A Digital Signature Scheme Secure Against Adaptive Chosen-Message Attacks. *SIAM Journal of Computing*, 17(2):281–308, April 1988.

[39] C. Hall, I. Goldberg, and B. Schneier. Reaction Attacks Against Several Public-Key Cryptosystems. In *Proc. of ICICS '99*, LNCS, pages 2–12. Springer-Verlag, 1999.

[40] J. Håstad. Solving Simultaneous Modular Equations of Low Degree. *SIAM Journal of Computing*, 17:336–341, 1988.

[41] A. Joux and R. Lercier. Improvements to the general Number Field Sieve for discrete logarithms in prime fields. *Mathematics of Computation*, 2000. to appear.

[42] M. Joye, J. J. Quisquater, and M. Yung. On the Power of Misbehaving Adversaries and Security Analysis of the Original EPOC. In *CT – RSA '01*, LNCS 2020, pages 208–222. Springer-Verlag, Berlin, 2001.

[43] KCDSA Task Force Team. The Korean Certificate-based Digital Signature Algorithm. Submission to IEEE P1363a. August 1998.
Available from http://grouper.ieee.org/groups/1363/.

[44] P. C. Kocher. Timing Attacks on Implementations of Diffie-Hellman, RSA, DSS, and Other Systems. In *Crypto '96*, LNCS 1109, pages 104–113. Springer-Verlag, Berlin, 1996.

[45] P. C. Kocher, J. Jaffe, and B. Jun. Differential Power Analysis. In *Crypto '99*, LNCS 1666, pages 388–397. Springer-Verlag, Berlin, 1999.

[46] A. Lenstra and H. Lenstra. *The Development of the Number Field Sieve*, volume 1554 of *Lecture Notes in Mathematics*. Springer-Verlag, 1993.

[47] A. Lenstra and E. Verheul. Selecting Cryptographic Key Sizes. In *PKC '00*, LNCS 1751, pages 446–465. Springer-Verlag, Berlin, 2000.

[48] H.W. Lenstra. On the Chor-Rivest Knapsack Cryptosystem. *Journal of Cryptology*, 3:149–155, 1991.

[49] J. Manger. A Chosen Ciphertext Attack on RSA Optimal Asymmetric Encryption Padding (OAEP) as Standardized in PKCS #1. In *Crypto '01*, LNCS 2139, pages 230–238. Springer-Verlag, Berlin, 2001.

[50] G. Miller. Riemann's Hypothesis and Tests for Primality. *Journal of Computer and System Sciences*, 13:300–317, 1976.

[51] D. M'Raïhi, D. Naccache, D. Pointcheval, and S. Vaudenay. Computational Alternatives to Random Number Generators. In *Fifth Annual Workshop on Selected Areas in Cryptography (SAC '98)*, LNCS 1556, pages 72–80. Springer-Verlag, Berlin, 1998.

[52] M. Naor and M. Yung. Public-Key Cryptosystems Provably Secure against Chosen Ciphertext Attacks. In *Proc. of the 22nd STOC*, pages 427–437. ACM Press, New York, 1990.

[53] V. I. Nechaev. Complexity of a Determinate Algorithm for the Discrete Logarithm. *Mathematical Notes*, 55(2):165–172, 1994.

[54] J. B. Nielsen. Separating Random Oracle Proofs from Complexity Theoretic Proofs: The Non-committing Encryption Case. In *Crypto '02*, LNCS 2442, pages 111–126. Springer-Verlag, Berlin, 2002.

[55] NIST. Digital Signature Standard (DSS). Federal Information Processing Standards PUBlication 186, November 1994.

[56] NIST. Secure Hash Standard (SHS). Federal Information Processing Standards PUB-lication 180–1, April 1995.

[57] NIST. Descriptions of SHA–256, SHA–384, and SHA–512. Available from http://www.nist.gov/sha/, October 2000.

[58] K. Ohta and T. Okamoto. On Concrete Security Treatment of Signatures Derived from Identification. In *Crypto '98*, LNCS 1462, pages 354–369. Springer-Verlag, Berlin, 1998.

[59] T. Okamoto and D. Pointcheval. REACT: Rapid Enhanced-security Asymmetric Cryptosystem Transform. In *CT – RSA '01*, LNCS 2020, pages 159–175. Springer-Verlag, Berlin, 2001.

[60] T. Okamoto and D. Pointcheval. The Gap-Problems: a New Class of Problems for the Security of Cryptographic Schemes. In *PKC '01*, LNCS 1992. Springer-Verlag, Berlin, 2001.

[61] D. Pointcheval. *Les Preuves de Connaissance et leurs Preuves de Sécurité*. PhD thesis, université de Caen, December 1996.

[62] D. Pointcheval. Chosen-Ciphertext Security for any One-Way Cryptosystem. In *PKC '00*, LNCS 1751, pages 129–146. Springer-Verlag, Berlin, 2000.

[63] D. Pointcheval. About Generic Conversions from any Weakly Secure Encryption Scheme into a Chosen-Ciphertext Secure Scheme. In *Proceedings of the Fourth Conference on Algebraic Geometry, Number Theory, Coding Theory and Cryptography*, pages 145–162, Tokyo, Japan, 2001.

[64] D. Pointcheval. Practical Security in Public-Key Cryptography. In *Proc. of ICISC '01*, LNCS 2288. Springer-Verlag, Berlin, 2001.

[65] D. Pointcheval. How to Encrypt Properly with RSA. *CryptoBytes*, 5(1):10–19, winter/spring 2002.

[66] D. Pointcheval. *Le chiffrement asymétrique et la sécurité prouvée*. PhD thesis, université de Paris VII, May 2002. Thèse d'habilitation.

[67] D. Pointcheval and J. Stern. Security Proofs for Signature Schemes. In *Eurocrypt '96*, LNCS 1070, pages 387–398. Springer-Verlag, Berlin, 1996.

[68] D. Pointcheval and J. Stern. Security Arguments for Digital Signatures and Blind Signatures. *Journal of Cryptology*, 13(3):361–396, 2000.

[69] D. Pointcheval and S. Vaudenay. On Provable Security for Digital Signature Algorithms. Technical Report LIENS-96-17, LIENS, October 1996.

[70] J. M. Pollard. Monte Carlo Methods for Index Computation (mod p). *Mathematics of Computation*, 32(143):918–924, July 1978.

[71] C. Rackoff and D. R. Simon. Non-Interactive Zero-Knowledge Proof of Knowledge and Chosen Ciphertext Attack. In *Crypto '91*, LNCS 576, pages 433–444. Springer-Verlag, Berlin, 1992.

[72] R. Rivest. The MD5 Message-Digest Algorithm. RFC 1321, The Internet Engineering Task Force, April 1992.

[73] R. Rivest, A. Shamir, and L. Adleman. A Method for Obtaining Digital Signatures and Public Key Cryptosystems. *Communications of the ACM*, 21(2):120–126, February 1978.

[74] RSA Data Security, Inc. Public Key Cryptography Standards – PKCS. Available from `http://www.rsa.com/rsalabs/pubs/PKCS/`.

[75] C. P. Schnorr. Efficient Identification and Signatures for Smart Cards. In *Crypto '89*, LNCS 435, pages 235–251. Springer-Verlag, Berlin, 1990.

[76] C. P. Schnorr. Efficient Signature Generation by Smart Cards. *Journal of Cryptology*, 4(3):161–174, 1991.

[77] C. P. Schnorr and M. Jakobsson. Security of Signed ElGamal Encryption. In *Asiacrypt '00*, LNCS 1976, pages 458–469. Springer-Verlag, Berlin, 2000.

[78] D. Shanks. Class Number, a Theory of Factorization, and Genera. In *Proceedings of the Symposium on Pure Mathematics*, volume 20, pages 415–440. AMS, 1971.

[79] H. Shimizu. On the Improvement of the Håstad Bound. In *1996 IEICE Fall Conference*, Volume A-162, 1996. In Japanese.

[80] V. Shoup. Lower Bounds for Discrete Logarithms and Related Problems. In *Eurocrypt '97*, LNCS 1233, pages 256–266. Springer-Verlag, Berlin, 1997.

[81] V. Shoup. A Proposal for an ISO Standard for Public-Key Encryption, december 2001. ISO/IEC JTC 1/SC27.

[82] V. Shoup. OAEP Reconsidered. In *Crypto '01*, LNCS 2139, pages 239–259. Springer-Verlag, Berlin, 2001. Also appeared in the Cryptology ePrint Archive 2000/060. November 2000.
Available from `http://eprint.iacr.org/`.

[83] V. Shoup. OAEP Reconsidered. *Journal of Cryptology*, 15(4):223–249, September 2002.

[84] J. Stern, D. Pointcheval, J. Malone-Lee, and N. Smart. Flaws in Applying Proof Methodologies to Signature Schemes. In *Crypto '02*, LNCS 2442, pages 93–110. Springer-Verlag, Berlin, 2002.

[85] Y. Tsiounis and M. Yung. On the Security of El Gamal based Encryption. In *PKC '98*, LNCS. Springer-Verlag, Berlin, 1998.

[86] S. Vaudenay. Cryptanalysis of the Chor-Rivest Scheme. In *Crypto '98*, LNCS 1462, pages 243–256. Springer-Verlag, Berlin, 1998.

[87] G. S. Vernam. Cipher Printing Telegraph Systems for Secret Wire and Radio Telegraphic Communications. *Journal of the American Institute of Electrical Engineers*, 45:109–115, 1926.

Efficient and Secure Public-Key Cryptosystems

Tsuyoshi Takagi

Abstract. Nowadays, RSA cryptosystem is used for practical security applications, e.g., SSL, IPSEC, PKI, etc. Elliptic curve cryptosystem has focused on the implementation on memory constraint environments due to its small key size. In this chapter we describe an overview of efficient algorithms applied to RSA cryptosystem and EC cryptosystem. On the other hand, novel attacks on the efficient implementation have been proposed, namely timing attack, side channel attacks, fault attack, etc. These attacks can break the secret key of the underlying cryptosystem, if the implementation method is not carefully considered. We also explain several attacks related to efficient implementation, and present countermeasures against them.

1. Efficient Integer Arithmetic

In this section we show several fast integer arithmetic used for cryptography.

Let Z be the integer ring. Let Z/nZ be the residue class ring modulo n, where n is a positive integer. In this article we set the following representative class $Z/nZ = \{0, 1, 2, \ldots, n-1\}$. We denote by $(Z/nZ)^*$ the multiplicative group of residue n, namely $\{g \in Z/nZ | gcd(g, n) = 1\}$, where $gcd(a, b)$ is the great common divisor of a and b. In cryptography we deal with quite large integers, e.g., 1024 bits for RSA cryptosystem, 160 bits for elliptic curve cryptosystem. Therefore the asymptotic complexity is useful for estimating the running time of cryptographic algorithms. Let $O(f(n))$ be a function $h(n)$ such that $|h(n)| \leq c|f(n)|$ for enough large n with some positive constant c. The basic operations in Z/nZ used for cryptography are modular addition $a+b$, modular subtraction $a-b$, modular multiplication ab, and modular inversion c^{-1}, where $a, b \in Z/nZ$ and $c \in (Z/nZ)^*$. Their asymptotic complexity are $O(\log n)$ for addition and subtraction, and $O((\log n)^2)$ for multiplication and inversion [MOV96].

1.1. Modular Exponentiation

The modular exponentiation is the core arithmetic for RSA cryptosystem. It computes $a^d \in Z/nZ$ for given integers a, d, and n. Let $d = \sum_{i=0}^{k-1} d[i]2^i$ be the binary representation of d, where k is the most non-zero bit and $d[i] = 0$ for $i > k - 1$. A standard algorithm of computing the modular exponentiation is the binary method, which repeatedly computes squares and multiplications based on the bits of exponent d. There are two different directions for implementing the modular multiplication, namely right-to-left and left-to-right. The binary methods are as follows:

Binary Exponentiation Method	
INPUT $d, n, a, (d[k-1], \dots, d[1], d[0]), d[k-1] = 1$	
OUTPUT $a^d \bmod n$	
(Left-to-Right)	(Right-to-Left)
1: $t \leftarrow a$	1: $t \leftarrow 1, s \leftarrow a$
2: for $i = k - 2$ down to 0	2: for $i = 0$ to $k - 1$
3: $t \leftarrow t^2 \bmod n$	3: $t \leftarrow t s^{d[i]} \bmod n$
4: $t \leftarrow t a^{d[i]} \bmod n$	4: if $i \neq k - 1$, then $s \leftarrow s^2 \bmod n$
5: return t	5: return t

The left-to-right computes from the most non-zero bit $(d[k-1] = 1)$ down to the least bit $(d[0])$. The squaring $s = s^2 \bmod n$ is always computed, and the multiplication $t = ta \bmod n$ with the base point a is computed if the i-th bit $d[i]$ is non-zero. The right-to-left algorithm prepares two registers s, t. It computes from the least bit $d[0]$ to the most bits $d[k-1]$. The register s is used for recursively computing the squaring $s = s^2 \bmod n$. The register t is multiplied with s if $d[i]$ is non-zero bit.

Both method require $(k-1)$ squaring and $(k-1)/2$ multiplications on average. For example a 1024-bit integer n requires about 1500 squaring and multiplications on average. The asymptotical running time of computing the modular exponentiation is $O((\log n)^3)$.

1.2. Window Methods

If we are allowed to use additional memory, the speed of modular multiplication can be improved by precomputing several points. Here we explain a 2^w-ary method and a sliding window method.

The 2^w-ary method represents a k-bit integer $d = \sum_{i=0}^{k-1} d[i] 2^i$ using 2^w-adic representation, namely

$$d = \sum_{j=0}^{\lfloor (k-1)/w \rfloor} (d_w[j])(2^w)^j, \quad d_w[j] = \sum_{h=0}^{w-1} (d[wj + h]) 2^h. \qquad (1.1)$$

In order to calculate a modular exponentiation $a^d \bmod n$, we precompute the following points $a^2, a^3, \dots, a^{2^w-1}$. Then it applies the left-to-right modular exponentiation to the 2^w-adic representation as follow.

2^w-Ary Exponentiation Method (Evaluation Stage)

INPUT $d, n, a, (d_w[m], \ldots, d_w[1], d_w[0]), m = \lfloor (k-1)/w \rfloor$

OUTPUT $a^d \bmod n$

1: $t \leftarrow d_w[m]$

2: for $i = m - 1$ to 0

3:    $t \leftarrow t^{2^w} \bmod n$

4:    $t \leftarrow t a^{d_w[i]} \bmod n$

5: return t

We estimate the efficiency of 2^w-ary method in the following. For the pre-computation stage we need $2^w - 2$ multiplications of Z/nZ. In the evaluation stage we always compute t^{2^w} which requires mw multiplications in total. If the digit by the base 2^w representation is not zero, we additionally compute $t = t a^{d_w[i]}$. The probability that a digit is not zero is $(1 - 1/2^w)$ and thus we compute $m(1 - 1/2^w)$ multiplications on average. In total the 2^k-ary exponentiation method requires $2^w - 2 + m(w + 1 - 1/2^w)$, where $m = \lfloor (k-1)/w \rfloor$. We show an example of 2^w-ary chain form as follows:

binary string	10011101111001110001011011110001101010111111001
$w = 2$	1001030103020103000101020303000102020203030201
$w = 3$	1001006007004007000005005007000006005003007001

Next we try to reduce the precomputed table size using a different exponent recording algorithm. The sliding window method is one of the most efficient window method with small table size for the general purposes. While the 2^w-ary method precomputes all positive integers smaller than 2^w, the width-w sliding window method precomputes only the odd integers smaller than 2^w, namely we represent an integer d as follows:

$$d = \sum_{i=0}^{k} d_{sw}[i] 2^i, \quad d_{sw}[i] = \{0, 1, 3, \ldots, 2^w - 1\}. \qquad (1.2)$$

We explain the exponent recording stage of sliding window method. The binary bit sequence of d is scanned from the most significant bit. If a zero bit appears, we skip to one lower bit. If a non-zero bit appears, we scan lower bits (at most w bits) and convert it to the largest odd integer smaller 2^w. The converted odd integer from the scanned bits is the digit of the sliding window method, and the other digits are assigned as zero. The conversion tables for small width w are $11 \rightarrow 03$ for $w = 2$ and $101 \rightarrow 005, 111 \rightarrow 007$ for $w = 3$. We show an example of the sliding window chain as follows:

binary string	10011101111001110001011011110001101010111111001
$w = 2$	1000310030300031000100300303000030101003031001
$w = 3$	1000070007100007000005005007000030005000703001

Then the sliding window method computes the modular multiplication using the left-to-right binary method.

Width-w Sliding Window Method (Evaluation Stage)

INPUT $d, n, a, (d_{sw}[k-1], \ldots, d_{sw}[1], d_{sw}[0])$
OUTPUT $a^d \bmod n$

1: $t \leftarrow a^{d_{sw}[k-1]}$
2: for $i = k - 2$ to 0
3:     $t \leftarrow t^2 \bmod n$
4:     $t \leftarrow ta^{d_{sw}[i]} \bmod n$
5: return t

The efficiency of width-w sliding window method is known as the following theorem.

Theorem 1.1. *The average density of non-zero bits of the width-w sliding window chain is asymptotically $1/(w+1)$.*

Proof. We assume that each bit of the binary string distributes with probability $1/2$. The width-w conversion table can be simulated by a finite automaton with two statuses (0) and (NZ) of binary strings, where (NZ) is the w-consecutive bits with non-zero leading bit. From the construction, the transition matrix of these statuses is as follows:

$$\begin{pmatrix} (0) & : & 1/2 & 1/2 \\ (NZ) & : & 1/2 & 1/2 \end{pmatrix}.$$

Therefore the statuses (0) and (NZ) asymptotically distribute with probability $1/2$. The average bit-length of the non-zero bits and the two statuses is $1 * \frac{1}{2}$ and $1 * \frac{1}{2} + w * \frac{1}{2}$, respectively. Thus the average non-zero density is asymptotically $(1 * \frac{1}{2})/(1 * \frac{1}{2} + w * \frac{1}{2}) = 1/(w+1)$. $\square$

1.3. Montgomery Multiplication

Let a, b be two elements in Z/nZ, where n is a positive integer. The straightforward implementation of modular multiplication $ab \bmod n$ requires a division with remainder, namely we compute the integer r such that $ab = qn + r, 0 \leq r < n$ for some integer q. The division of integer is an relatively expensive and complicated operation for implementation. The Montgomery multiplication is able to avoid the division in the modular multiplication. The general description of Montgomery multiplication is as follows:

Montgomery Multiplication

INPUT $a, b \in Z/nZ$, $R = 2^r$, (r is bit-length of n), $n' = -n^{-1}$ mod R.
OUTPUT abR^{-1} mod n

1: $t \leftarrow ab$ in Z
2: $u \leftarrow tn'$ mod R
3: $v \leftarrow (t + un)/R$ in Z
4: if $v > n$, then $v \leftarrow v - n$
5: return v

Montgomery multiplication still utilizes the reduction modulo R in Step 2 and the division by R in Step 3, where $R = 2^r$. However, these operations are quite efficient, because integers in computer system are usually represented as a binary representation. The reduction modulo $R = 2^r$ in Step 2 is a re-assignment of least r bits of integer tn'. The integer $t + un$ in Step 3 is divisible by R, and the division by R is a r-bit right shift operation.

In the following we explain the correctness of Montgomery multiplication. At first we claim that $(t + un)$ is divisible by R. There are integers k, l such that $u = tn' + kR$ and $n'n = -1 + lR$. Thus we obtain $t + un = R(tl + kn)$. Next note that $(t + un)/R = (t + un)R^{-1}$ mod $n = tR^{-1}$ mod $n = abR^{-1}$ mod n. Therefore v in Step 3 is contained in the same residue class of the output abR^{-1} mod n. Finally we show v is at most $2n$, namely $(t + un)/R < (n^2 + Rn)/R < 2n$.

Note that the output from Montgomery multiplication is different from the ab mod n. We describe how to apply the Montgomery multiplication to the modular exponentiation algorithm. Denote by $Mont(a, b)$ Montgomery multiplication for $a, b \in Z/nZ$ and $R = 2^r$, where $r = \lceil \log_2 n \rceil$. In the following we explain how to compute the modular exponentiation a^d mod n using the $Mont(\cdot, \cdot)$, where d is an integer. Let $d = \sum_{i=0}^{n-1} d[i]2^i$ be the bit representation of d. We apply the left-to-right binary method as follows:

Binary Method with Montgomery Multiplication

INPUT d, n, a, $(d[k-1], \ldots, d[1], d[0])$, R^2 mod n, $d[k-1] = 1$
OUTPUT a^d mod n

1: $t \leftarrow Mont(a, R^2)$
2: $s \leftarrow t$
3: for $i = k - 2$ down to 0
4:    $s \leftarrow Mont(s, s)$
5:    if $d[i] = 1$ then $s \leftarrow Mont(s, t)$
6: $s \leftarrow Mont(s, 1)$
7: return s

We assume that R^2 mod n is precomputed. In Step 1 we convert the integer a to $Mont(a, R^2) = aR$ mod n. In the main loop of the binary method, the integer in the register is represented by $s = a^{2k}R$ mod n for some $k \in Z$. Thus we obtain $Mont(s, s) = s^2R$ mod n in Step 4 and $Mont(s, t) = saR$ mod n in Step 5. After the main loop, the integer is still multiplied with R, namely $s = a^d R \bmod n$. Then we recover it to the standard representation by computing $Mont(s, 1) = s \bmod n$.

The algorithm calls only Montgomery multiplication as subroutine. The overheads from the standard binary method are Step 1 and Step 6, namely two Montgomery multiplications. Therefore, we can efficiently implement the binary exponentiation using the Montgomery multiplication.

2. Fast Variants of RSA Cryptosystem

The RSA cryptosystem is one of the most practical public key cryptosystems and is used throughout the world [RSA78]. In this section we show several efficient variants of RSA cryptosystems, namely RSA with Chinese remainder theorem, Multi-Prime RSA, and Multi-Exponent RSA.

The original RSA cryptosystem is as follows: Generate two random primes p, q, and let $n = pq$. Compute $L = \mathrm{LCM}\,(p-1, q-1)$, and find e, d which satisfy $ed = 1 \bmod L$. Then e, n are the public keys, and d is the secret key. Let $M \in Z/nZ$ be the plaintext. The algorithms of encryption and decryption consist of exponentiation to the eth and d^{th} powers modulo n, respectively. We encrypt the plaintext by the equation: $C = M^e \bmod n$. We decrypt the ciphertext by the equation: $M = C^d \bmod n$.

We can make e small, but the low exponent attacks should be considered ([CFPR96], [Cop96], [Has88]). The encryption process takes less computation and is fast. On the other hand, the decryption key d must be larger than $n^{1/2}$ to preclude Wiener's attack [Wie90] and its extensions ([VT97], [BD00]). Therefore, the cost of the decryption process is dominant for the RSA cryptosystem.

2.1. PKCS #1 Version 2.1

We review the RSA primitives described in the PKCS # 1 version 2.1, namely the RSA with Chinese Remainder Theorem (CRT) [QC82] and the Multi-Prime RSA [PKCS].

RSA with CRT. At first we describe the RSA primitive using the CRT [QC82]. The secret keys of this RSA variant are the primes p, q and d_p, d_q, where $n = pq$ and $d_p = d \bmod p - 1, d_q = d \bmod q - 1$. The value $M = C^d \bmod n$ can be computed from $M_p = C^{d_p} \bmod p$ and $M_q = C^{d_q} \bmod q$ using the CRT. We usually use the Garner's theorem:

$$M = M_p + pV, \quad V = (M_q - M_p)p^{-1} \bmod q.$$

The inverse value $p^{-1} \bmod q$ is also stored as a part of the secret key, and we do not have to compute the modular inversion, but the total secret key size becomes 1.5 times larger. In this case, the computation time of $C^d \bmod n$ using the CRT is about 4 time faster than the original one.

Multi-Prime RSA. We describe the Multi-Prime RSA [PKCS]. The public key (e, n) and the encryption function $f(x) = x^e \bmod n$ of the Multi-Prime RSA are equal to those of the RSA primitive, where e satisfies $\text{GCD}(e, \phi(n)) = 1$. We explain the decryption algorithm in the following.

At first we describe the simplest case of the Multi-Prime RSA, which uses the modulus $n = p_1 p_2 p_3$, where p_1, p_2, p_3 are primes with the same size. If we carefully choose the size of the primes, the modulus $n = p_1 p_2 p_3$ is secure for cryptographic purpose [Sil00]. For example, a 1024-bit Multi-Prime RSA modulus is as secure as a 1024-bit RSA modulus as we described in section 2.3. The plaintext M is encrypted by $C = M^e \bmod n$. The secret keys of the Multi-Prime RSA are the primes p_i and d_{p_i} for $i = 1, 2, 3$, where $d_{p_i} = d \bmod p_i - 1$. The message $M \bmod n$ can be computed from $M_{p_i} = C^{d_{p_i}} \bmod p_i$ for $i = 1, 2, 3$ using CRT. We use twice the Garner's algorithm for the CRT:

$$M = M_{p_1 p_2} + (p_1 p_2)V, \qquad V = (M_{p_3} - M_{p_1 p_2})(p_1 p_2)^{-1} \bmod p_3,$$
$$M_{p_1 p_2} = M_{p_1} + p_1 U, \qquad U = (M_{p_2} - M_{p_1})p_1^{-1} \bmod p_2.$$

The inverse values $((p_1 p_2)^{-1} \bmod p_3)$ and $(p_1^{-1} \bmod p_2)$ are stored as a part of the secret key, and we do not have to compute the modular inversion.

We describe the Multi-Prime RSA for general modulus $n = \Pi_i p_i$, where p_i are primes $i = 1, 2, \ldots, m$ as follows:

Decryption of Multi-Prime RSA

INPUT $C, d_{p_1}, \ldots, d_{p_m}, p_1, \ldots, p_m,$
$\quad p(1)\_inv\_p_2, p(2)\_inv\_p_3, \ldots, p(m-1)\_inv\_p_m$
OUTPUT M

1: for $i = 1$ to m
2: $\quad M_{p_i} = C^{d_{p_i}} \bmod p_i$
3: $\quad A = M_{p_1}$
4: for $i = 1$ to $m - 1$
5: $\quad p(i) = p(i-1)p_i$
6: $\quad F = M_{p_{i+1}} - A$
7: $\quad E = F(p(i)\_inv\_p_{i+1}) \bmod p_{i+1}$
8: $\quad A = A + p(i)E$
9: Return A

The plaintext M is encrypted by $C = M^e \bmod n$. The relation between the encryption exponent e and the decryption exponent d is $ed \equiv 1 \bmod \text{LCM}(\Pi_i(p_i - 1))$. Moreover, we denote $d_{p_i} = d \bmod p_i - 1$, $p(i) = p_1 \cdots p_i$ for $i = 1, 2, \ldots, m$ and $p(i)\_inv\_p_{i+1} = p(i)^{-1} \bmod p_{i+1}$ for $i = 1, 2, \ldots, m - 1$. Note that $p(1) = p_1$ and we define $p(0) = 1$.

2.2. Multi-Exponent RSA

In this section, we describe another variant of RSA cryptosystem, called Multi-Exponent RSA ([BS02, Tak98]).

Generation of the keys: Generate two random primes p, q, and let $n = p^k q$. Compute $L = \text{LCM}(p - 1, q - 1)$, and find e, d which satisfies $ed = 1 \bmod L$ and $\text{GCD}(e, p) = 1$. Let $d_p = d \bmod p - 1$ and $d_q = d \bmod q - 1$. Moreover, we pre-compute $(p^k)^{-1} \bmod q$ for the sake of efficiency. We denote $(p^k)\_inv\_q = (p^k)^{-1} \bmod q$. Then e, n are the public keys, and $d_p, d_q, p, q, (p^k)\_inv\_q$ are the secret keys.

Encryption: Let $M \in (Z/nZ)^*$ be the plaintext. We encrypt the plaintext by the equation:

$$C = M^e \bmod n. \tag{2.1}$$

Decryption: We decrypt $M_p = M \bmod p^k$ and $M_q = M \bmod q$ using the secret keys. The plaintext M can be recovered by the Chinese remainder theorem. Here, M_q is computed by $M_q = C^{d_q} \bmod q$ and M_p is computed by the Hensel lifting from $M \bmod p = C^{d_p} \bmod p$. The details of the decryption algorithm is as follows:

Multi-Exponent RSA Decryption

INPUT $C, e, d_p, d_q, p, q, (p^k)\_inv\_q, e\_inv\_p$
OUTPUT M

1: $M_q = C^{d_q} \bmod q$
2: $K = C^{d_p - 1} \bmod p$
3: $A = KC \bmod p$
4: for $i = 1$ to $k - 1$
5:     $p^{i+1} = p^i p$
6:     $F = A^e \bmod p^{i+1}$
7:     $E = C - F \bmod p^{i+1}$
8:     $B = EK(e\_inv\_p) \bmod p^{i+1}$
9:     $A = A + B$
10: $V = (M_q - A)((p^k)\_inv\_q) \bmod q$
11: $A = A + (p^k)V$
12: Return A

We explain that the decryption algorithm of Multi-Exponent RSA returns the correct value in the following. We prove that $M_i = M \bmod p^i$ can be lifted to $M_{i+1} = M \bmod p^{i+1}$ using the Multi-Exponent RSA decryption by the induction of i. We have proved it for $i = 1$ above. We assume that it is true for $i = j - 1$, which means the algorithm works correct up to $i = j - 1$ and we have obtained the correct $M_j = M \bmod p^j$. We will prove that $M_{j+1} \bmod p^{j+1}$ can be lifted from M_j using the Multi-Exponent RSA decryption. There is a unique positive integer $X_j < p$ such that $M_{j+1} = M_j + p^j X_j \bmod p^{j+1}$. If we find the value $X_j < p$, the M_{j+1} can be computed. From $C = (M_j + p^j X_j)^e = M_j^e + (p^j X_j) e M_j^{e-1} \bmod p^{j+1}$, we have the following relationship:

$$C - M_j^e = (p^j X_j) e M_j^{e-1} \bmod p^{j+1}. \tag{2.2}$$

The value $(M_j^{e-1})^{-1} \bmod p = M^{1-e} \bmod p$ is equal to $K = C^{d_p-1} = M^{1-e} \bmod p =$ in Step 2. Thus we obtain the formula of Step 7:

$$p^j X_j = (C - F)K(e^{-1} \bmod p) \bmod p^{j+1}. \qquad (2.3)$$

2.3. Size of Secret Primes

We discuss the size of the secret primes p and q. The RSA cryptosystem uses a composite number of the symmetry type pq, where p and q are the same bit size. The cryptosystem proposed in this paper bases its security on the difficulty of factoring the modulus $p^k q$. We have to carefully choose the size of p and q.

There are two types of fast factoring algorithm to consider: the number field sieve [LL91] and the elliptic curve method [Len87]. Other factoring algorithms have the same or slower running times, so the size of the RSA-modulus can be estimated by these two factoring algorithms ([KR95], [MOV96], [RS97], [Bre00]). Let $L_N[s, c] = exp((c+o(1)) \log^s(N) \log \log^{1-s}(N))$. The number field sieve is the fastest factoring algorithm, and the running time is estimated from the total bit size of the integer n to be factored, which is expected as $L_n[1/3, (64/9)^{1/3}]$. If we choose n to be larger than 1024 bits, the number field sieve becomes infeasible. In our case we have to make the modulus $n = p^k q$ larger than 1024 bits. The elliptic curve method is effective for finding primes which are divisors of the integer n to be factored. The running time is estimated in terms of the bit size of the prime divisor p. Its expected value is $L_p[1/2, 2^{1/2}]$. Note that the running time of the elliptic curve method is different from that of the number field sieve, and the order is much different. If we choose the primes to be larger than 342 bits, the elliptic curve method requires much more time in comparison with the NFS for factoring a 1024-bit composite number.

The factoring algorithm strongly depends on the implementation. The fastest implementation record for the number field sieve factored 512-bit RSA modulus [RSA155][1] and that for the elliptic curve method found a 183-bit prime factor [ECMNET]. Here again, we emphasize that there is a big difference in the cost between the number field sieve and the elliptic curve method. Therefore, if we choose the 1024-bit modulus $p^2 q$ with 342-bit primes p and q, neither of the factoring algorithms is feasible, so the Multi-Exponent RSA is secure for cryptographic purposes. Silverman discussed the key size based on the cost based analysis and he concluded that the 1024-bit modulus $p^2 q$ with p, q of the same size is secure against both the NFS and the ECM [Sil00].

We wonder if there exists factoring algorithms against the modulus with a square factor $p^2 q$. This factoring problem appeared in the list of open problems in number theoretic complexity by Adleman and McCurley [AM94], and it is unknown whether there exists $L_p[1/3]$-type sub-exponential algorithm which finds the primes of the composite number $p^2 q$. Peralta and Okamoto proposed a factoring algorithm against numbers of the form $p^2 q$ based on the elliptic curve method [PO96]. They focused on the fact that the Jacobi symbol modulo $p^2 q$ is equal to

---

[1] Recently, the RSA-160 (530 bits) was factored (See [BFKLB03]).

one modulo q, and the running time becomes a little bit faster than that for the original elliptic curve method. Recently, Ebinger and Teske reported that their algorithm does not improve the running time of the ECM [ET02].

2.4. Comparison

We compare the Multi-Prime RSA with the Multi-Exponent RSA. The running time and the size of secret key of both the Multi-Prime decryption and the Multi-Exponent decryption are discussed.

In order to estimate the running times we use the straight-forward algorithms described in book [MOV96]. An integer is represented $\sum_{i=0}^{n} u[i]b^i$ with base b and digit $u[i]$, where b is chosen suitable for computer architecture and $u[i] = 0, 1, \ldots, b-1$. The multiplication of two base digits is called single-precision multiplication (SPM). A multiplication of $(n+1)$ digits and $(t+1)$ digits requires $(n+1)(t+1)$ SPMs (Algorithm 14.12 and Note 14.15 of [MOV96]). A division of $(n+1)$ digits by $(t+1)$ digits requires $(n-t)(t+3)$ SPMs (Algorithm 14.20 and Note 14.25 of [MOV96]) We assume that a modular multiplication of $(n+1)$ digits requires $2n^2 + 5n + 1$ SPMs (Algorithm 14.28 of [MOV96]). Let $a^x \bmod b$ be a modular exponentiation of $(n+1)$ digits, where a, b and x are $(n+1)$-digit integers. If we compute the modular exponentiation using the standard binary method, it requires $1.5n(2n^2 + 5n + 1)$ SPMs on average. We assume that the computation times of an addition and a subtraction are negligible compared with those of the multiplication or the division.

At first we estimate the running time of Multi-Prime decryption. We assume that the secret primes p_i (for $i = 1, 2, \ldots, m$) are $(n + 1)$ digits. In the beginning of Step 2 we reduce the ciphertext C modulo p_i, which requires $(m - 1)n(n + 3)$ SPMs. Then m modular multiplications of $C^{d_{p_i}} \bmod p_i$ (for $i = 1, 2, \ldots, m$) are computed, which require $m(1.5n)(2n^2 + 5n + 1)$ SPMs. In Step 5 we compute multiplications of $(n + 1)$ digits and $(n + 1)i$ digits for $i = 1, 2, \ldots, m - 2$, which require $\frac{(m-2)(m-1)}{2}(n + 1)n + (m - 2)(n + 1)$ SPMs. In Step 7 we compute $m - 1$ modular multiplications of $(n+1)$ digits and $(n+1)i$ digits modulo $(n+1)$ digits for $i = 1, 2, \ldots, m-1$, which require $\frac{(m-1)m}{2}(2n^2+4n)+(m-1)(n+1)$ SPMs. In Step 8 we compute multiplication of $(n+1)$ digits and $i(n+1)$ digits for $i = 1, 2, \ldots, m-1$, which require $\frac{(m-1)m}{2}(n^2 + n) + (m - 1)(n + 1)$ SPMs. The size of the total secret key is $(3m - 1)(n + 1)$ digits. If we choose $m = 3$, then Multi-Prime decryption requires $9n^3 + 34.5n^2 + 31.5n + 5$ single-precision multiplications and the total size of the secret keys is $8(n + 1)$ digits.

Next we estimate the running time of Multi-Exponent decryption. Let c be the number of modular multiplications modulo n for computing $a^e \bmod n$ using some addition chain. For example, we can choose $c = 17$ for $e = 2^{16} + 1$ using the standard binary method. We assume that the secret primes p, q are $(n + 1)$ digits. ¿From Step 1 to Step 3, two modular multiplications of $C^{d_p} \bmod p$ and $C^{d_q} \bmod p$ are computed, which require $kn(n+3)+2(1.5n)(2n^2+5n+1)$ SPMs. In Step 5 we compute multiplications of $(n+1)$ digits and $(n+1)i$ digits for $i = 1, 2, \ldots, k - 1$, which require $\frac{(k-1)k}{2}(n+1)n+(k-1)(n+1)$ SPMs. The computation $A^e \bmod p^{i+1}$ for $i = 1, 2, \ldots, k-1$ in Step 6 requires $c((2n^2)(\frac{(k+1)(2k+1)k}{6} - 1) + (5n)(\frac{(k+1)k}{2} - 1) + (k-1))$ SPMs. The computation $E = C - F \bmod p^{i+1}$ for $i = 1, 2, \ldots, k-1$ in Step 7 requires $n((3+n)(k-1)k + (kn - n - 3)\frac{k(k-1)}{2} - n\frac{(k-1)(2k-1)k}{6})$ SPMs. The computation $EK(e\_inv\_p) \bmod p^{i+1}$ for $i = 1, 2, \ldots, k-1$ in Step 8 requires $2n(2n+1)\frac{k(k-1)}{2} + 2(2n^2 + 5n + 1)(k-1)$ SPMs. The CRT part of Step 10 and Step 11 requires $3kn^2 + (5k+2)n + 2$ SPMs. The size of the total secret key is $6(n+1)$ digits, which does not depend on the exponent k. If we choose $k = 2$, then Multi-Exponent decryption requires $6n^3 + (8c+34)n^2 + (10c+38)n + (c+5)$ single-precision multiplications. For the encryption exponent $e = 2^{16} + 1$ the Multi-Exponent decryption requires $6n^3 + 170n^2 + 208n + 22$ single-precision multiplications.

Here we choose the same bit length $n = 341, (b = 2)$ for the primes of both the Multi-Prime RSA with $m = 3$ and the Multi-Exponent RSA with $k = 2$. Then the decryption time of the Multi-Exponent RSA with $e = 2^{16} + 1$ is about 1.40 times faster than that of the Multi-Prime RSA.

TABLE 1. Comparison of efficiency for 1024-bit modulus

	PKCS #1 Multi-Prime RSA	Multi-Exponent RSA $(e = 2^{16} + 1)$
Key generation	880.12 ms	589.08 ms
Decryption	20.04 ms	14.13 ms
Secrete Keys	2736 bits	2052 bits

In order to demonstrate the efficiency of Multi-Exponent RSA, we implemented both the Multi-Prime RSA (Multi-Prime decryption with $m = 3$) and the (Multi-Exponent decryption with $k = 2, e = 2^{16} + 1$) on a Celeron 500 MHz using the LiDIA library version 2.0 [LiDIA] and TurboLinux 6.0. We also implemented the key generation of these schemes. In Table 1 we show the timings for 1024-bit modulus with 342-bit primes. The timings in the table are average values of 10,000 random instances. The improvements of the Multi-Exponent RSA over Multi-Prime RSA is as follows: the key generation is about 49% faster, the decryption is about 42% faster, and the key size is about 33% smaller .

3. Implementation Attack on RSA-CRT

Recently many attacks on the practical implementation of cryptography have been proposed. We describe some attacks on the RSA with Chinese remainder theorem (RSA-CRT). The algorithm and notation used in this section are same with those of the previous section.

Timing Attack. We explain the timing attack proposed by Kocher [Koc96]. The decryption algorithm of RSA-CRT computes $C^{d_p} \bmod p$ using the secret key d_p, p for a given ciphertext C. Before computing $C^{d_p} \bmod p$, we usually reduce the ciphertext C modulo p in order to achieve the faster decryption. However, if the ciphertext C is smaller than the secret prime p, then the ciphertext C is not reduced by modulo p. There is a difference of timing between $C < p$ and $C > p$ for computing $C^{d_p} \bmod p$ in the implementation. Let O_p be the oracle that answers 1 (or 0) if $C \leq p$ (or $C > p$) for a given ciphertext c. The attacker can recover p by the binary search as follows:

Timing Attack on RSA-CRT

Input: public key (n, e), bit-length B of p
Output: secret prime p such that $n = pq$.
---
1. Set $C \leftarrow 2^{B-2}$
2. For $i = B - 1$ down to 0
2.1. Set $A \leftarrow C + 2^i$
2.2. If $O_p(C) = 1$ holds, then set $C \leftarrow A$
3. Return C.

We assume that the most significant bit of prime p is one, namely $p \in [2^{B-2}, 2^{B-1}]$. In Step 1, we assign the lower bound of the secret prime p. In Step 2, the approximation of p is computed by adding C with 2^i for $i = B - 3, B - 4, \ldots, 1, 0$. If the oracle answers $O_p(C) = 1$, then we know $C \leq p$ and we assign the larger lower bound $C \leftarrow A$. In Step 3 we return the secret prime p. Recently Boneh et al. showed an experimental result of this timing attack in the server-client model — some implementation of SSL are vulnerable [BB03].

We explain a standard countermeasure against the timing attack, called the ciphertext blinding method. Before decrypting ciphertext $C = M^e \bmod n$, we randomize it by $C' = CR^e \bmod n$ with a random integer $R \in Z/nZ$. Then C' is decrypted by $M' = C'^d \bmod n = MR \bmod n$. Then the randomness R is removed by $M = M'R^{-1} \bmod n$. A drawback of this scheme is the expensive computation of the inverse $R^{-1} \bmod n$. While we can compute $R^{-1} \bmod n$ using the modular exponentiation $R^{\phi(n)-1} \bmod n$, it requires a large overhead.

Fault Attack. We explain the fault attack on RSA-CRT proposed by [JLQ99]. Let $C = M^e \bmod n$ be a ciphertext of message M. The fault attack tries to manipulate one bit of the message modulo q (we call M'_q) during the decryption of C (the message modulo p remains correct). Then the resulting message obtained by the Garner algorithm is

$$M' = M_p + pV, \quad V = (M'_q - M_p)p^{-1} \bmod q.$$

Note that $M' = M \bmod p$ and $M' \neq M \bmod q$, and thus the modulus can be factored by computing $gcd(M - M', n)$.

This attack was extended to more sophisticated fault attack ([BDL01, KR02]), etc. Aumüller et al. showed an experimental result of this attack [ABF+02]. They also proposed a countermeasure, which checks every process during the decryption, e.g. $M_p = M \bmod p$, $M^e = C \bmod p$, etc.

SPA/DPA. Kocher et al. proposed the power analysis against the cryptographic devices, namely the simple power analysis (SPA) and the differential power analysis (DPA) [KJJ99]. SPA tries to break the secret information by using a single power consumption as leaked data, and DPA additionally uses statistical analysis of the power consumption. The binary method for computing the modular multiplication $C^d \bmod n$ is vulnerable against SPA. The power consumption required for squaring and multiplication is not completely same, and the SPA can distinguish the two operations. Messerges et al. experimentally showed the binary method is vulnerable against SPA [MDS99]. An experimental DPA against the modular multiplication $C^{d_p} \bmod p$ was demonstrated by den Boer et al. [BLW02]. The ciphertext blinding method resists this type of attacks. The other countermeasure is the exponent blinding method, which randomizes the secret exponent by computing $d' = d + \phi(n)r$ for some integer r.

Novak Attack. Novak proposed an SPA against the Chinese remainder theorem part [Nov02]. He focused on the following implementation of $M_q - M_p \bmod q$; first compute $y = M_q - M_p$ and then $y = y + q$ if $y = M_q - M_p < 0$ holds. The experimental result shows the side channel information of $y = M_q - M_p < 0$ can be detected by SPA.

Novak developed a binary search algorithm of finding secret prime q using the oracle δ that answers $\delta(x) = 1$ for $x \le 0$ and $\delta(x) = 0$ otherwise. The characteristic function δ has the following property.

Lemma 3.1. *Let $q > p$. In ascending order of $x = 0, 1, 2, \ldots$, the sign $\delta(x)$ has the pattern*

$$\delta(x) = 1, 1, \ldots, 1, 0, 0, \ldots, 0, 1, \ldots, 1, 0, \ldots, 0, 1, \ldots.$$

If $\delta(x - 1) = 1$ and $\delta(x) = 0$ hold, then $q|x$ (q is a divisor of x).

Proof. We divide Z/nZ into two parts, namely $Z/nZ = LP \cup UP$, where $LP = \{0, 1, \ldots, p-1\}$, $UP = \{p, p+1, \ldots, n-1\}$. Note that $\delta(x) = 1$ holds for all $x \in LP$ due to $q > p$. Thus we assume that $x \in UP$. Let $f(x) = x \bmod q - x \bmod p$, then $\delta(x) = 1$ iff $f(x) \ge 0$. Next $\delta(kq) = 0, \delta(kq - 1) = 1$ holds for $0 < k < p$, because of $f(kq) < 0$ and $f(kq - 1) > (q - 1) - (p - 1) = 0$. Moreover, $\delta(k'p) = 1$ holds for $0 < k' < q$. Thus, two sets $x \bmod p$ and $x \bmod q$ have the following pattern:

$$x \bmod q = \{\ldots, q - 2, q - 1, 0, 1, 2, \ldots\},$$
$$x \bmod p = \{\ldots, l - 2, l - 1, l, l + 1, l + 2, \ldots\},$$

where l is an integer $0 < l < p$. Once $t \bmod q > t \bmod p$ holds for successive $t \bmod q$, then $\delta(x) = 1$ for $x = t, t + 1, \ldots, q - 2, q - 1$. Thus the corresponding δ sequence is $\delta(x) = \underbrace{0, \ldots, 0}_{q-s}, \underbrace{1, \ldots, 1}_{s}$ for $x \bmod q = 0, 1, 2, \ldots, q - 1$ and some integer s. Consequently we have proved the proposition. $\qquad\square$

From this lemma we can construct a binary search algorithm for secret prime q in the setting of the adaptive chosen ciphertext attack.

Novak Attack on RSA-CRT

Input: public key (n, e), bit-length B of p

Output: secret prime q such that $n = pq$

1. Choose $x_0, x_1 \in Z/nZ$ s.t. $x_0 > x_1$, $x_1 - x_0 < 2^B$, $\delta(x_0) = 1, \delta(x_1) = 0$
2. Set $LB = x_0, UB = x_1$
3. While $LB \neq UB$ do the following
3.1. $M = \lceil (LB + UB)/2 \rceil$
3.2. Compute $\delta(M)$ of $C = M^e \bmod n$
3.2. If $\delta(M) = 1$, then $LB = M$, otherwise $UB = M$
4. Compute $q = gcd(M, n)$
5. Return q

We should note that Novak's attack is effective for $M_q \approx M_p$ only, because y often takes different signs. A countermeasure against SPA is to always compute $y' = y + q$, and then we choose y' if and only if $M_q - M_p < 0$. Note that the exponent blinding method does not resist Novak attack.

Remark 3.2. The timing attack and Novak attack are effective on the chosen ciphertext attack setting. However, they are not feasible to the probabilistic signature, e.g., RSA-PSS [PKCS]. Even if the attacker chooses a message M, it is randomized by padding function ρ such that $\rho(M)$. The attacker cannot control the size of $\rho(M)$. Very recently, Fouque et al. proposed an extension of Novak attack on RSA with the randomly chosen messages, but this attack is restricted to the unbalanced modulus s.t. $p \not\approx q$ [FMP03].

4. EPOC Cryptosystem

EPOC-2 is a public-key cryptosystem that can be proved IND-CCA2 under the factoring assumption in the random oracle model. It was written into a standard specification P1363 of IEEE, and it has been a candidate of the public-key cryptosystem in several international standards (or portfolio) on cryptography, e.g. NESSIE, CRYPTREC, ISO, etc.

In this section we analyze a chosen ciphertext attack against EPOC-2 from NESSIE by observing the timing of the reject signs from the decryption oracle. We construct an algorithm, which can factor the public modulus using the difference of the reject symbols. For random 384-bit primes, the modulus can be factored with probability at least $1/2$ by invoking about 385 times to the decryption oracle.

4.1. EPOC-2 Cryptosystem

We review the EPOC-2 encryption scheme in the following. There are several different versions of EPOC-2 as scientific papers ([FO99b], [FO01]) or as specifications of international standards (or portfolio) ([IEEE], [NESSIE], [CRYPTREC]), etc. Here we consider the current specification and the notation of the self-evaluation report that were submitted to the 2nd phase of NESSIE project [NESSIE]. The specifications of EPOC-2 from IEEE and CRYPTREC are similar to that of NESSIE.

Key Generation
$pLen$, the bit length of prime p
$n = p^2 q$, the modulus, $g \in Z/nZ$ s.t. $p \mid ord_{p^2}(g)$
$g_p = g \bmod p^2$, $h = g^n \bmod n$
Public-key: $(n, g, h, pLen)$, Secret key: p, q, g_p
Encryption of m
$m \in \{0,1\}^*$, a message, $\sigma \in \{0,1\}^{pLen-1}$, a random integer
$c_2 = m \oplus G(\sigma)$, $c_1 = g^\sigma h^{H(m,\sigma,c_2)} \bmod n$
The ciphertext: (c_1, c_2)
Decryption of c
$\sigma^* = L(c_1^{p-1} \bmod p^2) L(g_p^{p-1} \bmod p^2)^{-1} \bmod p$ $(= [[c_1]]_g)$
If $\|\sigma^*\| \le pLen - 1$, then go to next step, otherwise return **Reject**,
$m^* = c_2 \oplus G(\sigma^*)$, if $c_1 = g^{\sigma^*} h^{H(m^*,\sigma^*,c_2)} \bmod q$ holds,
then output m^* as decryption of (c_1, c_2), otherwise return **Reject**.

FIGURE 1. EPOC-2 Cryptosystem

EPOC-2 is an probabilistic encryption scheme based on the hardness of the factoring problem of $n = p^2 q$, where p, q are distinct prime numbers. Let $pLen$ be the bit-length of the prime p. In the key generation, we additionally generate an integer g of Z/nZ such that $p \mid ord_{p^2}(g)$ (the order of $g \bmod p^2$ in group Z/p^2Z is divisible by p). Moreover, we compute $g_p = g \bmod p^2$ and $h = g^n \bmod n$. Then the public-key and the secret key of EPOC-2 are $(n, g, h, pLen)$ and (p, q, g_p), respectively. Let G be a mask generation function: $\{0,1\}^{pLen-1} \to \{0,1\}^*$ and let H be a hash function: $\{0,1\}^* \times \{0,1\}^{pLen-1} \times \{0,1\}^* \to \{0,1\}^{rLen}$, where $rLen$ is the bit-length of the output of the hash function H, defined by the security parameter for primes p, q. There are several variations of EPOC-2 in the key generation (e.g. h of CRYPTREC is chosen differently), but the proposed attack is not affected by its variations.

The encryption of EPOC-2 is computed as follows: $m \in \{0,1\}^*$ is a message with arbitrary bit length. For a random integer $\sigma \in \{0,1\}^{pLen-1}$, we encrypt the message m as follows: $c_2 = m \oplus G(\sigma)$, $c_1 = g^\sigma h^{H(m,\sigma,c_2)} \bmod n$. The ciphertext of m is $C = (c_1, c_2)$.

The decryption of EPOC-2 is as follows: At first the first component c_1 of the ciphertext C is decrypted by computing $\sigma^* = L(c_1^{p-1} \bmod p^2) L(g_p^{p-1} \bmod p^2)^{-1} \bmod p$, where $L(x) = (x-1)/p$. We also denote by $[[c_1]]_g = L(c_1^{p-1} \bmod p^2) L(g_p^{p-1} \bmod p^2)^{-1} \bmod p$. Here we have the first reject function based on the size of σ^*. Let $|\sigma^*|$ be the bit-length of σ^*. If $|\sigma^*| > pLen-1$, we stop the decryption procedure and return **Reject**. Otherwise we go to next step. This rejection function is necessary in order to prevent the attack proposed by Joye, Quisquater, and Yung [JQY01]. We denote by Reject 1 this reject symbol. Note that the ciphertext $C = (c_1, c_2)$ with $c_1 = g^r \bmod n$ for integer $r < 2^{pLen-1}$ and random integer $c_2 \in Z/nZ$ is not rejected by this test and go to next step, although C is an invalid ciphertext (it is rejected in the next step). The message m^* is decrypted by computing $m^* = c_2 \oplus G(\sigma^*)$. Here we have the second rejection function. If $c_1 = g^{\sigma^*} h^{H(m^*, \sigma^*, c_2)} \bmod q$ holds, then output m^* as decryption of (c_1, c_2), otherwise return Reject.

History of EPOC. We shortly review the history of the specifications of EPOC family. We mainly discuss how the reject symbol that is returned by the decryption oracle has been changed.

The cryptographic primitive of EPOC was proposed by Uchiyama and Okamoto at EUROCRYPT'98 [OU98]. The one-wayness and the semantic security (IND-CPA) of the primitive are as secure as factoring and p-subgroup problem in the standard model. The EPOC primitive has no reject symbol in the decryption oracle, so that it is insecure against the chosen ciphertext attack. Indeed, Joye, Quisquater, and Yung proposed a chosen ciphertext attack against the EPOC primitive at rump session of Eurocrypt'98 [JQY98]. Let c be the ciphertext of m, which is larger than the secret key p. If the attacker obtains the decrypted message m' of the ciphertext c, the modulus n of the EPOC primitive can be factored by computing $gcd(m - m', n) = p$.

At CRYPTO'99 Fujisaki and Okamoto proposed a conversion technique that enhances the EPOC primitive to be IND-CCA2 under factoring assumption in the random oracle model [FO99b]. In the decryption process the conversion checks the integrity of the ciphertext by re-encrypting the message. This version of EPOC was submitted to the IEEE P1363a on October 1998 [IEEE]. Joye et al. proposed a chosen ciphertext attack against the submission (ver. D6 of EPOC-2 in IEEE) [JQY01]. We call it the JQY attack. The JQY attack based on the chosen ciphertext attack against the EPOC primitive [JQY98], and the attack tries to find the approximation of the secret prime p by adaptively asking ciphertexts (whose message is as large as p) to the decryption oracle. In the paper [JQY01] they suggested that if the decryption oracle checks the size of the integer decrypted by the EPOC primitive, the JQY attack is no longer successful. The reject symbol arisen from this rejection function is called Reject 1 in Section 4.2. The current version of EPOC-2 from IEEE supports this reject function and the JQY attack does not work for it.

The security reduction from [FO99b] was evaluated for general cryptographic primitives and the advantage of the reduction was not so tight. Fujisaki and Okamoto proved the better security reduction in the paper [FO01]. In that paper they included the reject treatment proposed by Joye et al. (Reject 1).

EPOC-2 have been proposed at NESSIE 1st/2nd phase [NESSIE], at CRYP-TREC 2000/2001 [CRYPTREC]. These versions support the rejection function (Reject 1). We notice that the specification of the EPOC-2 from NESSIE 1st phase is different — the decryption oracle returns only one reject symbol after completing all steps of the decryption process. Although EPOC has not incorporated into the draft of ISO Standard, EPOC-2 will be included in the standard [Sho01].

We summarize these history of EPOC related to the reject function in the next table.

EPOC Version	Based Paper	Reject 1	JQY Attack
EPOC Primitive	[OU98]	NO	YES
IEEE (ver. D6)	[FO99b]	NO	YES
IEEE Version	[FO01]	YES	NO
CRYPTREC Version	[FO01]	YES	NO
NESSIE Version	[FO01]	YES	NO

4.2. Reject Timing Attack on EPOC-2

We describe the reject timing attack against the current version of EPOC-2. Dent initially proposed a reject timing attack against EPOC-2 cryptosystem [Den02a]. The attack is based on the JQY attack [JQY01]. Although the current version of EPOC-2 is secure against the JQY attack, the reject timing attack can break it using the timing of the two different rejection symbols.

At first we show an observation on the decryption algorithm of EPOC-2. In the decryption process, the calculation of the integrity check $c_1 = g^{\sigma^*} h^{H(m^*, \sigma^*, c_2)}$ mod q is executed if and only if $|\sigma^*| \leq pLen - 1$ holds. It has two modular exponentiations modulo q and their running time is relatively slow — several milliseconds in standard computation environments. The timing attack, which measures the timing of receiving Reject from the decryption oracle, can observe the calculation. Therefore we use the following assumption:

For any ciphertext $C = (c_1, c_2)$, the attacker can know that $\sigma^* = [[c_1]]_g$ satisfies $\sigma \in \{0, 1\}^{pLen-1}$ or not by asking the ciphertext C to the decryption oracle.

¿From this assumption, the attacker can tell the difference of two reject symbols: the error of the primitive decryption (Reject 1) and the error of the integrity check (Reject 2) in the decryption oracle. If the decrypted ephemeral integer σ^* by the EPOC primitive is large than 2^{pLen-1}, then Reject 1 is returned. The reject symbol Reject 2 is returned, if both $|\sigma^*| \leq pLen - 1$ and $c_1 \neq g^{\sigma^*} h^{H(m^*, \sigma^*, c_2)}$ mod q for $m^* = c_2 \oplus G(\sigma^*)$ hold.

Decryption of c
$\sigma^* = L(c_1^{p-1} \bmod p^2) L(g_p^{p-1} \bmod p^2)^{-1} \bmod p (= [[c_1]]_g)$
If $|\sigma^*| \leq k - 1$, then go to next step, otherwise return Reject 1,
$m^* = c_2 \oplus G(\sigma^*)$, if $c_1 = g^{\sigma^*} h^{H(m^*, \sigma^*, c_2)}$ mod q holds,
then output m^* as decryption of (c_1, c_2), otherwise return Reject 2.

We state this observation as the following lemma.

Lemma 4.1. *Let $C = (c_1, c_2)$ be a ciphertext of EPOC. Let $\sigma^* = [[c_1]]_g$ be the ephemeral integer decrypted by the EPOC primitive. We have the following conditions:*

(1) $\sigma^* > 2^{pLen-1} \Rightarrow$ Reject 1,

(2) $c_1 \neq g^{\sigma^*} h^{H(m^*,\sigma^*,c_2)} \bmod q$ for $m^* = c_2 \oplus G(\sigma^*) \Rightarrow$ Reject 2.

Main Idea. We describe the main idea of our attack. Let $C = (c_1, c_2)$ be a valid ciphertext of EPOC-2. Let $\sigma^* = [[c_1]]_g$. The attacker manipulates the ciphertext C by multiplying it with an integer $D = g^\alpha \bmod n$, namely $C' = (c_1/D \bmod n, c_2)$. The ciphertext C' is rejected in the decryption oracle with overwhelming probability, because the second integrity check fails $(c_1 \neq g^{\sigma^*} h^{H(m^*,\sigma^*,c_2)} \bmod q$ for $m^* = c_2 \oplus G(\sigma^*))$. However the attacker can know a relation of σ^* and α based on the rejection symbols: Reject 1 or Reject 2. Indeed we have the following lemma.

Lemma 4.2. *Assume* $p > 2^{pLen-1} + \alpha$ *for a positive integer* $\alpha < 2^{pLen-1}$. *Let* $C = (c_1, c_2)$ *be a ciphertext of EPOC-2. Let* $[[c_1]]_g = \sigma^*$. *The reject symbol against the ciphertext* $C' = (c_1/D, c_2)$ *with* $D = g^\alpha \bmod n$ *is equal to* Reject 2 *if and only if* $\sigma^* > \alpha$ *holds.*

Proof. Note that $[[c_1/D \bmod n]]_g = \sigma^* - \alpha \bmod p$. If $\sigma^* > \alpha$ holds, then we have $[[c_1/D \bmod n]]_g = \sigma^* - \alpha < 2^{pLen-1}$ and the reject symbol is Reject 2. If $\sigma^* < \alpha$ holds, then we have $[[c_1/D \bmod n]]_g = \sigma^* - \alpha + p$. Because of $\sigma^* - \alpha + p > \sigma^* + 2^{pLen-1} > 2^{pLen-1}$, the ciphertext C' is reject with Reject 1. $\qquad\square$

Therefore the difference of the reject symbols yields an oracle, which answers that the condition $\sigma^* > \alpha$ holds or not for a given ciphertext $C = (c_1, c_2)$ and an integer α, where $[[c_1]]_g = \sigma^*$. If we ask the ciphertext C with different many α to the decryption oracle, the attacker can find the approximation of σ^*.

Once we know an algorithm which answers $\sigma^* = [[c_1]]_g$ for a given ciphertext $C = (c_1, c_2)$, we can factor the modulus n. We have the following lemma.

Lemma 4.3. *Let* $c_1 = g^\sigma \bmod n$ *with* $\sigma > p$. *If we know the decryption* $\sigma^* = L(c_1^{p-1} \bmod p^2)L(g_p^{p-1} \bmod p^2)^{-1} \bmod p = [[c_1]]_g$, *then we can factor the modulus by computing* $\gcd(\sigma - \sigma^*, n) = p$.

Proof. Because $\sigma^* = [[c_1]]_g = \sigma \bmod p$ holds, we have $p|(\sigma - \sigma^*)$. $\qquad\square$

This lemma is used for the security proof of the EPOC primitive [OU98] and the chosen ciphertext attack on the EPOC primitive (JQY attack) [JQY01].

In the following we will construct an algorithm that finds σ^* for a given ciphertext c_1 and an integer σ using the oracle above. We show the high level description of the attack as follows.

1. Choose an integer σ such that $\sigma > 2^{pLen} > p$. Compute $c_1 = g^\sigma \bmod n$. Let $C = (c_1, c_2)$ be a ciphertext for random $c_2 \in \{0,1\}^*$.

2. The attacker asks the manipulated ciphertext $C' = (c_1/D, c_2)$ to the decryption oracle, where $D = g^\alpha \bmod n$ for some integers $0 < \alpha < 2^{pLen-1}$. He/She analyzes the reject symbols for the ciphertexts C'.

3. The attacker outputs $\sigma^*(= \sigma \bmod p)$ and factors n by $\gcd(\sigma - \sigma^*, n)$.

Initialization. In the beginning of the attack, we require a ciphertext $c_1 = g^\sigma \bmod n$ with $\sigma > p$ and $\sigma^* = \sigma \bmod p < 2^{pLen-1}$. This condition is easily tested by asking the ciphertext $C = (c_1, c_2)$ to the decryption oracle.

If we choose the σ from the interval $[2^{pLen}, 2^{pLen+1}]$, then $\sigma \bmod p < 2^{pLen-1}$ is satisfied with probability at least $1/2$. Thus we have the following initialization for our attack.

Initialization

Input: $n, g, pLen$

Output: $C = (c_1, c_2)$ with $\sigma > p, \sigma^* = \sigma \bmod p < 2^{pLen-1}$

---

1. Generate $\sigma \in_R [2^{pLen}, 2^{pLen+1}]$
2. Compute $C = (c_1, c_2)$, where $c_1 = g^\sigma \bmod n$, $c_2 \in_R \{0,1\}^*$
3. Ask C to the decryption oracle. If we receive Reject 1, goto step 1
4. Return C

Outline of Attack. We explain the outline of the reject timing attack. The attack guesses the bits of $\sigma^* = \sigma \bmod p$ from the most significant bit. From Lemma 4.2, the attack can guess σ^* is larger or smaller than a given bound. Let UB, LB be the upper bound and lower bound of σ^* known by the oracle call, respectively. UB and LB are stored as temporary values. The attacker tries to shrink the distance $LB - UB$ by asking the oracle. ¿From the initialization, we have $LB = 0$ and $UB = 2^{pLen-1}$ in the beginning. Moreover we assume that $p > 2^{pLen-1} + 2^{pLen-2}$, which is satisfied with probability at least $1/2$ for randomly chosen $pLen$-bit primes.

We explain how to guess whether $\sigma^* > 2^{pLen-2}$ or not. We assume that the ciphertext is already initialized. Let $D = g^\alpha \bmod n$ for $\alpha = 2^{pLen-2}$. If we ask the ciphertext $C' = (c_1/D \bmod n, c_2)$ to the decryption oracle, from Lemma 4.2 we have following relationship:

(1) $\sigma^* > 2^{pLen-2} \Leftrightarrow$ Reject 2
(2) $\sigma^* < 2^{pLen-2} \Leftrightarrow$ Reject 1

Therefore we know the σ^* is in intervals $[0, 2^{pLen-2}]$ or $[2^{pLen-2}, 2^{pLen-1}]$. Indeed we assign $LB = Av$ if Reject 2, otherwise, $UB = Av$, where $Av = (LB + UB)/2$.

In order to guess the next most bits, the following normalization of the ciphertext is executed. If σ^* is in the upper interval $[2^{pLen-2}, 2^{pLen-1}]$, then the ciphertext is normalized by calculating $c_1/D \bmod n$ with integer $D = g^\beta \bmod n$ for $\beta = 2^{pLen-1}$. Here $c_1/D \bmod n$ was already computed in the previous step, and we just assign $c_1 = c_1/D \bmod n$ if integer σ in the upper interval $[2^{pLen-2}, 2^{pLen-1}]$.

Then we manipulate the ciphertext $c_1/D = g^\alpha \bmod n$ for $\alpha = 2^{pLen-3}$. ¿From $p > 2^{pLen-1} + 2^{pLen-2}$, the prime p satisfies the assumption of Lemma 4.2 for $\alpha = 2^{pLen-3}$, namely $p > 2^{pLen-1} + 2^{pLen-3}$. By asking $C' = (c_1/D, c_2)$ to the oracle, we know σ^* is in the intervals $[0, 2^{pLen-3}]$, $[2^{pLen-3}, 2^{pLen-2}]$, and thus σ^* is in one of intervals $[(i-1)2^{pLen-3}, i2^{pLen-3}]$ for $i = 1, 2$. Consequently we assign the new upper/lower bound of σ^* by selecting $LB = Av$ if Reject 2 or $UB = Av$ otherwise, where $Av = (LB + UB)/2$.

If we iterate these steps, the lower bits of integer σ^* can be found. We eventually find the approximation of σ^* with a small error bound.

Details of Algorithm. We describe the algorithm that factors the modulus n using the reject timing attack.

Reject Timing Attack on EPOC
Input: $n, g, pLen$ (Public Key)
Output: p, q (Secret Primes)
1. $\sigma, C = (c_1, c_2) \leftarrow$ Initialization$(n, g, pLen)$
2. $LB = 0, UB = 2^{pLen-1}$
3. For $i = 2$ to $pLen$
4. $\alpha = 2^{pLen-i}, A = c_1/g^\alpha \bmod n$
5. $Av = (LB + UB)/2$
6. Ask $C = (A, c_2)$ to the decryption oracle for random c_2
7. If Reject 2, then $c_1 = A, LB = Av$
8. If Reject 1, then $UB = Av$
9. If $n > p = gcd(\sigma - \sigma^*, n) > 1$ for $\sigma^* \in [LB, UB]$, then compute $q = n/p^2$.
8. Return p, q

In step 1, the first component c_1 of the ciphertext satisfies $\sigma^* = [[c]]_g < 2^{pLen-1}$. The difference $UB - LB$ in step 9 is at most 2 because we iterate $pLen - 1$ times the approximation finding algorithm. The gcd computation in step 9 is performed at most twice.

If $gcd(\sigma - \sigma^*, n) = 1$ or n holds, the algorithm fails to factor the modulus. If the prime p satisfies the condition $p > 2^{pLen-1} + 2^{pLen-2}$, the algorithm always outputs the prime p due to Lemma 4.2. If we chose randomly the prime from $2^{pLen-1} < p < 2^{pLen}$, this requirement is satisfied with probability at least $1/2$. Thus we have the following theorem.

Theorem 4.4. *Algorithm* RTA_EPOC *can factor the modulus n with probability at least $1/2$ if the secret prime p is randomly chosen from $pLen$-bit primes.*

Note that our attack is not restricted to these above conditions. The algorithm works in general situations, although the probability of success may change.

An Example. We demonstrate an example of the reject timing attack against EPOC-2. A key from the test vector distributed by NTT [EPOC] is examined, namely the public key we tested is as follows:

$$g = 2$$

$$n = 4152082246314238505355867044990543688751999781554451624701106598380392$$
$$15424048181304933087306526022590055923617205805726379994358837338676663$$
$$89399817044374374516393502103692694950685397085324359599936584125928819$$
$$41150432040813228433987742010304682227696157664293649691342062932597079$$
$$1087072520403087020944100627497661376574278795207514968894743015333$$

The initial integer σ should satisfy both $\sigma > 2^{pLen}$ and $\sigma^* = \sigma \bmod p < 2^{pLen-1}$. The criteria $\sigma^* < 2^{pLen-1}$ is examined by asking $C = (c_1, c_2)$ to the decryption oracle, where $c_1 = g^\sigma \bmod n$ and c_2 is an random integer. We chose the following value:

$$\sigma = 45967310160463599521985689654286761916183121570558985158546512 6859$$
$$6009452061357008908408225953085289537652667149 45265$$

Then we compute the main loop of the reject timing attack. At step i the ciphertext $C = (c_1, c_2)$ is manipulated by computing $c_1/D \bmod n$ with $D = g^\alpha \bmod n$ for $\alpha = 2^{pLen-i}$ for some integers. The manipulated ciphertext is asked to the decryption oracle and the attacker knows the lower bound LB and the upper bound UB of the approximation of integer σ^*. The difference $UB - LB$ is shrinking for each iteration. We list up several first and last values of the lower bound LB and the upper bound UB of integer σ^* from our experiment.

i	Rejection	LB (Lower Bound) / UB (Upper Bound)
2	Reject 2	$LB[2] = 0$
		$UB[2] = 9850501549098619803069760025035903451269934817 61636166$
		$6987073351061430442874302652853566563721228910201656997576704$
3	Reject 1	$LB[3] = 49252507745493099015348800125179517256349674088 0818083$
		$3493536675530715221437151326426783281860614455100828498788352$
		$UB[3] = UB[2]$
4	Reject 2	$LB[4] = LB[3]$
		$UB[4] = 7387876161823964852302320018776927588452451 11321227125$
		$0240305013296072832155726989640174922790921682651242748182528$
...	...	...
		...
382	Reject 1	$LB[382] = 5067002360797088795877355807121865756660222 7664593135$
		$5609664755149218764337332512491852217190886534297 7731283270068$
		$UB[382] = LB[381]$
383	Reject 2	$LB[383] = LB[382]$
		$UB[383] = 50670023607970887958773558071218657566602227664593135$
		$56096647551492187643373325124918522171908865342977731283270070$
384	Reject 1	$LB[384] = 5067002360797088795877355807121865756660222 7664593135$
		$5609664755149218764337332512491852217190886534297 7731283270069$
		$UB[384] = LB[383]$

At the end of the main loop, we know $UB - LB = 1$. Finally we compute $gcd(\sigma - \sigma^*, n)$ for integer $\sigma^* \in [LB[384], UB[384]]$. If $0 < gcd(\sigma - \sigma^*, n) < n$ holds, we obtain the secret prime $p = gcd(\sigma - \sigma^*, n)$ and the other factor by computing $q = n/p^2$. In our example, we have successfully obtained the secret prime p.

$$gcd(\sigma - \sigma^*, n) = 378838416036532422019982950613121461170975827449275 4483578$$
$$070660900906313023019798049352503528330530089896128597 2933$$

How to Repair EPOC-2. The reject timing attack against EPOC is effective, because there are two different rejection processes. One possibility to resist the attack is to use only one rejection function.

Modified Decryption
$\sigma^* = L(c^{p-1} \bmod p^2) L(g_p^{p-1} \bmod p^2)^{-1} \bmod p$,
$m^* = c_2 \oplus G(\sigma^*)$, $\quad c_1^* = g^{\sigma^*} h^{H(m^*, \sigma^*, c_2)} \bmod q$.
Event 1 = $\{|\sigma^*| \le k - 1\}$, Event 2 = $\{c_1 = c_1^*\}$.
Set $\Gamma = \{$Event 1 $\wedge$ Event 2$\}$.
If $\Gamma = 1$, output m^* as decryption of (c_1, c_2), otherwise, return Reject.

The decryption oracle always computes both σ^* and c_1^*. Then the Boolean logic functions Event 1 = $\{|\sigma^*| \leq k - 1\}$ and Event 2 = $\{c_1 = c_1^*\}$ are evaluated. Then the control bit Γ = {Event 1 $\wedge$ Event 2} is assigned. If Γ = 1 holds, m^* is output as the decryption of (c_1, c_2), otherwise Reject is returned. Because the timings for computing the values of Event 1 and the control bit Γ are negligible, the attacker can not know the value of Event 1.

On the other hands, the implementer also has to care the treatment of Event 1 and Event 2. If the history of Event 1 is stored in a log file, then the attacker can perform the reject timing attack by knowing the log file. This was discussed by Manger [Man01] and was extended to the memory dump attack [KCJ+01]. As described in the current version of PKCS #1, the implementer should make efforts not to correlate Event 1 with the decrypted ciphertexts.

4.3. Relation to Other Cryptosystems

In this section we discuss how the reject timing attack can be extended to other provably secure cryptosystems.

EPOC-2 consists of the encryption primitive from Okamoto and Uchiyama [OU98] and the conversion technique from Fujisaki and Okamoto [FO99b] that makes the encryption primitive semantically secure against the chosen ciphertext attack. We can consider two possible variations of EPOC-2: (1) to replace the conversion technique to others. (2) to replace the encryption primitive to others. We discuss how these variations are secure against the reject timing attack.

Other Conversion Techniques. We can convert the EPOC primitive to be secure against the chosen ciphertext attack using different conversions. Fujisaki and Okamoto proposed a conversion technique that converts an IND-CPA scheme to be IND-CCA [FO99a]. The Fujisaki-Okamoto conversion with the EPOC primitive is called EPOC-1. The EPOC primitive is IND-CPA under a non-standard assumption, e.g. the p-subgroup assumption [OU98], and there is no significant advantage for EPOC to use this conversion. Pointcheval proposed a general conversion technique that can convert a one-way function to be IND-CCA2 [Poi00]. However the security reduction is not so tight. A conversion technique that has the tight security reduction from the encryption primitive is the REACT conversion [OP01], which is based on the conversion proposed Bellare and Rogaway [BR93]. The REACT conversion with the EPOC primitive is called EPOC-3. In Figure 2, we show a construction of EPOC using REACT conversion, which is modified – the original description in [OP01] does not support two different rejection symbols – in order to compare the security of the converted scheme against the reject timing attack with that of EPOC-2.

Here h is a hash function that tests the integrity check in the decryption oracle. In this construction there are two different reject functions. If the timing of calculating $m^* = c_2 \oplus G(\sigma^*)$ and $c_3 = H(m^*, \sigma^*, c_1, c_2)$ are relative slow, then the attacker have a possibility to tell the difference between two reject symbols. However, the computation time of hash functions is generally very fast. On the

Encryption of m
$m \in \{0,1\}^*$, a message, $\sigma \in \{0,1\}^{pLen-1}$, a random integer
$c_2 = m \oplus G(\sigma)$, $c_1 = g^\sigma h^r \bmod n$ for random $r \in \{0,1\}^{rLen}$
$c_3 = H(m, \sigma, c_1, c_2)$
The ciphertext: (c_1, c_2, c_3)

Decryption of c
$\sigma^* = L(c_1^{p-1} \bmod p^2) L(g_p^{p-1} \bmod p^2)^{-1} \bmod p$
If $|\sigma^*| \leq pLen - 1$, then go to next step, otherwise return Reject,
$m^* = c_2 \oplus G(\sigma^*)$, if $c_3 = H(m^*, \sigma^*, c_1, c_2)$ holds,
then output m^* as decryption of (c_1, c_2, c_3), otherwise return Reject.

FIGURE 2. EPOC using REACT Conversion

other hand, the integrity check of EPOC-2 computes two modular exponentiations. The attacker has a larger chance to break EPOC-2 using the reject timing attack. Similarly the EPOC-1 using the Fujisaki-Okamoto conversion is vulnerable against the reject timing attack, because it utilizes the re-encryption technique.

Coron et al. proposed the GEM family ([CHJPPT02a], [CHJPPT02b]). The construction of their conversion technique is based on hash functions and a symmetric key cryptosystem — the invalid ciphertexts are rejected by the integrity test using the hash functions and the symmetric key cryptosystem. The computation time of these integrity test are much faster than that of the re-encryption test of EPOC-2, and the reject timing attack on GEM family is more difficult.

4.4. Other Encryption Primitives

The conversion technique by Fujisaki and Okamoto is designed for converting any one-way function to be IND-CCA2 [FO99b]. The Fujisaki-Okamoto conversion is applicable to other cryptographic primitives. We discuss the possibility of adapting our attack to other primitives.

We shortly describe their conversion technique in the following. We do not describe the hybrid version using symmetric key system, but the scheme using hash functions. Let (pk, sk) be the public key for a given security parameter k. Let MSP be the message space and let k_1 be the size of message space. E_{pk} is the encryption function that encrypts a message in MSP with k_2-bit random integer. D_{sk} is the decryption function that satisfies $D_{sk}(E_{pk}(\sigma, r)) = \sigma$ for $\sigma \in$ MSP and a random k_2-bit integer r. We use a hash function $h : \{0,1\}^{k_1} \to \{0,1\}^*$, and a mask generation function $g : \{0,1\}^* \to \{0,1\}^{k_2}$. In Figure 3 we describe the Fujisaki-Okamoto conversion technique.

Here we have two different rejection functions. The first one is arisen from checking $D_{sk}(c_1) \in$ MSP. In the case of EPOC-2, the message space MSP is equal to $\{0,1\}^{pLen-1}$, which is strictly smaller than the space of $D_{sk}(c_1)$. Here, most

Encryption of m
$m \in \{0,1\}^*$, a message, $\sigma \in_R$ MSP,
$c_2 = m \oplus G(\sigma)$, $c_1 = E_{pk}(\sigma, H(m,\sigma))$
The ciphertext: $c = (c_1, c_2)$
Decryption of c
$\sigma^* = D_{sk}(c_1)$,
If $\sigma^* \in$ MSP, then go to next step, otherwise return Reject,
$m^* = c_2 \oplus G(\sigma^*)$, if $c_1 = E_{pk}(\sigma^*, H(m^*, \sigma^*))$ holds,
then output m^* as decryption of (c_1, c_2), otherwise return Reject.

FIGURE 3. Fujisaki-Okamoto Conversion

standard cryptographic primitives like RSA, ElGamal-type encryption are permutation and they satisfy the following condition:

$$\text{MSP} = \{D_{sk}(c_1) | c_1 = E_{pk}(\sigma, r) \text{ for } \sigma \in \text{MSP}, r \in \{0,1\}^{k_2}\}. \qquad (4.1)$$

The message space MSP is not smaller than the space of the decrypted messages. Therefore, any ciphertexts are not rejected by the first test. However, when we design a new cryptographic primitive, we have to care the treatment of the reject function.

The cryptographic primitives that have the degenerated MSP are the Rabin-type cryptosystem ([Bon01], [NSS01]) or the NICE cryptosystem [BST01]. It is an interesting problem to investigate the security against the reject timing attack. Note that Manger's attack [Man01] is not effective on the Rabin-type cryptosystem because the Rabin primitive has no reject function based on the size of the integer-to-octet conversion. On the other hand, Paillier primitive is known as an extension of EPOC to the ring Z/n^2Z where n is the RSA modulus [Pai99]. The message space of the Paillier primitive is Z/nZ, which is equal to that of the decrypted messages, and thus the Paillier primitive has no reject function based on checking $D_{sk}(c_1) \in$ MSP. We can not break the cryptosystem based on the Paillier primitive using the reject timing attack.

5. Elliptic Curve Cryptosystem

In this section we explain several efficient algorithms used for elliptic curve cryptosystems.

We assume that $K = \mathbb{F}_p$ $(p > 3)$ be a finite field with p elements. Elliptic curves over K can be represented by the equation

$$E(K) := \{(x,y) \in K \times K | y^2 = x^3 + ax + b \ (a, b \in K, \ 4a^3 + 27b^2 \neq 0)\} \cup O, \quad (5.1)$$

where O is the point of infinity. Every elliptic curve is isomorphic to a curve of this form, and we call it the Weierstrass form. An elliptic curve $E(K)$ has an additive group structure. Let $P_1 = (x_1, y_1)$, $P_2 = (x_2, y_2)$ be two elements of $E(K)$ that are different from O and satisfy $P_2 \neq \pm P_1$. Then the sum $P_1 + P_2 = (x_3, y_3)$ is defined as follows:

$$x_3 = \lambda^2 - x_1 - x_2, \quad y_3 = \lambda(x_1 - x_3) - y_1, \qquad (5.2)$$

where $\lambda = (y_2 - y_1)/(x_2 - x_1)$ for $P_1 \neq P_2$, and $\lambda = (3x_1^2 + a)/(2y_1)$ for $P_1 = P_2$. We call $P_1 + P_2 (P_1 \neq P_2)$ the elliptic curve addition (ECADD) and $P_1 + P_2 (P_1 = P_2)$, that is $2P_1$, the elliptic curve doubling (ECDBL). Let d be an integer and P be a point on the elliptic curve $E(K)$. The scalar multiplication is to compute the point dP. There are several enhancements of the scalar multiplication. The first one is to represent the elliptic curve $E(K)$ with a different coordinate system, whose scalar multiplication is more efficient. For examples, a projective coordinate and a class of Jacobian coordinate has been studied [CMO98]. The second one is to use an efficient addition chain. The addition-subtraction chain is an example [MO90]. We can also apply the addition chains developed for the ElGamal cryptosystem over finite fields [Gor98].

5.1. Scalar Multiplication

Let d be an n-bit integer and P be a point of the elliptic curve $E(K)$. A standard way for computing the scalar multiplication dP is to use the binary expression of $d = d_{n-1}2^{n-1} + d_{n-2}2^{n-2} + \cdots + d_1 2 + d_0$, where $d_{n-1} = 1$ and $d_i = 0, 1$ ($i = 0, 1, \ldots, n-2$). Then the following binary method computes $d*P$ efficiently. We call these methods the binary methods (or the add-and-double methods). On average they require $(n - 1)$ ECDBLs + $(n - 1)/2$ ECADDs.

Scalar Multiplication using Binary Method

INPUT d, P, $(d[k-1], \ldots, d[1], d[0])$, $d[k-1] = 1$	
OUTPUT dP	
(Left-to-Right)	(Right-to-Left)
1. $Q[0] = P$	1. $Q[0] = P, Q[1] = 0$
2. for $i = k - 2$ down to 0	2. for $i = 0$ to $k - 1$
2.1. $Q[0] = ECDBL(Q[0])$	2.1. $Q[1] = ECADD(Q[1], d[i]Q[0])$
2.2. $Q[0] = ECADD(Q[0], d[i]P)$	2.2. $Q[0] = ECDBL(Q[0])$
3: return $Q[0]$	3. return $Q[1]$

The main difference between right-to-left and left-to-right algorithm is the treatment of ECADD. The left-to-right algorithm utilizes the ECADD with the base point P, so that the Z-coordinate of ECADD is always one. On the contrary, the Z-coordinate of the input point used for the ECADD of right-to-left algorithm is not one. Therefore, the running time using the left-to-right algorithm achieves faster computation time.

Width-w Non-Adjacent Form. The fastest method with less memory is the width-w non-adjacent form (NAF). The width-w NAF represents an n-bit integer $d = \sum_{i=0}^{n} d_w[i]2^i$, where $d_w[i]$ are odd integers with $|d_w[i]| < 2^{w-1}$ and there are at most one non-zero digit among w-consecutive digits. In order to compute the scalar multiplication we pre-compute the table with points $P, 3P, \ldots, (2^{w-1}-1)P$, which has 2^{w-2} points including base point P. The points with the opposite sign are generated on the fly during the scalar multiplication.

Generating Width-w NAF	Scalar Multiplication with Width-w NAF
INPUT An n-bit d, a width w OUTPUT $\qquad d_w[n], d_w[n-1], \ldots, d_w[0]$	INPUT $d_w[i]$, P, $(|d_w[i]|)P$ OUTPUT dP
1. $i \leftarrow 0$ 2. While $d > 0$ do the following 2.1. if d is odd then do following 2.1.1. $d_w[i] \leftarrow d$ mods 2^w 2.1.2. $d \leftarrow d - d_w[i]$ 2.2. else $d_w[i] \leftarrow 0$ 2.3. $d \leftarrow d/2$, $i \leftarrow i+1$ 3: Return $d_w[n], d_w[n-1], \ldots, d_w[0]$	1. $Q \leftarrow d_w[c]P$ $\quad$ for the largest c with $d_w[c] \neq 0$ 2. For $i = c - 1$ to 0 2.1. $Q \leftarrow \text{ECDBL}(Q)$ 2.2. $Q \leftarrow \text{ECADD}(Q, d_w[i]P)$ 3. Return Q

Several methods for generating the width-w NAF have been proposed ([KT92], [MOC97], [BSS99], [Sol00]). Generating_Width-w_NAF is an algorithm that generates the width-w NAF proposed by Solinas [Sol00]. Notation "mods 2^w" at Step 2.1.1 stands for the signed residue modulo 2^w, namely $\pm 1, \pm 3, \ldots, \pm(2^{w-1} - 1)$. Note that the next $(w-1)$ consecutive bits of non-zero bits in the width-w NAF are always zero. It is known that the density of the non-zero bits of the width-w NAF is asymptotically equal to $1/(1 + w)$. We show an example of non-adjacent form as follows:

binary string	10011101111001110001011011110001101010111111001
$w = 2$	$1010001000\bar{1}0100\bar{1}0010\bar{1}00\bar{1}000\bar{1}00100\bar{1}0\bar{1}0\bar{1}00001\bar{1}001$
$w = 3$	$1003000\bar{1}0000030 0\bar{1}00003 00\bar{1}000\bar{1}00100030030000\bar{1}001$

Scalar_Multiplication_with_Width-w_NAF is an algorithm of computing the scalar multiplication using the width-w NAF. It is calculated from the most significant bit — elliptic curve doubling (ECDBL) at Step 2.1 is executed for each bit and elliptic curve addition (ECADD) at Step 2.2 is executed if and only if $d_w[i]$ is non-zero. Therefore we have to compute $(c+1)$-time ECDBLs and $(c+1)/(1+w)$-time ECADDs, where c is the largest integer with $d_w[c] \neq 0$. If we choose larger width w, then the scalar multiplication becomes faster, but with more memory.

5.2. Efficient Coordinate System

There are several ways to represent a point on an elliptic curve. The costs of computing an ECADD and an ECDBL depend on the representation of the coordinate system. The detailed description of the coordinate systems is given in [CMO98].

TABLE 2. Computing times of ECADD and ECDBL

Coordinate	ECADD		ECDBL	
System	$Z \neq 1$	$Z = 1$	$a \neq -3$	$a = -3$
A	$2M + 1S + 1I$	—	$2M + 2S + 1I$	
P	$12M + 2S$	$9M + 2S$	$7M + 5S$	$7M + 3S$
J	$12M + 4S$	$8M + 3S$	$4M + 6S$	$4M + 4S$
J^C	$11M + 3S$	$8M + 3S$	$5M + 6S$	$5M + 4S$
J^m	$13M + 6S$	$9M + 5S$	$4M + 4S$	

The major coordinate systems are as follows: the affine coordinate system (A), the projective coordinate system (P), the Jacobian coordinate system (J), the Chudonovsky coordinate system (J^C), and the modified Jacobian coordinate system (J^m). We summarize the costs in Table 2, where M, S, I denotes the computation time of a multiplication, a squaring, and an inverse in the definition field K, respectively. The speed of ECADD or ECDBL can be enhanced when the third coordinate is $Z = 1$ or the coefficient of the definition equation is $a = -3$. We show the concrete algorithms for computing $ECDBL^J$, $ECDBL^{J,a=-3}$, $ECADD^J$, $ECADD^{J,Z=1}$.

$ECDBL^J$, $4M + 6S + 11A$

Input (X_1, Y_1, Z_1, a)
Output (X_2, Y_2, Z_2)

$R_4 \leftarrow X_1, R_5 \leftarrow Y_1, R_6 \leftarrow Z_1$

$R_1 \leftarrow R_4^2$
$R_2 \leftarrow R_5^2$
$R_2 \leftarrow R_2 + R_2$
$R_4 \leftarrow R_4 \times R_2$
$R_4 \leftarrow R_4 + R_4$
$R_2 \leftarrow R_2^2$
$R_2 \leftarrow R_2 + R_2$
$R_3 \leftarrow R_6^2$
$R_3 \leftarrow R_3^2$
$R_6 \leftarrow R_5 \times R_6$
$R_6 \leftarrow R_6 + R_6$
$R_5 \leftarrow R_1 + R_1$
$R_1 \leftarrow R_1 + R_5$
$R_3 \leftarrow a \times R_3$
$R_1 \leftarrow R_1 + R_3$
$R_3 \leftarrow R_1^2$
$R_5 \leftarrow R_4 + R_4$
$R_5 \leftarrow R_3 - R_5$
$R_4 \leftarrow R_4 - R_5$
$R_1 \leftarrow R_1 \times R_4$
$R_4 \leftarrow R_1 - R_2$

$X_2 \leftarrow R_5, Y_2 \leftarrow R_4, Z_2 \leftarrow R_6$

$ECDBL^{J,a=-3}$, $4M + 4S + 13A$

Input (X_1, Y_1, Z_1)
Output (X_2, Y_2, Z_2)

$R_4 \leftarrow X_1, R_5 \leftarrow Y_1, R_6 \leftarrow Z_1$

$R_2 \leftarrow R_5^2$
$R_2 \leftarrow R_2 + R_2$
$R_3 \leftarrow R_4 \times R_2$
$R_3 \leftarrow R_3 + R_3$
$R_2 \leftarrow R_2^2$
$R_2 \leftarrow R_2 + R_2$
$R_5 \leftarrow R_5 \times R_6$
$R_5 \leftarrow R_5 + R_5$
$R_6 \leftarrow R_6^2$
$R_4 \leftarrow R_4 + R_6$
$R_6 \leftarrow R_6 + R_6$
$R_6 \leftarrow R_4 - R_6$
$R_4 \leftarrow R_4 \times R_6$
$R_6 \leftarrow R_4 + R_4$
$R_4 \leftarrow R_4 + R_6$
$R_6 \leftarrow R_4^2$
$R_6 \leftarrow R_6 - R_3$
$R_6 \leftarrow R_6 - R_3$
$R_3 \leftarrow R_3 - R_6$
$R_4 \leftarrow R_4 \times R_3$
$R_4 \leftarrow R_4 - R_2$

$X_2 \leftarrow R_5, Y_2 \leftarrow R_4, Z_2 \leftarrow R_5$

ECADD^J, $12M + 4S + 7A$
Input $(X_1, Y_1, Z_1, X_2, Y_2, Z_2)$
Output (X_3, Y_3, Z_3)

$R_2 \leftarrow X_1, R_3 \leftarrow Y_1, R_4 \leftarrow Z_1$
$R_5 \leftarrow X_2, R_6 \leftarrow Y_2, R_7 \leftarrow Z_2$

$R_1 \leftarrow R_7^2$
$R_2 \leftarrow R_2 \times R_1$
$R_3 \leftarrow R_3 \times R_7$
$R_3 \leftarrow R_3 \times R_1$
$R_1 \leftarrow R_4^2$
$R_5 \leftarrow R_5 \times R_1$
$R_6 \leftarrow R_6 \times R_4$
$R_6 \leftarrow R_6 \times R_1$
$R_5 \leftarrow R_5 - R_2$
$R_7 \leftarrow R_4 \times R_7$
$R_7 \leftarrow R_5 \times R_7$
$R_6 \leftarrow R_6 - R_3$
$R_1 \leftarrow R_5^2$
$R_4 \leftarrow R_6^2$
$R_2 \leftarrow R_2 \times R_1$
$R_5 \leftarrow R_1 \times R_5$
$R_4 \leftarrow R_4 - R_5$
$R_1 \leftarrow R_2 + R_2$
$R_4 \leftarrow R_4 - R_1$
$R_2 \leftarrow R_2 - R_4$
$R_6 \leftarrow R_6 \times R_2$
$R_1 \leftarrow R_3 \times R_5$
$R_1 \leftarrow R_6 - R_1$

$X_3 \leftarrow R_4, Y_3 \leftarrow R_1, Z_3 \leftarrow R_7$

ECADD$^{J, Z_1=1}$, $8M + 3S + 7A$
Input $(X_1, Y_1, X_2, Y_2, Z_2)$
Output (X_3, Y_3, Z_3)

$R_2 \leftarrow X_1, R_3 \leftarrow Y_1, R_5 \leftarrow X_2$
$R_6 \leftarrow Y_2, R_7 \leftarrow Z_2$

$R_1 \leftarrow R_7^2$
$R_2 \leftarrow R_2 \times R_1$
$R_3 \leftarrow R_3 \times R_7$
$R_3 \leftarrow R_3 \times R_1$
$R_5 \leftarrow R_5 - R_2$
$R_7 \leftarrow R_5 \times R_7$
$R_6 \leftarrow R_6 - R_3$
$R_1 \leftarrow R_5^2$
$R_4 \leftarrow R_6^2$
$R_2 \leftarrow R_2 \times R_1$
$R_5 \leftarrow R_1 \times R_5$
$R_4 \leftarrow R_4 - R_5$
$R_1 \leftarrow R_2 + R_2$
$R_4 \leftarrow R_4 - R_1$
$R_2 \leftarrow R_2 - R_4$
$R_6 \leftarrow R_6 \times R_2$
$R_1 \leftarrow R_3 \times R_5$
$R_1 \leftarrow R_6 - R_1$

$X_3 \leftarrow R_4, Y_3 \leftarrow R_1, Z_3 \leftarrow R_7$

6. Side Channel Attacks on ECC

In this section we describe several side channel attacks on ECC.

6.1. SPA on ECC

The SPA observes the power consumption of devices, and detects the difference of operations using the secret key. The scalar multiplication using binary method (ECC binary method) is vulnerable to the SPA. The scalar multiplication is computed by the addition formula, namely ECDBL and ECADD, based on the bit of the secret scalar. The operation ECADD in ECC binary method is computed if and only if the underlying bit is 1, although the operation ECDBL is always computed. The addition formula is assembled by the basic operations of the definition field (See Section 5.2). There are differences between the basic operations of ECDBL and those of ECADD. Thus the SPA attacker can detect the secret bit. In order to resist the SPA, we have to eliminate the relations between the bit information and their addition formula.

Double-and-add-always method. Coron proposed a simple countermeasure, which is called as the double-and-add-always method [Cor99]. The double-and-add-always method is described as follows:

Double-And-Add-Always Method

Input: $d = (d_{n-1} \cdots d_1 d_0)_2$, $P \in E(K)$ $(d_{n-1} = 1)$

Output: dP

1. $Q[0] \leftarrow P$
2. For $i = (n - 2)$ down to 0 do:
2.1. $Q[0] \leftarrow \text{ECDBL}(Q[0])$
2.2. $Q[1] \leftarrow \text{ECADD}(Q[0], P)$
2.3. $Q[0] \leftarrow Q[d_i]$
3. Return($Q[0]$)

The double-and-add-always method always computes ECADD whether $d_i = 0$ or 1. Therefore, attackers cannot guess the bit information of d using SPA.

6.2. DPA and Countermeasures

The differential power analysis (DPA) observes many power consumptions and analyze these information together with statistic tools. Even if a method is secure against SPA, it might not secure against the DPA. The DPA attacker tries to guess that the computation cP for an integer c is performed during the exponentiation. He/She gathers many power consumptions cP_i for $i = 1, 2, 3, \ldots$, and detects the spike arisen from the correlation function based on the specific bit of cP_i. The DPA can break the binary method, because the sequence of points generated by the binary method is deterministic and the DPA can find correlation for a specific bit.

Coron pointed out that it is necessary to insert random numbers during the computation of dP to prevent DPA [Cor99]. The randomization eliminates the correlation between the secret bit and the sequence of points. The standard randomization methods are Coron's 3rd [Cor99] and Joye-Tymen countermeasures [JT01]. The main idea of these countermeasures is to randomize the base point before starting the scalar multiplication. If the base point is randomized, there is no correlation among the power consumptions of each scalar multiplication. The DPA cannot obtain the spike of the power consumption derived from the statistical tool. We describe the two standard randomization in the following. There are other DPA countermeasures (e.g. randomized window methods [Wal02a, IYTT02], etc), but in this paper we aim at investigating the security of Coron's 3rd and Joye-Tymen countermeasures.

Coron's 3rd Countermeasure. Coron proposed three countermeasures against DPA for elliptic curve cryptosystems [Cor99]. But, Okeya and Sakurai pointed out that only his 3rd countermeasure was secure against DPA [OS00]. This countermeasure is based on randomization of Jacobian coordinates. To prevent DPA we transform $P = (x, y)$ in affine coordinate to $P = (r^2 x : r^3 y : r)$ in Jacobian coordinates for a random value $r \in K^*$. This randomization produce the randomization in each representation of point and the randomization of power consumption during scalar multiplication dP.

Joye-Tymen Countermeasure. Joye-Tymen countermeasure uses an isomorphism of an elliptic curve [JT01]. For a random value $r \in K^*$, an elliptic curve $E : y^2 = x^3 + ax + b$ and the point $P = (x, y)$ can be transformed to its isomorphic class like $E' : y'^2 = x'^3 + a'x' + b'$ for $a' = r^4 a$, $b' = r^6 b$ and $P' = (x', y') = (r^2 x, r^3 y)$. Instead of computing dP, we compute $Q' = dP' = (x_{Q'}, y_{Q'})$ on E' and then pull back $Q = (x_Q, y_Q)$ by computing $x_Q = r^{-2} x_{Q'}$ and $y_Q = r^{-3} y_{Q'}$. This countermeasure can hold the Z-coordinate equal to 1 during the computation of dP' and it enables good efficiency.

6.3. Goubin's Power-Analysis Attack

Goubin proposed a new power analysis using a point that cannot be randomized by neither Coron's 3rd nor Joye-Tymen countermeasure [Gou03]. Goubin focused on the following two points: $(x, 0)$ and $(0, y)$. The points $(x, 0)$ and $(0, y)$ are represented by $(X : 0 : Z)$ and $(0 : Y : Z)$ in Jacobian coordinates. Even these points are randomized by Coron's 3rd countermeasure, one of the coordinate remains zero, namely $(rX : 0 : rZ)$ and $(0 : rY : rZ)$ for some random integer $r \in K^*$. Similarly Joye-Tymen randomization cannot randomize these points. Therefore, the attacker can detect whether the points $(x, 0)$ or $(0, y)$ are used in the scalar multiplication using the DPA.

The attacker can break the secret scalar using these points as follows: We can compute $P = (c^{-1} \bmod \#E)(0, y)$ for give scalar c, because the order of the curve $\#E$ is prime. If the scalar multiplication computes the point $cP = (0, y)$, the power consumption of the next step is always significantly different from the others. Thus the DPA can detect whether cP is computed or not for the scalar c during the scalar multiplication. The attacker can obtain the whole secret scalar by recursively applying this process.

Goubin's attack is effective on the curves that have points $(x, 0)$ or $(0, y)$. The point $(x, 0)$ is not on the curves with prime order ($\neq 2$), because the order of the point $(x, 0)$ is 2. The point $(0, y)$ appears on the curve if b is quadratic residue modulo p, which is computed by solving $y^2 = b$.

7. Zero-Value Point Attack on ECC

In this section, we explain the zero-value point attack (ZVP attack). The ZVP attack is an extension of Goubin's attack, and it utilizes the auxiliary register which takes the zero-value in the definition field. We investigate the zero-value registers that are randomized by neither Coron's 3rd nor Joye-Tymen countermeasure.

The addition formula is assembled by the operations of the base field, namely the multiplication and the addition. We have about 20 different operations of the auxiliary registers for both ECDBL and ECADD (See the addition formula in Appendix A WHERE IS?). There are a lot of possibilities that the value of the auxiliary registers become zero. The zero-value registers of the ECDBL and those of the ECADD are quite different. We examine all possible operations that take zero in the auxiliary registers.

We show several criteria, with which the ZVP attack is effective — the attack is strongly depending on the implementation of the addition formula. We list up all possible security conditions and we discuss their effectiveness. Moreover, we demonstrate the attack is effective on several standard curves.

Outline of Attack. We describe the outline of the zero-value point attack in the following. The goal of the zero-value point attack is to break the secret scalar by adaptively choosing the base point Q. We assume that the scalar multiplication is computed by the binary method. But, we can apply our zero-value point attack to the SPA countermeasures using the deterministic addition chain described in section 3.1. The attacker breaks the secret key from the most significant bits. The second most significant bit d_{n-2} can be broken by checking whether one of addition formulae ECDBL($2Q$), ECADD($2Q, Q$), ECDBL($3Q$), and ECADD($3Q, Q$) is computed. If we can generate the zero-value register for these addition formulae, we can detect the second most bit — $d_{n-2} = 0$ holds if ECDBL($2Q$) or ECADD($2Q, Q$) has the zero-value register, and $d_{n-2} = 1$ holds if ECDBL($3Q$) or ECADD($3Q, Q$) has the zero-value register.

Next, we assume that $(n - i - 1)$ most significant bits $(d_{n-1}, \cdots, d_{i+1})_2$ of d are known. We can break the i-th bit d_i by checking whether one of ECDBL($2kQ$), ECADD($2kQ, Q$), ECDBL($(2k + 1)Q$), and ECADD($(2k + 1)Q, Q$) is computed, where $k = \sum_{j=i+1}^{n-1} d_j 2^{j-i-1}$. We know that $d_i = 0$ holds if ECDBL($2kQ$) or ECADD($2kQ, Q$) has the zero-value register, and $d_i = 1$ holds if ECDBL($(2k + 1)Q$) or ECADD($(2k + 1)Q, Q$) has the zero-value register. Therefore if we find a point P that takes the zero-value register at ECDBL, we can use the base point $Q = (c^{-1} \mod \#E)P$ for some integer c for this attack. On the other hand, in order to use the zero-value register at ECADD, the base point Q that causes the zero-value register at ECADD(cQ, Q) must be found.

Thus the attacker has to find the points Q which cause the zero-value register at ECDBL(cQ) or ECADD(cQ, Q) for given integer c. The ECDBL causes the zero-value register for a given one point Q, but the zero-value register for the ECADD depends on the two points Q and cQ. In this paper we call these points *zero-value point (ZVP)*.

Possible Zero-Value Points from ECDBL. We investigate the ZVP for addition formulae in Jacobian coordinates, but the same arguments apply to addition formulae in projective coordinates. We search the zero-value points in the following. We examine all auxiliary registers of the ECDBL in Jacobian coordinates. There are 21 intermediate values for ECDBL^J, as described in Appendix A. We prove the following theorem.

Theorem 7.1. *Let E be an elliptic curve over a prime field $\mathbf{F}_p$ defined by $y^2 = x^3 + ax + b$. The elliptic curve E has the zero-value point $P = (x, y)$ of ECDBL$^J(P)$ if and only if one of the following five conditions is satisfied:*
(ED1) $3x^2 + a = 0$,
(ED2) $5x^4 + 2ax^2 - 4bx + a^2 = 0$,

(ED3) *the order of P is equal to 3,*
(ED4) $x(P) = 0$ *or* $x(2P) = 0$, *and*
(ED5) $y(P) = 0$ *or* $y(2P) = 0$.
Moreover, the zero-value points are not randomized by Coron's 3rd or Joye-Tymen randomization.

Conditions (ED4) and (ED5) are exactly those of Goubin's attack.
We will prove this theorem in the following. Let $P_1 = (X_1 : Y_1 : Z_1)$, and $P_3 = (X_3 : Y_3 : Z_3) = \text{ECDBL}^J(P_1)$. The intermediate values of ECDBL can be zero if and only if one of the following value is zero.

$$X_1, Y_1, Z_1, X_3, Y_3, M, -S + M^2, S - T$$

Here $Z_1 = 0$ implies $P = O$, which never appears for input of $\text{ECDBL}^J(P)$. The conditions $X_1 = 0$, $Y_1 = 0$, $X_3 = 0$, and $Y_3 = 0$ are equivalent to $x(P) = 0$, $y(P) = 0$, $x(2P) = 0$, and $y(2P) = 0$ which are exactly the points discussed by Goubin. Next $M = 3X_1^2 + aZ_1^4 = 0$ implies the condition $3x^2 + a = 0$, which is the condition (ED1). Note that neither Coron's 3rd nor Joye-Tymen randomization can randomize the points. Indeed the randomized point $(X_1' : Z_1') = (r^2 X_1 : r Z_1)$ by Coron's 3rd randomization satisfies $3X_1'^2 + aZ_1'^4 = r^4(3X_1^2 + aZ_1^4) = 0$, where $r \in K^*$. The randomized point $(X_1'' : Z_1'') = (s^2 X_1 : Z_1)$ and curve parameter $a'' = s^4 a$ by Joye-Tymen randomization satisfies $3X_1''^2 + a'' Z_1''^4 = s^4(3X_1^2 + aZ_1^4) = 0$, where $s \in K^*$. The condition $-S + M^2 = 0$ implies $-4X_1Y_1^2 + (3X_1^2 + aZ_1^4)^2 = 0$, which is equivalent to $-4xy^2 + (3x^2 + a)^2 = 0$, namely condition (ED2). The condition $S - T = 0$ implies $x_1 = x_3$. This occurs only if $2P = \pm P$, which means $P = O$ or the order of P equals to 3, namely condition (ED3).

Remark 7.2. In order to obtain $T = -2S + M^2$ we computed with the following ordered additions $W = -S + M^2$ and then $T = W - S$. If we compute $-2S$ and then $-2S + M^2$, condition (ED2) does not appear in the ECDBL. Thus we should avoid the former order of the two additions for the implementation of ECDBL.

Possible Zero-Value Points from ECADD. We investigate the possible zero-value points from ECADD, namely all possible zero-value points P which satisfies ECADD (cP, P) for some integer c. There are 23 auxiliary values in the ECADD. We examine the addition formula in Jacobian coordinates. We prove the following theorem.

Theorem 7.3. *Let E be an elliptic curve over prime field $\mathbf{F}_p$ defined by $y^2 = x^3 + ax + b$. The elliptic curve E has the zero-value point $P = (x, y)$ of $\text{ECADD}^J(cP, P)$ for some $c \in \mathbf{Z}$ if and only if one of the following seven conditions is satisfied:*
(EA1) *P is a y-coordinate self-collision point,*
(EA2) $x(cP) + x(P) = 0$,
(EA3) $x(P) - x(cP) = \lambda(P, cP)^2$,
(EA4) $2x(cP) = \lambda(P, cP)^2$, $x(cP) = \lambda(P, cP)^2$, *or* $x(P) = \lambda(P, cP)^2$,
(EA5) *the order of P is odd,*
(EA6) $x(cP) = 0$, $x(P) = 0$, *or* $x((c + 1)P) = 0$, *and*

(EA7) $y(cP) = 0$, $y(P) = 0$, or $y((c+1)P) = 0$.
Moreover, the zero-value points are not randomized by Coron's 3rd or Joye-Tymen randomization.

A point $P = (x, y)$ is called the y-coordinate self-collision point if there is a positive integer c such that the y-coordinate of the point cP is equal to y. Conditions (EA6) and (EA7) are those of Goubin's attack.

Let $P_1 = (X_1 : Y_1 : Z_1)$, $P_2 = (X_2 : Y_2 : Z_2)$, and $\text{ECADD}^J(P_1, P_2) = (X_3 : Y_3 : Z_3)$. Here we can set $P_1 = cP_2$ for some integer c. If one of the following values is zero, at least one of the intermediate values must be zero.

$$X_1, Y_1, Z_1, X_2, Y_2, Z_2, X_3, Y_3, H, R, U_1 H^2 - X_3.$$

Here if one of $X_1, Y_1, X_2, Y_2, X_3, Y_3$ is zero, this provides conditions (EA6) and (EA7). $Z_1 = 0$, $Z_2 = 0$ and $H = 0$ imply $P_1 = O$, $P_2 = O$, and $P_1 = \pm P_2$, respectively, which never appear for input of $\text{ECADD}^J(P_1, P_2)$. Next, $R = Y_1 Z_2^3 - Y_2 Z_1^3 = 0$ implies $y_1 = y_2$, where $y_1 = Y_1/Z_1^3$ and $y_2 = Y_2/Z_2^3$, namely condition (EA1). This is equal to the y-coordinate collision point. Note that neither Coron's 3rd nor Joye-Tymen randomization can randomize the points. Indeed the randomized point $(Y_1' : Z_1') = (r^3 Y_1 : r Z_1), (Y_2' : Z_2') = (s^3 Y_2 : s Z_2)$ by Coron's 3rd randomization satisfies $Y_1' Z_2'^3 - Y_2' Z_1'^3 = r^3 s^3 (Y_1 Z_2^3 - Y_2 Z_1^3) = 0$, where $r, s \in K^*$. The randomized point $(Y_1'' : Z_1'') = (t^3 Y_1 : Z_1), (Y_2'' : Z_2'') = (t^3 Y_2 : Z_2)$ by Joye-Tymen randomization satisfies $Y_1'' Z_2''^3 - Y_2'' Z_1''^3 = t^3 (Y_1 Z_2^3 - Y_2 Z_1^3) = 0$, where $t \in K^*$. Finally $U_1 H^2 - X_3 = 0$ implies $3U_1 H^2 + H^3 - R^2 = 0$, which is $x_1 - x_3 = 0$. This occurs only if $(c+1)P = \pm cP$, which means $P = O$ or the order of P equals to $2c + 1$, namely condition (EA5).

The other possible intermediate values appear only at the computation of $X_3 = -H^3 - 2U_1 H^2 + R^2$. For ECADD^J in Appendix A, we compute $-H^3 + R^2$, but we can differently implement it. Indeed, we have 6 possible intermediate values:

(a1) $-H^3 - 2U_1 H^2$,
(a2) $-2U_1 H^2 + R^2$,
(a3) $-H^3 + R^2$,
(a4) $-H^3 - U_1 H^2$,
(a5) $-U_1 H^2 + R^2$,
(a6) $(-H^3 - U_1 H^2) + R^2$.

We examine these conditions in the following. These above points are randomized by neither Coron 3rd nor Joye-Tymen randomization. Condition (a1) implies $H(X_2 Z_1^2 + X_1 Z_2^2) = 0$, namely $H = 0$ or $x_1 + x_2 = 0$ in affine coordinate. The condition $H = 0$ has already appeared in the multiplicative ZVP. $x_1 + x_2 = 0$ implies $x(cP) + x(P) = 0$, which is equal to condition (EA2). Condition (a2) implies $-2X_1 Z_2^2(X_2 Z_1^2 - X_1 Z_2^2)^2 + (Y_2 Z_1^3 - Y_1 Z_2^3)^2 = 0$, which is $2x_1 = \lambda^2$ in affine coordinate. It is condition (EA4). Condition (a3) implies $-(X_2 Z_1^2 - X_1 Z_2^2)^3 + (Y_2 Z_1^3 - Y_1 Z_2^3)^2 = 0$, which is $x_2 - x_1 = \lambda(P_2, P_1)^2$, namely condition (EA3). Condition (a4) implies $H = 0$ or $U_2 = 0$, which was discussed in the multiplication case. Condition (a5) is converted to $-X_1 Z_2^2(X_2 Z_1^2 - X_1 Z_2^2)^2 + (Y_2 Z_1^3 - Y_1 Z_2^3)^2 = 0$, which is $x_1 = \lambda^2$ in affine coordinate. It is equal to condition (EA4). Condition (a6) implies $-(X_2Z_1^2 - X_1Z_2^2)^3 - X_1Z_2^2(X_2Z_1^2 - X_1Z_2^2)^2 + (Y_2Z_1^3 - Y_1Z_2^3)^2 = 0$, which is $x_2 = \lambda(P_2, P_1)^2$ in affine coordinate. It is equal to condition (EA4).

Remark 7.4. If we implement the addition $-H^3 - 2U_1H^2 + R^2$ with either condition (a1), (a2), or (a3), then conditions (a4), (a5), and (a6) never appear in the ECADD. Condition (a1), (a2), and (a3) never simultaneously are satisfied — only one of them can be occurred. For example, the implementation of ECADD in Appendix A uses (a3), and thus the other conditions will never appear. The security of ECADD against the zero-value point attack strongly depends on its implementation, and we should care how to implement it.

How to Find the ZVP. We discuss how to find the ZVP described in the previous sections. A zero-value point is called as non-trivial, if the order of the point is smaller than that of the curve. The standard curves over prime fields have prime order, i.e., the orders of these elliptic curves are always prime and there are no non-trivial ZVP on them. We know that the Goubin's point can be easily computed. In the following we discuss the non-trivial ZVP that is different from the Goubin's points.

First we discuss the non-trivial ZVP from the ECDBL. There are two non-trivial points (x, y) such that

(ED1) $3x^2 + a = 0$,

(ED2) $5x^4 + 2ax^2 - 4bx + a^2 = 0$.

The solution of these polynomials over finite fields can be easily computed using the Cantor-Zassenhaus algorithm [Coh94].

Next we discuss the non-trivial ZVP from the ECADD. The existence conditions of these points are determined by not only one base point P but also the exponent c. In order to find these ZVP we have to know how to represent the relation between P and cP, for example, $x(cP) + x(P) = 0$. Izu and Takagi discussed a similar self-collision for Brier-Joye addition formula [IT03]. Here we can similarly apply their approach for finding the ZVP. We explain it in the following. Let $P = (x, y)$ be the point on the elliptic curve. The division polynomial $\psi(P)$, $\phi(P)$, $\omega(P)$ is a useful tool for representing these relationships as the polynomials over definition field K. The point cP can be represented as follows:

$$cP = \left(\frac{\phi_c(P)}{\psi_c^2(P)}, \frac{\omega_c(P)}{\psi_c^3(P)} \right)$$

where c is a scalar value (See for example, [Sil86]). For small c, we know $\psi_1(P) = 1$, $\psi_2(P) = 2y$, and $\psi_3(P) = 3x^4 + 6ax^2 + 12bx - a^2$, where $P = (x, y)$. We define $\phi_c = x\psi_c^2 - \psi_{c-1}\psi_{c+1}$ and $4y\omega_c = \psi_{c+2}\psi_{c-1}^2 - \psi_{c-2}\psi_{c+1}^2$.

For example, the points $P = (x, y)$ which satisfy $x(cP) + x(P) = 0$ are the solutions of $\phi_c(P) + x(P)\psi_c^2(P) = 0$. The points $P = (x, y)$ with $x(P) - x(cP) = \lambda(P, cP)^2$ are the solutions of polynomial $(x(P)\psi_c^2(P) - \phi_c(P))^3 = (y(P)\psi_c^3(P) - \omega_c(P))^2$. Similarly we can construct the equations whose solutions imply the ZVP.

The polynomials $\psi_c(P), \omega_c(P), \phi_c(P)$ have degree with order $O(c^2)$, which increases exponentially in $\log c$. Therefore, it is a hard problem to find the solutions of these equations for large c — we can find the ZVP only for small c using the division polynomials. It is an open problem to find a more efficient algorithm of computing the ZVP.

ZVP on Standard Curves. We have examined the existence of several ZVP over the SECG [SECG] random curves over prime fields. Especially we discuss the non-trivial conditions from ECDBL^J, namely (ED1) $3x^2 + a = 0$, (ED2) $5x^4 + 2ax^2 - 4bx + a^2 = 0$. These conditions are most effectively used for the zero-value point attack. We have found enough curves which have the points with condition (ED1) or (ED2). In Table 1 we summarize the existence of these points. Notation 'o' means that the curve has the point with one of the aforementioned conditions. For comparison we also show point $(0, y)$ used in Goubin's attack in Table 1. Some curves, e.g., secp112r1, secp224r1, are secure against the Goubin's attack, but not against ours. SECG secp224r1 is insecure only against condition (ED2).

TABLE 3. The existence of non-trivial ZVP of ECDBL^J

	$(0, y)$	(ED1)	(ED2)
SECG secp112r1	-	o	o
SECG secp128r1	o	-	-
SECG secp160r1	o	-	-
SECG secp160r2	o	-	o
SECG secp192r1	o	o	o
SECG secp224r1	-	-	o
SECG secp256r1	o	-	o
SECG secp384r1	o	o	-
SECG secp521r1	o	o	-

Countermeasure using Isogeny. In order to resist Goubin's attack, Smart proposed a countermeasure using isogeny of elliptic curve [Sma03].

Let $\Phi_l(X, Y)$ be a modular polynomial of degree l. Two elliptic curves $E_1(a_1, b_1)$ and $E_2(a_2, b_2)$ are called l-isogenous if and only if $\Phi_l(j_1, j_2) = 0$ satisfies, where j_i are j-invariant of curve E_i for $i = 1, 2$. Isogenous curves have the same order. The isogeny is given by

$$\psi : \begin{cases} E_1 & \longrightarrow & E_2 \\ (x, y) & \longmapsto & (\frac{f_1(x)}{g(x)^2}, \frac{y \cdot f_2(x)}{g(x)^3}) \end{cases} ,$$

where f_1, f_2 and g are polynomials of degree l, $(3l-1)/2$ and $(l-1)/2$ respectively (see details in [BSS99, Chapter VII]). By Horner's rule, the computational cost of this mapping is estimated as $(l + (3l-2)/2 + (l-1)/2 + 5)M + I = (3l+4)M + I$.

Smart proposed that if the original curve E has the point $(0, y)$, the isogenous curve E' to E could have no point $(0, y)$. If we can find E' which has no point

$(0, y)$, we transfer the base point $P \in E$ to $P' \in E'$ using the isogeny $\psi : E \to E'$. Instead of computing scalar multiplication $Q = dP$, we compute $Q' = dP'$ on E' and then pull back $Q \in E$ from $Q' \in E'$ by the mapping $\psi^{-1} : E' \to E$. The mappings ψ, ψ^{-1} require $(3l + 4)M + I$ respectively, so that the additional cost for this countermeasure is $(6l + 8)M + 2I$. This countermeasure with a small isogeny degree is faster than randomizing the secret scalar d with the order of the curve.

It is a further research topic to investigate the isogenous curve that are secure against ZVP attack.

7.1. Non-Zero Digit Methods

In this section we explain three approaches that resist the SPA. The first one is the Montgomery-type method, which always computes both ECADD and ECDBL for bit information d_i. It was originally proposed by Montgomery [Mon87], and enhanced the Weierstrass form of elliptic curves over K ([IT02, IBT02, BJ02, FGKS02]). The second one is to use an indistinguishable addition formula, with which we can compute both ECDBL and ECADD ([BJ02, CJ01]). The third one is to use the addition chain with fixed pattern with pre-computed points ([Möl01, OT03a]).

7.2. Montgomery Ladder Method

We explain the scalar multiplication using Montgomery ladder in the following. The algorithm improved on the addition chain and the addition formula. Both improvements are based on the scalar multiplication by Montgomery [Mon87]. However, we firstly point out that the addition chain is applicable for not only Montgomery form curves but any type of curves. We also establish the addition formulas, which only use the x-coordinate of the points, for the Weierstrass form curves.

Scalar Multiplication using Montgomery Ladder. We describe the scalar multiplication using Montgomery ladder in the following:

Scalar Multiplication using Montgomery Ladder
Input: $d = (d_{n-1} \cdots d_1 d_0)_2$, $P \in E(K)$ $(d_{n-1} = 1)$
Output: dP
1. $Q[0] \leftarrow P, Q[1] \leftarrow 2P$
2. For $i = (n - 2)$ down to 0 do:
2.1. $Q[2] = ECDBL(Q[d[i]])$
2.2. $Q[1] = ECADD(Q[0], Q[1])$
2.3. $Q[0] = Q[2 - d[i]]$
2.4. $Q[1] = Q[1 + d[i]]$
3. return $Q[0]$

For each bit $d[i]$, we compute $Q[2] = ECDBL$ $(Q[d[i]])$ in Step 2.1 and $Q[1] = ECADD(Q[0], Q[1])$ in Step 2.2. Then the values are assigned $Q[0] = Q[2], Q[1] = Q[1]$ if $d[i] = 0$ and $Q[0] = Q[1], Q[1] = Q[2]$ if $d[i] = 1$. We prove the correctness of the Montgomery ladder algorithm in the following.

Theorem 7.5. *The scalar multiplication using Montgomery ladder, on input a point P and an integer $d > 2$, outputs the correct value of the scalar multiplication $d * P$.*

Proof. When we write $Q[0], Q[1]$, it means that $Q[0]$ in Step 2.3 and $Q[1]$ in Step 2.4 of Montgomery ladder in the following. The loop of Step 2 generates a sequence

$$(Q[0], Q[1])_{n-2}, (Q[0], Q[1])_{n-3}, \ldots, (Q[0], Q[1])_1, (Q[0], Q[1])_0, \qquad (7.1)$$

from the bit sequence $d[n-2], d[n-3], \ldots, d[1], d[0]$. At first we prove $Q[1] = Q[0] + P$ for each $(Q[0], Q[1])_i, i = 0, 1, \ldots, n-2$, by the induction for the number of the sequence. For $n = 2$ we have only one loop in Step 2 and we have two cases $d[0] = 0$ or 1. Then we obtain $Q[0] = 2 * P, Q[1] = 3 * P$ for $d[0] = 0$, and $Q[0] = 3 * P, Q[1] = 4 * P$ for $d[0] = 1$. The fact $Q[1] = Q[0] + P$ is correct for $n = 2$. Next, we assume that $Q[1] = Q[0] + P$ up to $n = k$. In this case we have $R[1] = R[0] + P$, where $(Q[0], Q[1])_1 = (R[0], R[1])$. For $n = k + 1$ we also have two cases $d[0] = 0$ or 1. Then we obtain $Q[0] = 2 * R[0], Q[1] = 2 * R[0] + P$ for $d[0] = 0$, and $Q[0] = 2 * R[0] + P, Q[1] = 2 * R[0] + 2 * P$ for $d[0] = 1$. The fact $Q[1] = Q[0] + P$ is correct for $n = k + 1$. Thus we proved that $Q[1] = Q[0] + P$ for each $(Q[0], Q[1])_i, i = 0, 1, \ldots, n-2$.

Next, we prove that $Q[0]$ is equivalent to $Q[0]$ in Step 3 of left-to-right binary method ($Q[0]$ in Step 2.1 of the double-and-add-always method) for each loop of $d[i], (i = 0, 1, \ldots, n-2)$. In each loop of $d[i]$, for given $Q[0], Q[1]$, the new $Q[0]$ is computed as follows: ECDBL($Q[0]$) for $d[i] = 0$ and ECADD($Q[0], Q[1]$) $= Q[0] + (Q[0] + P) = 2 * Q[0] + P = $ ECADD(ECDBL($Q[0]$), P) for $d[i] = 1$. On the other hand, in each loop of $d[i]$ in the left-to-right binary method, for given $Q[0]$, the new $Q[0]$ is computed as follows: ECDBL($Q[0]$) for $d[i] = 0$ and ECADD(ECDBL($Q[0]$), P) for $d[i] = 1$. They are completely the same computations. Thus we can conclude that the output $d * P$ is correct. $\square$

Montgomery ladder requires one ECDBL in the initial Step 1, and $(n-1)$ ECDBLs and $(n-1)$ ECADDs in the loop. The computation time of the loop is same as that of double-and-add-always method.

Remark 7.6. Scalar multiplication using Montgomery ladder does not depend on the representation of elliptic curves, and it is applicable to execute a modular exponentiation in any abelian group. Therefore the RSA cryptosystem, the DSA, the ElGamal cryptosystem can use the Montgomery ladder.

Addition formula. Let E be an elliptic curve defined by the standard Weierstrass form (5.1) and $P_1 = (x_1, y_1)$, $P_2 = (x_2, y_2)$, $P_3 = P_1 + P_2 = (x_3, y_3)$ be points on $E(K)$. Moreover, let $P_3' = P_1 - P_2 = (x_3', y_3')$. Then we obtain the following relations:

$$x_3 \cdot x_3' = \frac{(x_1 x_2 - a)^2 - 4b(x_1 + x_2)}{(x_1 - x_2)^2}, \quad x_3 + x_3' = \frac{2(x_1 + x_2)(x_1 x_2 + a) + 4b}{(x_1 - x_2)^2}. \qquad (7.2)$$

On the other hand, letting $P_4 = 2 * P_1 = (x_4, y_4)$ leads to the relation

$$x_4 = \frac{(x_1^2 - a)^2 - 8bx_1}{4(x_1^3 + ax_1 + b)}. \qquad (7.3)$$

Thus the x-coordinates of both P_3 and P_4 can be computed just from the x-coordinates of the points P_1, P_2, P_3'. We call this method the multiplicative (additive) x-coordinate-only method. The x-coordinate-only methods for a scalar multiplication were originally introduced by Montgomery [Mon87]. However, his main interest was to find a special form of elliptic curves on which the computing times are optimal. The additive method was not discussed in his paper.

In the projective coordinate system, equations (7.2) and (7.3) turn to be

$$\frac{X_3}{Z_3} = \frac{Z_3'}{X_3'} \frac{(X_1 X_2 - a Z_1 Z_2)^2 - 4 b Z_1 Z_2 (X_1 Z_2 + X_2 Z_1)}{(X_1 Z_2 - X_2 Z_1)^2}, \tag{7.4}$$

$$\frac{X_3}{Z_3} = \frac{2(X_1 Z_2 + X_2 Z_1)(X_1 X_2 + a Z_1 Z_2) + 4 b Z_1^2 Z_2^2}{(X_1 Z_2 - X_2 Z_1)^2} - \frac{X_3'}{Z_3'}, \tag{7.5}$$

$$\frac{X_4}{Z_4} = \frac{(X_1^2 - a Z_1^2)^2 - 8 b X_1 Z_1^3}{4(X_1 Z_1 (X_1^2 + a Z_1^2) + b Z_1^4)}. \tag{7.6}$$

The computing times for (7.4),(7.5),(7.6) are $\text{ECADD}_m^{(x)} = 9M + 2S$, $\text{ECADD}_a^{(x)} = 10M + 2S$, $\text{ECDBL}^{(x)} = 6M + 3S$. If $Z_3' = 1$, the computing times are reduced to $\text{ECADD}_{m(Z_3'=1)}^{(x)} = \text{ECADD}_{a(Z_3'=1)}^{(x)} = 8M + 2S$.

When we use the x-coordinate-only methods, we need the difference of two points $P_3' = P_1 - P_2$. This may be a problem in general, but not in Montgomery ladder. In each loop of Montgomery ladder, the two points $(Q[0], Q[1])$ are simultaneously computed and they satisfy the equation $Q[1] - Q[0] = P$, where P is a base point of the scalar multiplication. Therefore, we can assume that the difference $P_2 - P_1$ for input values of $\text{ECADD}(P_1, P_2)$ of Montgomery ladder are always known. On the contrary, in order to know that of double-and-add-always method we need extra computation. The x-coordinate-only methods for double-and-add-always method have no computational advantage.

Y-Coordinate Recovering. When we apply the x-coordinate-only methods to the Montgomery ladder, the output is only the x-coordinate of $d * P$. This is enough for some cryptographic applications such as a key exchange scheme and an encryption/decryption scheme [SECG]. But other applications also require the y-coordinate of $d * P$ in the verification of a signature scheme [SECG]. However, the y-coordinate of $d * P$ is easily obtained in the following way: The final values of $Q[0], Q[1]$ in Montgomery ladder are related by $Q[1] = Q[0] + P$. Let $P = (x_1, y_1), Q[0] = (x_2, y_2), Q[1] = (x_3, y_3)$. Here known values are x_1, y_1, x_2, x_3 and the target is y_2. Using a standard addition formula (2), we obtain the equation $y_2 = (2y_1)^{-1}(y_1^2 + x_2^3 + a x_2 + b - (x_1 - x_2)^2 (x_1 + x_2 + x_3))$. This y-recovering technique was originally introduced by Agnew et al. for curves over $\mathbb{F}_{2^m}$ [AMV93]. In the projective coordinate, we show an algorithm that computes $dP = (X_d' : Y_d' : Z_d')$ for input $X_d, Z_d, X_{d+1}, Z_{d+1}, P = (x, y)$, where $x(dP) = X_d/Z_d, x((d+1)P) = X_{d+1}/Z_{d+1}$. It requires $11M + 2S + 7A$ and 7 auxiliary variables.

YRecovering, $11M + 2S + 7A$
Input $(X_d, Z_d, X_{d+1}, Z_{d+1}, x, y, a, b)$
Output (X'_d, Y'_d, Z'_d)
$R_1 \leftarrow X_d, R_2 \leftarrow Z_d, R_3 \leftarrow X_{d+1}, R_4 \leftarrow Z_{d+1}$
$R_5 \leftarrow x \times R_2$
$R_6 \leftarrow R_5 - R_1$
$R_6 \leftarrow R_6^2$
$R_6 \leftarrow R_3 \times R_6$
$R_5 \leftarrow R_5 + R_1$
$R_7 \leftarrow x \times R_1$
$R_1 \leftarrow R_1 \times R_2$
$R_3 \leftarrow a \times R_2$
$R_2 \leftarrow R_2^2$
$R_7 \leftarrow R_3 + R_7$
$R_7 \leftarrow R_5 \times R_7$
$R_5 \leftarrow y \times R_4$
$R_5 \leftarrow R_5 + R_5$
$R_3 \leftarrow R_5 \times R_2$
$R_1 \leftarrow R_5 \times R_1$
$R_2 \leftarrow b \times R_2$
$R_2 \leftarrow R_2 + R_2$
$R_7 \leftarrow R_7 + R_2$
$R_7 \leftarrow R_4 \times R_7$
$R_7 \leftarrow R_7 - R_6$
$X'_d \leftarrow R_1, Y'_d \leftarrow R_7, Z'_d \leftarrow R_3$

Formula xECADDDBL. In the above ladder, ECADD and ECDBL are computed separately. For performing SCA-resistant scalar multiplication efficiently, Izu et al. [IBT02] encapsulated these formulae into one formula xECADDDBL, which outputs x-coordinate values of $P_3 = P_1 + P_2$ and $P_4 = 2P_1$ on inputs P_1, P_2. In fact, with a projective version of the x-coordinate-only formulae, we can compute X_3, Z_3, X_4, Z_4 with $13M + 4S + 18A$ for $a \neq -3$ and $11M + 4S + 23A$ for $a = -3$. The number of auxiliary variables for the formulae is 7.

The scalar multiplication used for formula xECADDDBL is as follows:

Improved Scalar Multiplication using Montgomery Ladder
Input: $d = (d_{n-1} \cdots d_1 d_0)_2$, $P \in E(K)$ $(d_{n-1} = 1)$
Output: dP
1. $Q[0] \leftarrow P, Q[1] \leftarrow ECDBL(P)$
2. for $i = n - 2$ down to 0
2.1. $(Q[d[i] \oplus 1], Q[d[i]]) = xECADDDBL(Q[d[i]], Q[d[i] \oplus 1])$
3. return $Q[0]$

Timing. In order to demonstrate the efficiency of xECADDDBL, we implemented the 160-bit scalar multiplication using xECADDDBL and the previously fastest algorithm on a Celeron 500 MHz using the LiDIA library [LiDIA]. It should be emphasized here that our implementation was not optimized for cryptographic purposes — it is only intended to provide a comparison. The improvement is about 15%. The results are as follows:

TABLE 4. Computing times of 160-bit ECC on a Celeron 500 MHz

Double-and-Add-Always/Joye-Tymen	25.5 ms
xECADDDBL /Joye-Tymen	21.5 ms

xECADDDBL, $13M + 4S + 18A$	xECADDDBL$^{a=-3}$, $11M + 4S + 23A$
Input $(X_1, Z_1, X_2, Z_2, x, a, b)$	Input $(X_1, Z_1, X_2, Z_2, x, b)$
Output (X_3, Z_3, X_4, Z_4)	Output (X_3, Z_3, X_4, Z_4)
$R_1 \leftarrow X_1, R_2 \leftarrow Z_1, R_3 \leftarrow X_2$	$R_1 \leftarrow X_1, R_2 \leftarrow Z_1, R_3 \leftarrow X_2$
$R_4 \leftarrow Z_2$	$R_4 \leftarrow Z_2$
$R_6 \leftarrow R_1 \times R_4$	$R_6 \leftarrow R_1 \times R_4$
$R_1 \leftarrow R_1 \times R_3$	$R_1 \leftarrow R_1 \times R_3$
$R_4 \leftarrow R_2 \times R_4$	$R_4 \leftarrow R_2 \times R_4$
$R_2 \leftarrow R_3 \times R_2$	$R_2 \leftarrow R_3 \times R_2$
$R_3 \leftarrow R_6 - R_2$	$R_3 \leftarrow R_6 - R_2$
$R_3 \leftarrow R_3^2$	$R_3 \leftarrow R_3^2$
$R_5 \leftarrow x \times R_3$	$R_5 \leftarrow x \times R_3$
$R_7 \leftarrow a \times R_4$	$R_1 \leftarrow R_1 - R_4$
$R_1 \leftarrow R_1 + R_7$	$R_1 \leftarrow R_1 - R_4$
$R_2 \leftarrow R_2 + R_6$	$R_1 \leftarrow R_1 - R_4$
$R_1 \leftarrow R_1 \times R_2$	$R_2 \leftarrow R_2 + R_6$
$R_2 \leftarrow R_4^2$	$R_1 \leftarrow R_1 \times R_2$
$R_7 \leftarrow b \times R_2$	$R_2 \leftarrow R_4^2$
$R_1 \leftarrow R_1 + R_7$	$R_7 \leftarrow b \times R_2$
$R_1 \leftarrow R_1 + R_1$	$R_1 \leftarrow R_1 + R_7$
$R_5 \leftarrow R_1 - R_5$	$R_1 \leftarrow R_1 + R_1$
$R_5 \leftarrow R_7 + R_5$	$R_5 \leftarrow R_1 - R_5$
$R_5 \leftarrow R_7 + R_5$	$R_5 \leftarrow R_7 + R_5$
$R_2 \leftarrow a \times R_2$	$R_5 \leftarrow R_7 + R_5$
$R_1 \leftarrow R_6^2$	$R_1 \leftarrow R_2 + R_2$
$R_1 \leftarrow R_1 + R_2$	$R_1 \leftarrow R_1 + R_1$
$R_2 \leftarrow R_2 + R_2$	$R_2 \leftarrow R_2 - R_1$
$R_2 \leftarrow R_1 - R_2$	$R_1 \leftarrow R_6^2$
$R_2 \leftarrow R_2^2$	$R_1 \leftarrow R_1 + R_2$
$R_1 \leftarrow R_6 \times R_1$	$R_2 \leftarrow R_2 + R_2$
$R_7 \leftarrow R_4 \times R_7$	$R_1 \leftarrow R_1 - R_2$
$R_1 \leftarrow R_1 + R_7$	$R_2 \leftarrow R_2^2$
$R_7 \leftarrow R_6 \times R_7$	$R_1 \leftarrow R_6 \times R_1$
$R_7 \leftarrow R_7 + R_7$	$R_7 \leftarrow R_4 \times R_7$
$R_7 \leftarrow R_7 + R_7$	$R_1 \leftarrow R_1 + R_7$
$R_7 \leftarrow R_7 + R_7$	$R_7 \leftarrow R_6 \times R_7$
$R_7 \leftarrow R_2 - R_7$	$R_7 \leftarrow R_7 + R_7$
$R_6 \leftarrow R_4 \times R_1$	$R_7 \leftarrow R_7 + R_7$
$R_6 \leftarrow R_6 + R_6$	$R_7 \leftarrow R_7 + R_7$
$R_6 \leftarrow R_6 + R_6$	$R_7 \leftarrow R_2 - R_7$
	$R_6 \leftarrow R_4 \times R_1$
	$R_6 \leftarrow R_6 + R_6$
	$R_6 \leftarrow R_6 + R_6$
$X_3 \leftarrow R_5, Z_3 \leftarrow R_3$	$X_3 \leftarrow R_5, Z_3 \leftarrow R_3$
$X_4 \leftarrow R_7, Z_4 \leftarrow R_6$	$X_4 \leftarrow R_7, Z_4 \leftarrow R_6$

7.3. Non-Zero Window Method

Okeya and Takagi proposed an SPA-resistant addition chain with small memory, which is based on the width-w NAF [OT03a]. The algorithm is as follows:

SPA-resistant_Width-w_NAF_with_Odd_Scalar
INPUT An odd n-bit d
OUTPUT $d_w[n], d_w[n-1], \ldots, d_w[0]$
1. $r \leftarrow 0$, $i \leftarrow 0$, $r_0 \leftarrow w$
2. While $d > 1$ do the following
2.1. $u[i] \leftarrow (d \bmod 2^{w+1}) - 2^w$
2.2. $d \leftarrow (d - u[i])/2^{r_i}$
2.3. $d_w[r + r_i - 1] \leftarrow 0, d_w[r + r_i - 2] \leftarrow 0, \ldots, d_w[r+1] \leftarrow 0, d_w[r] \leftarrow u[i]$
2.4. $r \leftarrow r + r_i$, $i \leftarrow i + 1$, $r_i \leftarrow w$
3. $d_w[n] \leftarrow 0, \ldots, d_w[r+1] \leftarrow 0, d_w[r] \leftarrow 1$
4. Return $d_w[n], d_w[n-1], \ldots, d_w[0]$

The algorithm generates the SPA-resistant chain only for odd scalar, and the treatment for even scalar was discussed in [OT03a]. We assume that the scalar d is odd in the following. At Step 2.1, the integer $u[i]$ is assigned as $(d \bmod 2^{w+1}) - 2^w$. The computation assures that $u[i]$ is odd whenever d is odd. Since $d - u[i] = d - (d \bmod 2^{w+1}) + 2^w = 2^w \bmod 2^{w+1}$, the resultant $(d - u[i])/2^w$ is odd. Thus, each integer $u[i]$ is odd. Note that d terminates with $d = 1$. Hence we can achieve the SPA-resistant chain, e.g., the fixed pattern

$$| \underbrace{0..0}_{w-1} \, x | \underbrace{0..0}_{w-1} \, x | \cdots | \underbrace{0..0}_{w-1} \, x | \text{ with odd integers } |x| < 2^w.$$

The number of the pre-computed points is 2^{w-1}, and the density of the non-zero bit is $1/w$. The scalar multiplication using this chain is computed as same for the scalar multiplication with width-w NAF.

We show an example of non-adjacent form as follows:

binary string	100111011110011100010110111100011010101111001
$w = 2$	$10\bar{1}030301030303\bar{3}030103\bar{0}1030\bar{1}03010303\bar{0}1\bar{0}1\bar{0}1030303\bar{0}3$
$w = 3$	$10010070\bar{0}100500\bar{1}001003005007001001\bar{0}0\bar{3}003007001$

Note that this scheme is optimal in respect of the memory, and the table size takes $2, 4, 8, \ldots$ for $w = 2, 3, 4, \ldots$.

References

[ABF+02]   C. Aumüller, P. Bier, W. Fischer, P. Hofreiter, and J.-P. Seifert, "Fault Attacks on RSA with CRT: Concrete Results and Practical Countermeasures," CHES 2002, LNCS 2523, pp.260-275, 2003.

[AM94]   L. Adleman and K. McCurley, "Open problems in number theoretic complexity, II" proceedings of ANTS-I, LNCS 877, pp.291-322, 1994.

[AMV93]   G. Agnew, R. Mullin and S. Vanstone, "An implementation of elliptic curve cryptosystems over $F_{2^{155}}$," IEEE Journal on Selected Areas in Communications, vol.11, pp.804-813, 1993.

[AT03]   T. Akishita and T. Takagi, "Zero-Value Point Attacks on Elliptic Curve Cryptosystem", ISC 2003, LNCS2851, pp. 218-233, 2003.

[BS02]   D. Boneh and H. Shacham, "Fast Variants of RSA," CRYPTOBYTES, Vol.5, No.1, pp.1-9, 2002.

[BD00]   D. Boneh and G. Durfee, "Cryptanalysis of RSA with private key d less than $N^{0.292}$," IEEE Transactions on Information Theory, Vol.46, No.4, pp.1339-1349, 2000.

[BDL01]   D. Boneh, R. DeMillo, R. Lipton, "On the Importance of Eliminating Errors in Cryptographic Computations." Journal of Cryptology 14(2), pp.101-119, 2001.

[BDPR98]   M. Bellare, A. Desai, D. Pointcheval, and P. Rogaway, "Relations among notions of security for public-key encryption schemes," CRYPTO'98, LNCS 1462, pp.26-45, 1998.

[BFKLB03] F. Bahr, J. Franke, T. Kleinjung, M. Lochter and M. Böhm, RSA-160, http://www.loria.fr/~zimmerma/records/rsa160.

[BJ02]   É. Brier and M. Joye, "Weierstrass Elliptic Curve and Side-Channel Attacks", PKC 2002, LNCS 2274, pp. 335-345, Springer-Verlag, 2002.

[BLW02]   B. den Boer, K. Lemke, and G. Wicke, "A DPA Attack against the Modular Reduction within a CRT Implementation of RSA," CHES 2002, LNCS 2523, pp.228-243, 2003.

[Bon01]   D. Boneh, "Simplified OAEP for the RSA and Rabin Functions," CRYPTO 2001, LNCS 2139, pp.275-291, 2001.

[BR93]   M. Bellare and P. Rogaway, "Random oracles are practical: a paradigm for designing efficient protocols," First ACM Conference on Computer and Communications Security, (1993), pp.62-73.

[Bre00]   R. Brent, "Recent Progress and Prospects for Integer Factorisation Algorithms," COCOON 2000, LNCS 1858, pp.3-22, 2000.

[BSS99]   I. Blake, G. Seroussi, and N. Smart, *Elliptic Curve in Cryptography*, Cambridge University Press, 1999.

[BB03]   D. Boneh and D. Brumley, "Remote Timing Attacks are Practical," http://crypto.stanford.edu/~dabo/

[BST01]   J. Buchmann, K. Sakurai, and T. Takagi, "An IND-CCA2 Public-Key Cryptosystem with Fast Decryption," Information Security and Cryptology - ICISC 2001, LNCS 2288, pp.51-71, 2001.

[CJ01]   C. Clavier and M. Joye, "Universal exponentiation algorithm", CHES 2001, LNCS 2162, pp.300-308, Springer-Verlag, 2001.

[Coh94]   H. Cohen, *Course in Computational Algebraic Number Theory*, Graduate Texts in Mathematics, Vol. 138, Springer-Verlag, 1994.

[CMO98]   H. Cohen, A. Miyaji, and T. Ono, "Efficient Elliptic Curve Exponentiation Using Mixed Coordinates", LNCS 1514, pp. 51-65, 1998.

[Com]   MultiPrime^{TM}, Compaq AXL300 Accelerator. http://www.compaq.com/
            products/servers/security/axl300/

[Cop96]   D. Coppersmith "Finding a Small Root of a Bivariate Integer Equa-
            tion; Factoring with High Bits Known," EUROCRYPT '96, LNCS 1070,
            pp.178-189, 1996.

[CFPR96]   D. Coppersmith, M. Franklin, J. Patarin and M. Reiter, "Low-exponent
            RSA with related messages," EUROCRYPT '96, LNCS 1070, pp.1-9, 1996.

[Cor99]   J. Coron, "Resistance against Differential Power Analysis for Elliptic
            Curve Cryptosystems," CHES'99, LNCS1717, pp.292-302, 1999.

[CHJPPT02a] J. -S. Coron, H. Handschuh, M. Joye, P. Paillier, D. Pointcheval, and
            C. Tymen, "Optimal Chosen-Ciphertext Secure Encryption of Arbitrary-
            Length Messages," Public Key Cryptography 2002, LNCS 2274, pp.17-33,
            2002

[CHJPPT02b] J. -S. Coron, H. Handschuh, M. Joye, P. Paillier, D. Pointcheval, and
            C. Tymen, "GEM: A Generic Chosen-Ciphertext Secure Encryption
            Method," Topics in Cryptology - CT-RSA 2002, LNCS2271, pp.263-276,
            2002.

[CRYPTREC] CRYPTREC, Evaluation of Cryptographic Techniques, IPA. http://www.
            ipa.go.jp/security/enc/CRYPTREC/

[Dav82]   G. Davida, "Chosen Signature Cryptanalysis of the RSA (MIT) Public
            Key Cryptosystem," TR-CS-82-2, University of Wisconsin, 1982.

[Den02a]   A. Dent, "An implementation attack against the EPOC-2 public-key cryp-
            tosystem," Electronics Letters, 38(9), pp.412, 2002.

[ECMNET]   ECMNET Project; http://www.loria.fr/~zimmerma/records/ecmnet.
            html

[ET02]   P. Ebinger and E. Teske "Factoring $N = pq^2$ with the elliptic curve
            method," Technical Report, CORR 2002-02, CACR, the University of
            Waterloo, 2002.

[EPOC]   EPOC, Efficient Probabilistic Public-Key Encryption. http://info.isl.
            ntt.co.jp/epoc/

[FGKS02]   W. Fischer, C. Giraud, E. Knundsen, and J. Seifert, "Parallel Scalar
            Multiplication on General Elliptic Curves over F_p Hedged against Non-
            Differential Side-Channel Attacks", IACR Cryptology ePrint Archive
            2002/007.

[FMP03]   P. Fouque, G. Martinet, G. Poupard, "Attacking Unbalanced RSA-CRT
            using SPA," CHES 2003, LNCS 2779, 2003, to appear.

[FO99a]   E. Fujisaki and T. Okamoto, "How to Enhance the Security of Public-Key
            Encryption at Minimum Cost," 1999 International Workshop on Practice
            and Theory in Public Key Cryptography, LNCS 1560, (1999), pp.53-68.

[FO99b]   E. Fujisaki and T. Okamoto, "Secure Integration of Asymmetric and Sym-
            metric Encryption Schemes," Advances in Cryptology – CRYPTO'99,
            LNCS 1666, (1999), pp.537-554.

[FO01]   E. Fujisaki and T. Okamoto, "A Chosen-Cipher Secure Encryption Scheme
            Tightly as Secure as Factoring," IEICE Trans. Fundamentals, Vol. E84-A,
            No.1, (2001), pp.179-187.

[Gar59]   H. Garner, "The residue number system," IRE Transactions on Electronic Computers, EC-8 (6), pp.140-147, 1959.

[Gor98]   D. Gordon, "A survey of fast exponentiation methods", J. Algorithms, vol.27, pp.129-146, 1998.

[Gou03]   L. Goubin, "A Refined Power-Analysis Attack on Elliptic Curve Cryptosystems", PKC 2003, LNCS 2567, pp. 199-211, 2003.

[Has88]   J. Håstad, "Solving simultaneous modular equations of low degree," SIAM Journal of Computing, 17, pp.336-341, 1988.

[IEEE]   IEEE P1363, Standard Specifications for Public-Key Cryptography, 2000. Available from http://grouper.ieee.org/groups/1363/

[IIT02]   K. Itoh, T. Izu, and M. Takenaka, "Address-bit Differential Power Analysis on Cryptographic Schemes OK-ECDH and OK-ECDSA", CHES 2002, LNCS 2523, pp.129-143, 2002.

[IYTT02]   K. Itoh, J. Yajima, M. Takenaka, and N. Torii, "DPA Countermeasures by Improving the Window Method", CHES 2002, LNCS 2523, pp.303-317, 2002.

[IBT02]   T. Izu, B, Möller, and T. Takagi, "Improved Elliptic Curve Multiplication Methods Resistant against Side Channel Attacks", INDOCRYPT 2002, LNCS 2551, pp. 296-313, 2002.

[IT02]   T. Izu and T. Takagi, "A Fast Parallel Elliptic Curve Multiplication Resistant against Side Channel Attacks", PKC 2002, LNCS 2274, pp.280-296, 2002.

[IT03]   T. Izu and T. Takagi, "Exceptional Procedure Attack on Elliptic Curve Cryptosystems", PKC 2003, LNCS 2567, pp. 224-239, 2003.

[JCA]   Java Cryptography Architecture, http://java.sun.com/products/jdk/1.2/docs/guide/security/CryptoSpec.html

[JQY98]   M. Joye, J. -J. Quisquater, and M. Yung, "The Policeman in the Middle Attack," presented at rump session of Eurocrypt'98, 1998.

[JQY01]   M. Joye, J. -J. Quisquater, and M. Yung, "On the Power of Misbehaving Adversaries and Security Analysis of the Original EPOC," Topics in Cryptology - CT-RSA 2001, LNCS 2020, pp.208-222, 2001.

[JQ01]   M. Joye and J. Quisquater, "Hessian elliptic curves and side-channel attacks," CHES2001, LNCS 2162, pp.402-410, Springer-Verlag, 2001.

[JLQ99]   M. Joye, A.K. Lenstra, and J.-J. Quisquater, "Chinese Remaindering Based Cryptosystems in the Presence of Faults," Journal of Cryptology 12(4), pp.241-245, 1999.

[JT01]   M. Joye and C. Tymen, "Protection against Differential Analysis for Elliptic Curve Cryptography", CHES 2001, LNCS 2162, pp. 377-390, 2001.

[KR95]   B. Kaliski and M. Robshaw, "Secure use of RSA," CRYPTOBYTES, Vol.1, No.3, pp.7-13, 1995.

[Kal96]   B. Kaliski, "Timing Attacks on Cryptosystems," RSA Laboratories Bulletin, No.2, 1996.

[KCJ+01]   S. Kim, J. Cheon, M. Joye, S. Lim, M. Mambo, D. Won, and Y. Zheng, "Strong Adaptive Chosen-Ciphertext Attacks with Memory Dump (or:

The Importance of the Order of Decryption and Validation)," Cryptography and Coding, 8th IMA Int. Conf., LNCS 2260, pp.114-127, 2001.

[KR02]   V. Klíma and T. Rosa, "Further Results and Considerations on Side Channel Attacks on RSA," CHES 2002, LNCS 2523, pp.244-259, 2003.

[Koc96]   C. Kocher, "Timing attacks on Implementations of Diffie-Hellman, RSA, DSS, and other Systems," CRYPTO '96, LNCS 1109, pp.104-113, 1996.

[KJJ99]   C. Kocher, J. Jaffe, and B. Jun, "Differential Power Analysis," CRYPTO '99, LNCS 1666, pp.388-397, 1999.

[KT92]   K. Koyama and Y. Tsuruoka, "Speeding Up Elliptic Curve Cryptosystems using a Signed Binary Windows Method," CRYPTO '92, LNCS740, pp. 345-357, 1992.

[Len87]   H. Lenstra, Jr., "Factoring integers with elliptic curves", Annals of Mathematics, 126, pp.649-673, 1987.

[LL91]   A. K. Lenstra and H. W. Lenstra, Jr. (Eds.), "The development of the number field sieve," Lecture Notes in Mathematics, 1554, Springer, 1991.

[LS01]   P. Liardet and N. Smart, "Preventing SPA/DPA in ECC Systems Using the Jacobi Form," CHES 2001, LNCS2162, pp.391-401, 2001.

[LiDIA]   LiDIA, A C++ Library For Computational Number Theory, Technische Universtät Darmstadt, http://www.informatik.tu-darmstadt.de/TI/LiDIA/

[Man01]   J. Manger, "A Chosen Ciphertext Attack on RSA Optimal Asymmetric Encryption Padding (OAEP) as Standardized in PKCS #1 v2.0," CRYPTO 2001, LNCS 2139, pp.230-238, 2001.

[MDS99]   T. Messerges, E. Dabbish, R. Sloan, "Power Analysis Attacks of Modular Exponentiation in Smartcards," CHES'99, LNCS 1717, pp.144-157, 1999.

[MO90]   F. Morain and J. Olivos, "Speeding Up the Computation on an Elliptic Curve Using Addition-Subtraction Chains," Inform. Theory Appl. 24, pp.531-543, 2000.

[MOV96]   A. Menezes, P. van Oorschot, and S. Vanstone, Handbook of Applied Cryptography, CRC Press, 1997.

[MOC97]   A. Miyaji, T. Ono, and H. Cohen, "Efficient Elliptic Curve Exponentiation," ICICS 1997, LNCS 1334, pp.282-291, 1997.

[Mon87]   P. L. Montgomery, "Speeding the Pollard and Elliptic Curve Methods of Factorization", Mathematics of Computation, vol. 48, pp. 243-264, 1987.

[Möl01]   B. Möller, "Securing Elliptic Curve Point Multiplication against Side-Channel Attacks", ISC 2001, LNCS 2200, pp.324-334, 2001.

[Möl02]   B. Möller, "Improved Techniques for Fast Exponentiation," ICISC 2002, LNCS 2587, pp.298-312, 2003.

[NSS01]   M. Nishioka, H. Satoh, and K. Sakurai, "Design and Analysis of Fast Provably Secure Public-Key Cryptosystems Based on a Modular Squaring," ICISC 2001, LNCS 2288, pp.81-102, 2001.

[NESSIE] NESSIE, New European Schemes for Signatures, Integrity, nd Encryption, IST-1999-12324. https://www.cosic.esat.kuleuven.ac.be/nessie/

[Nov02]   R. Novak, "SPA-Based Adaptive Chosen-Ciphertext Attack on RSA Implementation," PKC 2002, LNCS 2274, pp.252-262, 2002.

[OP01]   T. Okamoto and D. Pointcheval, "REACT: Rapid Enhanced-security Asymmetric Cryptosystem Transform," In Proceedings of the Cryptographers' Track at RSA Conference '2001, LNCS 2020, (2001), pp.159-175.

[OU98]   T. Okamoto and S. Uchiyama; "A New Public-Key Cryptosystem as Secure as Factoring," Eurocrypt'98, LNCS 1403, pp.308-318, 1998.

[OS00]   K. Okeya and K. Sakurai, "Power Analysis Breaks Elliptic Curve Cryptosystems even Secure against the Timing Attack", INDOCRYPT 2000, LNCS 1977, pp.178-190, Springer-Verlag, 2000.

[OS02a]   K. Okeya and K. Sakurai, "On Insecurity of the Side Channel Attack Countermeasure using Addition-Subtraction Chains under Distinguishability between Addition and Doubling," ACISP 2002, LNCS2384, pp.420-435, 2002.

[OS02b]   K. Okeya and K. Sakurai, "A Second-Order DPA Attack Breaks a Window-method based Countermeasure against Side Channel Attacks," ISC 2002, LNCS 2433, pp.389-401, 2002.

[OT03a]   K. Okeya, and T. Takagi, "The Width-w NAF Method Provides Small Memory and Fast Elliptic Scalar Multiplications Secure against Side Channel Attacks", CT-RSA 2003, LNCS 2612, pp.328-342, 2003.

[OT03b]   K. Okeya, and T. Takagi, "A More Flexible Countermeasure against Side Channel Attacks Using Window Method," CHES 2003, LNCS 2779, pp.397-410, 2003.

[Osw02]   E. Oswald, "Enhancing Simple Power-Analysis Attacks on Elliptic Curve Cryptosystems," CHES 2002, LNCS 2523, pp.82-97. 2002.

[OA01]   E. Oswald and M. Aigner, "Randomized Addition-Subtraction Chains as a Countermeasure against Power Attacks," CHES 2001, LNCS2162, pp.39-50, 2001.

[Pai99]   P. Paillier, "Public-Key Cryptosystems based on Composite Degree Residuosity Classes," Eurocrypt'99, LNCS 1592, pp.223-238, 1999.

[PKCS]   Public-Key Cryptography Standards, PKCS # 1, Amendment 1: Multi-Prime RSA, RSA Laboratories. http://www.rsasecurity.com/rsalabs/pkcs/

[PO96]   R. Peralta and E. Okamoto, "Faster factoring of integers of a special form," IEICE Trans. Fundamentals, Vol.E79-A, No.4, pp.489-493, 1996.

[Poi00]   D. Pointcheval, "Chosen-ciphertext security for any one-way cryptosystem," Public Key Cryptography 2000, LNCS 1751, pp.129-146, 2000.

[QC82]   J. -J. Quisquater and C. Couvreur, "Fast decipherment algorithm for RSA public-key cryptosystem," Electronic Letters, 18, pp.905-907, 1982.

[RSA78]   R. Rivest, A. Shamir, and L. Adleman, "A method for obtaining digital signatures and public-key cryptosystems," Communications of the ACM, 21(2), pp.120-126, 1978.

[RS97]   R. Rivest and R. D. Silverman, "Are 'strong' primes needed for RSA," The 1997 RSA Laboratories Seminar Series, Seminars Proceedings, 1997.

[RSA155]   S. Cavallar, B. Dodson, A. Lenstra, W. Lioen, P. Montgomery, B. Murphy, H. te Riele, K. Aardal, J. Gilchrist, G. Guillerm, P. Leyland, J. Marchand, F. Morian, A. Muffett, C. Putnam, C. Putnam, and P. Zimmermann, "Factorization of a 512-Bit RSA Modulus," EUROCRYPT 2000, LNCS1807, pp.1-18, 2000.

[Sho01]   V. Shoup, "A Proposal for an ISO Standard for Public-Key Encryption (version 2.1)," http://www.shoup.net.

[Sil86]   J. Silverman, *The Arithmetic of Elliptic Curves*, GMT 106, Springer-Verlag, 1986.

[Sil00]   R. Silverman, "A cost-based security analysis of symmetric and asymmetric key lengths," RSA Laboratories Bulletin, No.13, 2000. http://www.rsasecurity.com/rsalabs/bulletins/bulletin13.html

[Sma01]   N. Smart, "The Hessian form of an elliptic curve," CHES2001, LNCS 2162, pp.118-125, 2001.

[Sma03]   N. Smart, "An Analysis of Goubin's Refined Power Analysis Attack," CHES 2003, LNCS 2779, pp. 281-290, Springer-Verlag, 2003.

[Sol00]   J. Solinas, "Efficient Arithmetic on Koblitz Curves," Design, Codes and Cryptography, 19, pp.195-249, 2000.

[SECG]   Standard for Efficient Cryptography (SECG), *SEC2: Recommended Elliptic Curve Domain Parameters*, Version 1.0, 2000. http://www.secg.org/

[Tak98]   T. Takagi, "Fast RSA-type cryptosystem modulo $p^k q$," CRYPTO '98, LNCS 1462, pp.318-326, 1998.

[VT97]   E. Verheul and H. van Tilborg, "Cryptanalysis of 'less short' RSA secret exponents," Applicable Algebra in Engineering, Communication and Computing, 8, pp.425-435, 1997.

[Wal02a]   C. Walter, "MIST: An Efficient, Randomized Exponentiation Algorithm for Resisting Power Analysis", CT-RSA 2002, LNCS 2271, pp.53-66, 2002.

[Wal02b]   C. Walter, "Some Security Aspects of the Mist Randomized Exponentiation Algorithm," CHES 2002, LNCS 2523, pp.564-578, 2002.

[Wie90]   M. J. Wiener, "Cryptanalysis of short RSA secret exponents," IEEE Transactions on Information Theory, IT-36, pp.553-558, 1990.

Printed in the United States
By Bookmasters